VOLUME SIX HUNDRED AND SIXTY ONE

METHODS IN ENZYMOLOGY

The DNA Replication-Repair Interface

METHODS IN ENZYMOLOGY

Editors-in-Chief

ANNA MARIE PYLE
Departments of Molecular, Cellular and Developmental Biology and Department of Chemistry
Investigator, Howard Hughes Medical Institute
Yale University

DAVID W. CHRISTIANSON
Roy and Diana Vagelos Laboratories
Department of Chemistry
University of Pennsylvania
Philadelphia, PA

Founding Editors

SIDNEY P. COLOWICK and NATHAN O. KAPLAN

VOLUME SIX HUNDRED AND SIXTY ONE

METHODS IN ENZYMOLOGY

The DNA Replication-Repair Interface

Edited by

BRANDT F. EICHMAN

Department of Biological Sciences and Department of Biochemistry, Vanderbilt University, Nashville, TN, United States

ELSEVIER

ACADEMIC PRESS

An imprint of Elsevier

Academic Press is an imprint of Elsevier
50 Hampshire Street, 5th Floor, Cambridge, MA 02139, United States
525 B Street, Suite 1650, San Diego, CA 92101, United States
The Boulevard, Langford Lane, Kidlington, Oxford OX5 1GB, United Kingdom
125 London Wall, London, EC2Y 5AS, United Kingdom

First edition 2021

Notices
Knowledge and best practice in this field are constantly changing. As new research and experience broaden our understanding, changes in research methods, professional practices, or medical treatment may become necessary.

Practitioners and researchers must always rely on their own experience and knowledge in evaluating and using any information, methods, compounds, or experiments described herein. In using such information or methods they should be mindful of their own safety and the safety of others, including parties for whom they have a professional responsibility.

To the fullest extent of the law, neither the Publisher nor the authors, contributors, or editors, assume any liability for any injury and/or damage to persons or property as a matter of products liability, negligence or otherwise, or from any use or operation of any methods, products, instructions, or ideas contained in the material herein.

ISBN: 978-0-323-90733-0
ISSN: 0076-6879

Publisher: Zoe Kruze
Developmental Editor: Federico Paulo S. Mendoza
Production Project Manager: James Selvam
Cover Designer: Alan Studholme

Typeset by STRAIVE, India

Contents

Contributors

Andrés Aguilera
Centro Andaluz de Biología Molecular y Medicina Regenerativa (CABIMER), Universidad de Sevilla-CSIC-Universidad Pablo de Olavide, Seville, Spain

Zamal Ahmed
Department of Cancer Biology; Department of Molecular & Cellular Oncology, University of Texas MD Anderson Cancer Center, Houston, TX, United States

Antoine Aze
Institut de Génétique Humaine, Université de Montpellier, Genome Surveillance and Stability, CNRS-UMR9002, Montpellier, France

Aakash Basu
Johns Hopkins University School of Medicine, Baltimore, Maryland, United States

Lou Baudrier
The University of Calgary, Cumming School of Medicine, Department of Biochemistry and Molecular Biology; Robson DNA Science Centre; Arnie Charbonneau Cancer Institute, Calgary, AB, Canada

Marina A. Bellani
Laboratory of Molecular Biology and Immunology, National Institute on Aging, National Institutes of Health, Baltimore, MD, United States

Orléna Benamozig
The University of Calgary, Cumming School of Medicine, Department of Biochemistry and Molecular Biology; Robson DNA Science Centre; Arnie Charbonneau Cancer Institute, Calgary, AB, Canada

Pierre Billon
The University of Calgary, Cumming School of Medicine, Department of Biochemistry and Molecular Biology; Robson DNA Science Centre; Arnie Charbonneau Cancer Institute, Calgary, AB, Canada

Matthew L. Bochman
Molecular & Cellular Biochemistry Department, Indiana University, Bloomington, IN, United States

Chris A. Brosey
Department of Cancer Biology; Department of Molecular & Cellular Oncology, University of Texas MD Anderson Cancer Center, Houston, TX, United States

Peter Chovanec
Department of Human Genetics, University of California, Los Angeles, CA, Unites States

Rajashree A. Deshpande
The Department of Molecular Biosciences, The University of Texas at Austin, Austin, TX, United States

John F.X. Diffley
Chromosome Replication Laboratory, The Francis Crick Institute, London, United Kingdom

Devon M. Fitzgerald
Department of Molecular and Human Genetics; Department of Biochemistry and Molecular Biology; Department of Molecular Virology and Microbiology; Dan L Duncan Comprehensive Cancer Center, Baylor College of Medicine, Houston, TX, United States

Himabindu Gali
Frederick National Laboratory for Cancer Research, Frederick, MD, United States

Julia Gichimu
Laboratory of Molecular Biology and Immunology, National Institute on Aging, National Institutes of Health, Baltimore, MD, United States

Belén Gómez-González
Centro Andaluz de Biología Molecular y Medicina Regenerativa (CABIMER), Universidad de Sevilla-CSIC-Universidad Pablo de Olavide, Seville, Spain

Eric C. Greene
Department of Biochemistry & Molecular Biophysics, Columbia University, New York, NY, United States

Guillaume Guilbaud
Division of Protein & Nucleic Acid Chemistry, MRC Laboratory of Molecular Biology, Cambridge, United Kingdom

Dipika Gupta
Department of Biochemistry and Molecular Pharmacology, New York University School of Medicine, New York, NY, United States

Taekjip Ha
Department of Biomedical Engineering; Department of Biophysics and Biophysical Chemistry; Department of Biophysics, Johns Hopkins University; Howard Hughes Medical Institute, Baltimore, MA, United States

James E. Haber
Department of Biology, Rosenstiel Basic Medical Sciences Research Center, Brandeis University, Waltham, MA, United States

Thanos D. Halazonetis
Department of Molecular Biology, University of Geneva, Geneva, Switzerland

Jing Huang
Institute of Chemical Biology and Nanomedicine, State Key Laboratory of Chemo/ Biosensing and Chemometrics, College of Biology, Hunan University, Changsha, China

Ryan C. James
Department of Molecular Biology and Genetics, Cornell University, Ithaca, NY, United States

Andrzej Joachimiak
Center for Structural Genomics of Infectious Diseases, Consortium for Advanced Science and Engineering; Department of Biochemistry and Molecular Biology, University of Chicago, Chicago, IL, United States

Erik Johansson
Department of Medical Biochemistry and Biophysics, Umeå University, Umeå, Sweden

Darin E. Jones
Department of Pharmaceutical Sciences, The University of Arkansas for Medical Sciences, Little Rock, AR, United States

Panagiotis Katsonis
Department of Molecular and Human Genetics, Baylor College of Medicine, Houston, TX, United States

In-Kwon Kim
Department of Chemistry, University of Cincinnati, Cincinnati, OH, United States

Youngchang Kim
Center for Structural Genomics of Infectious Diseases, Consortium for Advanced Science and Engineering; Department of Biochemistry and Molecular Biology, University of Chicago, Chicago, IL, United States

Youngho Kwon
Department of Biochemistry and Structural Biology, University of Texas Health Science Center at San Antonio, TX, United States

Wei Ting C. Lee
Department of Biochemistry and Molecular Pharmacology, New York University School of Medicine, New York, NY, United States

Olivier Lichtarge
Department of Molecular and Human Genetics, Baylor College of Medicine, Houston, TX, United States

Todd M. Link
Department of Cancer Biology; Department of Molecular & Cellular Oncology, University of Texas MD Anderson Cancer Center, Houston, TX, United States

Jingjing Liu
Department of Molecular and Human Genetics, Baylor College of Medicine, Houston, TX, United States

Liping Liu
Department of Biology, University of Iowa, Iowa City, IA, United States

Elena Lo Furno
Institut de Génétique Humaine, Université de Montpellier, Genome Surveillance and Stability, CNRS-UMR9002, Montpellier, France

Zhijun Ma
Department of Chemistry, University of Cincinnati, Cincinnati, OH, United States

Jonathan Mailler
Department of Molecular Biology, University of Geneva, Geneva, Switzerland

Domenico Maiorano
Institut de Génétique Humaine, Université de Montpellier, Genome Surveillance and Stability, CNRS-UMR9002, Montpellier, France

Anna Malkova
Department of Biology, University of Iowa, Iowa City, IA, United States

Qian Mei
Department of Molecular and Human Genetics; Department of Biochemistry and Molecular Biology; Department of Molecular Virology and Microbiology; Dan L Duncan Comprehensive Cancer Center, Baylor College of Medicine; Systems, Synthetic, and Physical Biology Program, Rice University, Houston, TX, United States

Davide Moiani
Department of Cancer Biology; Department of Molecular & Cellular Oncology, University of Texas MD Anderson Cancer Center, Houston, TX, United States

David G. Nickens
Molecular & Cellular Biochemistry Department, Indiana University, Bloomington, IN, United States

Sadeieh Nimer
Department of Molecular and Human Genetics; Department of Biochemistry and Molecular Biology; Department of Molecular Virology and Microbiology; Dan L Duncan Comprehensive Cancer Center, Baylor College of Medicine, Houston, TX, United States

Pedro Ortega
Centro Andaluz de Biología Molecular y Medicina Regenerativa (CABIMER), Universidad de Sevilla-CSIC-Universidad Pablo de Olavide, Seville, Spain

Laura Padayachy
Department of Molecular Biology, University of Geneva, Geneva, Switzerland

Tanya T. Paull
The Department of Molecular Biosciences, The University of Texas at Austin, Austin, TX, United States

Durga Pokharel
Horizon Discovery, Lafayette, CO, United States

Viktor Posse
Chromosome Replication Laboratory, The Francis Crick Institute, London, United Kingdom; Department of Medical Biochemistry and Cell Biology, University of Gothenburg, Gothenburg, Sweden

Suresh S. Pujari
Department of Medicinal Chemistry and Masonic Cancer Center, University of Minnesota, Minneapolis, MN, United States

Bénédicte Recolin
Institut de Génétique Humaine, Université de Montpellier, Genome Surveillance and Stability, CNRS-UMR9002, Montpellier, France

Susan M. Rosenberg
Department of Molecular and Human Genetics; Department of Biochemistry and Molecular Biology; Department of Molecular Virology and Microbiology; Dan L Duncan Comprehensive Cancer Center, Baylor College of Medicine; Systems, Synthetic, and Physical Biology Program, Rice University, Houston, TX, United States

Eli Rothenberg
Department of Biochemistry and Molecular Pharmacology, New York University School of Medicine, New York, NY, United States

Upasana Roy
Department of Biochemistry & Molecular Biophysics, Columbia University, New York, NY, United States

Julian E. Sale
Division of Protein & Nucleic Acid Chemistry, MRC Laboratory of Molecular Biology, Cambridge, United Kingdom

Michael M. Seidman
Laboratory of Molecular Biology and Immunology, National Institute on Aging, National Institutes of Health, Baltimore, MD, United States

Althaf Shaik
Laboratory of Molecular Biology and Immunology, National Institute on Aging, National Institutes of Health, Baltimore, MD, United States

Neal Sugawara
Department of Biology, Rosenstiel Basic Medical Sciences Research Center, Brandeis University, Waltham, MA, United States

Patrick Sung
Department of Biochemistry and Structural Biology, University of Texas Health Science Center at San Antonio, TX, United States

John A. Tainer
Department of Cancer Biology; Department of Molecular & Cellular Oncology, University of Texas MD Anderson Cancer Center, Houston, TX; Biophysics and Integrated Bioimaging, Lawrence Berkeley National Laboratory, Berkeley, CA, United States

Arun K. Thazhathveetil
Nitto Denko Avecia Inc., Cincinnati, OH, United States

Natalia Tretyakova
Department of Medicinal Chemistry and Masonic Cancer Center, University of Minnesota, Minneapolis, MN, United States

Susan E. Tsutakawa
Biophysics and Integrated Bioimaging, Lawrence Berkeley National Laboratory, Berkeley, CA, United States

Siem van der Laan
Institut de Génétique Humaine, Université de Montpellier, Genome Surveillance and Stability, CNRS-UMR9002, Montpellier, France

Yi Yin
Department of Human Genetics, University of California, Los Angeles, CA, Unites States

Jing Zhang
Department of Neurosurgery, Shanghai East Hospital, Tongji University School of Medicine, Shanghai, China

Roger S. Zou
Department of Biomedical Engineering, Johns Hopkins University, Baltimore, MA, United States

Preface

Replication machinery frequently encounters aberrant DNA structures and other impediments that inhibit progression of the replication fork. Replication stress and stalled forks can lead to double-strand breaks (DSBs) and chromosomal rearrangements, which are a source of genomic instability and underlie multiple diseases including cancer. DNA repair and recombination enzymes work together at stalled forks to restart replication, clear the impediment, and/or mitigate further instability resulting from harmful DNA structures generated as the fork is processed.

Much progress has been made in the last 10 years to understand the causes, consequences, and resolution of replication stress. This progress has been driven largely by new methods to visualize or map the products and frequency of replisome conflicts across the genome at sufficient resolution to obtain a mechanistic understanding of the proteins and DNA structures involved. In this volume of *Methods in Enzymology*, leading investigators describe state-of-the-art techniques to probe various aspects of replication stress and genome maintenance. These methods incorporate a wide array of technologies from chemical and computational biology, biochemistry, biophysics, cell biology, and genetics, providing unprecedented insight into how replication, recombination, and repair function together to ensure completion of DNA synthesis.

Chapters 1–5 focus on monitoring DNA replication and its progression in the face of fork stalling lesions. In Chapter 1, John Diffley and colleagues describe purification protocols for the 24 yeast proteins used to reconstitute replication *in vitro*. Julian Sale and Guillaume Guilbaud detail in Chapter 2 an assay for mapping sites of replisome stalling by changes in histone modification and transcription. In Chapter 3, Michael Seidman and coworkers describe preparation of psoralen interstrand crosslinks in cells, as well as DNA fiber and proximity ligation assays to image the effects of impediments to the replisome. Chapter 4 describes the use of high-resolution, multicolor single-molecule localization microscopy to detect the association of the replisome with G-quadraplexes in cells by Eli Rothenberg and colleagues. In Chapter 5, Domenico Maiorano and coauthors outline two experimental systems to analyze the DNA damage response in early *Xenopus laevis* embryos and mouse embryonic stem cells.

Chapters 6–14 deal with detection and/or mapping the locations of replication-repair intermediates and products in the genome. Andrés Aguilera and colleagues detail protocols to quantify the efficiency and genetic products of the recombination reactions during repair of replication-associated DSBs in Chapter 6. This theme is continued in Chapter 7, in which James Haber and coauthors describe a method for rapid detection of DNA synthesis during recombination repair of strand breaks at the replication fork. In Chapter 8, Susan Rosenberg and colleagues discuss chemical biology methods for genomic mapping of DNA replication-repair intermediates including Holliday junctions and 3′-ends of single-stranded DNA. Yi Yin and Peter Chovanec detail in Chapter 9 high-throughput, whole-genome sequencing methods for mapping mitotic crossovers formed during recombination repair of DSBs. Keeping with the theme of DSB repair, Tanya Paull and Rajashree Deshpande describe in Chapter 10 a protocol for mapping sites of nonhomologous end joining. In Chapter 11, Taekjip Ha and Roger Zou provide descriptions of their approach for targeted generation of DSB using rapid light regulation of CRISPR-Cas9, as well as methods to detect DSB levels, repair factors, and insertion-deletion repair outcomes. Pierre Billon and colleagues detail in Chapter 12 a protocol to identify mutations resulting from genetic variation or introduced by CRISPR-based genome editing. In Chapter 13, Thanos Halazonetis and coauthors describe a method to map the genomic regions that are replicated during mitosis in mammalian cells. Aakash Basu outlines a method to measure a mechanical property of DNA—its propensity to cyclize—on a genome-wide scale in Chapter 14.

Chapters 15–17 focus on generation of purified reagents and their uses in biochemical replication-repair studies. Matthew Bochman and David Nickens describe in Chapter 15 their purification of DNA helicases Pif1 and Hrq1 along with *in vitro* methods to assess their regulation of telomerase activity. In Chapter 16, Eric Greene, Patrick Sung, and coworkers detail purification and single-molecule analysis of Rad51 paralogs important for recombination repair of DSBs. This is followed in Chapter 17 by Natalia Tretyakova and Suresh Pujari, who describe the chemical synthesis of DNA-protein and DNA-peptide crosslink substrates and their use in biochemical studies of translesion synthesis. Finally, in Chapter 18, John Tainer and colleagues discuss their *in silico* methods for efficient inhibitor screening of enzymes involved in replication-repair.

I am grateful to all the authors for their hard work and timely submission of these chapters. I am also indebted to Paulo Mendoza and the Elsevier staff for their administrative help throughout assembly of this volume. Finally, I wish to thank David Christianson and Anna Pyle, the Editors-in-Chief of *Methods in Enzymology*, for the invitation to serve as editor of what I hope will be a useful resource to propel the replication-repair field forward.

BRANDT F. EICHMAN
Editor

CHAPTER ONE

Eukaryotic DNA replication with purified budding yeast proteins

Viktor Posse[a,b], Erik Johansson[c], and John F.X. Diffley[a,*]
[a]Chromosome Replication Laboratory, The Francis Crick Institute, London, United Kingdom
[b]Department of Medical Biochemistry and Cell Biology, University of Gothenburg, Gothenburg, Sweden
[c]Department of Medical Biochemistry and Biophysics, Umeå University, Umeå, Sweden
*Corresponding author: e-mail address: john.diffley@crick.ac.uk

Contents

Abstract

The in vitro reconstitution of origin firing was a key step toward the biochemical reconstitution of eukaryotic DNA replication in budding yeast. Today the basic replication assay involves proteins purified from 24 separate protocols that have evolved since their first publication, and as a result, the efficiency and reliability of the in vitro replication system has improved. Here we will present protocols for all 24 purifications together with a general protocol for the in vitro replication assay and some tips for troubleshooting problems with the assay.

Methods in Enzymology, Volume 661
ISSN 0076-6879
https://doi.org/10.1016/bs.mie.2021.08.018

1. Introduction

Origin firing with purified budding yeast proteins was first reconstituted in 2015 (Yeeles et al., 2015) and opened the possibility to study origin-dependent eukaryotic DNA replication with purified proteins. Since the initial publication, the assay has evolved to reproduce the replication rate observed in vivo, to support Okazaki fragment synthesis and maturation, as well as to replicate chromatinized DNA templates (Deegan et al., 2019; Devbhandari et al., 2017; Kurat et al., 2017; Yeeles et al., 2017). This chapter includes all steps required to set up the basic DNA replication system on naked DNA, reconstituting both leading and lagging strand synthesis, including the maturation of Okazaki fragments. First, we describe the approach used to over-express the proteins needed for the DNA replication assay. This includes the codon optimization for genes expressed in *Saccharomyces cerevisiae* as well as generalized protocols for protein expression in *E. coli* and *S. cerevisiae*. Second, we describe the strategies for the 24 different purifications that are needed for the reconstitution of origin dependent DNA replication. We describe the general protocols for each purification method followed by protocols for each protein, containing details both for their expression and purification. The buffers are displayed in Table 1 and referred to as buffer 1–28 in each protocol. Third, we describe the setup of the assay including some insights on optimization and troubleshooting.

The origin dependent in vitro replication system has proven to be very stable, despite the apparent complexity when combining so many purified proteins. The system is still young and will allow the field to continue to develop a deep mechanistic understanding of how the eukaryotic genome is duplicated.

2. Expression of proteins

2.1 Codon optimizations and strains

Codon optimization was done as previously described (Sharp et al., 1987), with additional data on protein abundance (de Godoy et al., 2008; Ghaemmaghami et al., 2003), using the script in Supplementary Materials available at https://doi.org/10.1016/bs.mie.2021.08.018. The codon-optimized open reading frame under the control of a GAL1–10 promotor (plasmids pRS303-pRS306) was stably integrated in the yJF1 yeast strain (MATa, ade2–1, ura3–1, his3–11,15, trp1–1, leu2–3112, can1–100, bar1::Hyg, pep4::KanMX). The name of each over-expression strain is specified in the protocols for the purifications of proteins (see below). Strains and plasmids are available upon request to J.F.X. Diffley.

Table 1 Buffer compositions for protein purification.

Buffer	Hepes-KOH pH 7.6 (mM)	Tris-HCl pH 7.2 (mM)	Tris-HCl pH 7.5 (mM)	Potassium phosphate pH 7.6 (mM)	Glycerol (%)	NP-40-S (%)	Triton X (%)	Tween20 (%)	EDTA (mM)	EGTA (mM)	DTT (mM)	ATP (mM)	Sucrose (%)	$Mg(OAc)_2$ (mM)	NaCl	KCl	KOAc	K-Glutamate	Protein
1	25				10	0.05					1					x			Orc
2	25				10	0.05					1						x		Orc
3	45				10	0.02					1			5			x		Cdt1/Mcm2-7
4				50			1				1	2		5			x		Cdc6
5				50			0.1				1	2		5			x		Cdc6
6				50	15		0.1				1			5			x		Cdc6
7	25				10	0.02					1			10			x		Cdc6
8	25				10	0.02					1				x				DDK
9	25				10	0.02					1							x	DDK
10	25				10	0.02			1		1					x			Sld3/7, Dpb11
11	25				10	0.02		0.1	1		1				x				Sld2
12	25				40	0.02			1		1				(x)		(x)		Sld2(Dialysis with NaCl or KOAc)
13	25				10				1		1						x		Cdc45
14	50				10	0.05			1	1	1			10		x			GINS
15	25				10	0.02			1		1						x		GINS
16	40				10	0.02					1						x		CDK
17	25				10						1						x		Polε
18		25			10	0.02					1				x				Polα, TopoI, Topo2, Csm3/Tof1, Polδ
19		25			10	0.02			1		1				x				Polα, Mrc1, Csm3/Tof1, Polδ
20		25			10						1				x				RPA, Ctf4
21		25			10				1		1				x				RPA, Ctf4, PCNA
22	25				10	0.05					1				x				Mcm10
23	25				10	0.05			1		1				x				Mcm10
24		25			40	0.02			1		1				x				Mrc1
25	25				10						1				x				RFC
26		50											10						PCNA
27			50		10				1		1				x				Ligase 1
28			25		10	0.02					1				x				Fen1

Salt concentration example: Buffer 1 with 400 mM KCl = Buffer 1^(400). Protease inhibitors are added to all lysis buffers (see protein expression) and initial binding and washing step for the first affinity chromatography step (see above). Sld2 is dialyzed with buffer 12 containing either NaCl or KOAc, see comments on Sld2 and salt in the replication reaction below. Buffers are also supplemented with additional chemicals for binding and elution in affinity chromatography steps (see details for specific proteins below).

2.2 Protocols for protein expression

2.2.1 *Expression in* S. cerevisiae

2.2.1.1 Chemicals

YPA (YP with 4 mg/L Adenine)
Glucose (20%, sterile)
Raffinose (20%, sterile)
Galactose (20%, sterile)
Alpha factor
Nocodazole (for S-CDK expression)
Protease inhibitors (final concentration in buffer: 1 mM AEBSF, 10 μg/mL leupeptin and 10 μg/mL pepstatin A)

2.2.1.2 Protocol

1. Grow a 50 mL YPA + 2% glucose starting culture of the protein expression strain at 30 °C to saturation (>24 h)
2. Inoculate 4–8 L of YPA + 2% raffinose with 3×10^8 cells/L and grow o/n at 30 °C to $1\text{–}3 \times 10^7$ cells/mL
3. If the expressed protein requires alpha factor arrest (see specific protein details below) add alpha factor to 100 ng/mL and continue grow for 2–3 h until all cells are arrested. If no alpha factor arrest is required, go to step 4
4. Induce protein expression by addition of 2% galactose and grow for 3 h at 30 °C
5. Harvest cells by centrifugation
6. Transfer cells to a 50 mL Falcon tube and wash once with lysis buffer without protease inhibitors and DTT (see specific protein details below)
7. Collect cells by centrifugation and resuspend in 0.3 volumes of lysis buffer with protease inhibitors and DTT (see specific protein details below)
8. Freeze dropwise in liquid nitrogen. Store the frozen cells at −80 °C or continue with step 9
9. Crush cells with a freezer mill (SPEX CertiPrep 6850 Freezer/Mill) with 6 cycles of 2 min at a crushing rate of 15. Store the frozen cell powder at −80 °C or continue with step 10
10. Thaw cell powder and roughly double the volume with cold lysis buffer

11. Spin at 45000 rpm in a Ti45 rotor (158,000 × *g*) for 45 min at 4 °C
12. Take the middle layer (avoiding lipid layer at the top and pelleted material at the bottom) and continue the purification with affinity chromatography (see below)

2.2.2 *Expression in* E. coli

2.2.2.1 Chemicals

LB
Antibiotics (Ampicillin, Kanamycin, Chloramphenicol)
IPTG
Lysozyme
cOmplete protease inhibitor cocktail tablets (Roche)

2.2.2.2 Protocol

1. Transform the plasmid for expression of the protein of interest to the expression strain of choice (see protocol for each protein below). Plate on LB-Agar plates containing the correct antibiotics for the plasmid and expression strain and grow at 37 °C o/n
2. Pick one colony and grow in a LB starter culture with the correct antibiotics at 37 °C o/n
3. Inoculate the selected volume (see specific protein details below) and grow at 37 °C to an OD_{600} of 0.6
4. Cool the culture to the temperature of protein expression and add IPTG (see specific protein details below)
5. Continue growth at the temperature for expression of the specific protein for indicated time (see specific protein details below)
6. Harvest by centrifugation
7. Transfer the pellet to a 50 mL Falcon tube on ice and resuspend in lysis buffer without protease inhibitors and DTT (see specific protein details below)
8. Collect cells by centrifugation (at 4 °C) and freeze in liquid nitrogen. Store the frozen pellet at −80 °C or continue with step 9
9. Thaw the cells and add up to 30 mL of cold lysis buffer (see specific protein details below)
10. Add Lysozyme to 100 μg/mL and incubate rotating in cold room for 20 min to 1 h
11. Sonicate cells on ice

12. Spin down in JA-25.50 at 15000 rpm (27,000 × *g*) for 15 min at 4 °C
13. Take supernatant and continue the purification with affinity chromatography as specified for each individual protein (see below)

3. Purification of proteins

3.1 General strategy and buffer compositions

All protein purification steps are performed on ice or in the cold room unless other is stated. For analysis purposes, save all flow-through and wash fractions as well as beads after elution for SDS-PAGE analysis. General chemicals and equipment are listed here as well as the description of common purification strategies. Unique chemicals and equipment for each specific protein are listed at the top of that proteins purification protocol (see below).

3.1.1 Common equipment and tubes needed for most purification methods

50 mL Falcon tubes
Cooled benchtop centrifuge for Falcon tubes
1.5 mL Eppendorf tubes
Cooled benchtop centrifuge for Eppendorf tubes
Tube rotator for Falcon tubes
Gravity chromatography column suitable for 0.5–2 mL bed volume
Liquid nitrogen
Chromatography system
SDS-PAGE system

3.1.2 Chromatography columns

MonoQ
SP FF
Heparin
DEAE
Hiload Superdex 16/600 200 pg
Superdex 200 Increase 10/300 GL

3.2 Affinity chromatography

In all protocols except for PCNA, a tag is fused to the overexpressed protein to allow for affinity purification. Protease inhibitors are used during binding and initial washing step for all purifications. For yeast expressed proteins, use final concentrations of 1 mM AEBSF, 10 μg/mL leupeptin and

10 μg/mL pepstatin A. For *E. coli* expressed proteins, use cOmplete protease inhibitor cocktail tablets (Roche).

3.2.1 Flag

3.2.1.1 Chemicals

Anti-FLAG M2 resin
$Mg(OAc)_2$
ATP
3×-FLAG peptide

1. Take 1–2 mL of Anti-FLAG M2 resin slurry (final bed volume 0.5–1 mL)
2. Resuspend in 10 volumes of lysis buffer in a 50 mL Falcon tube
3. Collect resin by centrifugation, 500 × *g* at 4 °C
4. Repeat steps 2 and 3 once
5. Add clarified cell lysate to the washed Anti-FLAG M2 resin and incubate for 1–2 h rotating in cold room
6. Collect resin by centrifugation, 500 × *g* at 4 °C
7. Resuspend the resin in 50 mL lysis buffer and collect by centrifugation, 500 × *g* at 4 °C
8. Resuspend in 25 mL of wash buffer with added $MgCl_2$ (10 mM) and ATP (1 mM) and incubate for 15 min rotating in cold room (ATP wash)
9. Collect resin by centrifugation, 500 × *g* at 4 °C
10. Resuspend in a small volume of wash buffer and transfer resin to a gravity chromatography column. Wash the Falcon tube with additional buffer to retain all of the resin
11. Wash the resin in the column with an additional 25 mL of wash buffer
12. Stop the flow with a bottom cap and add 2 mL elution buffer (wash buffer with 0.5 mg/mL FLAG peptide)
13. Incubate 10 min on ice before eluting into a 15 mL Falcon tube
14. Repeat steps 12 and 13 twice
15. Wash the column with 1 mL elution buffer into the same Falcon tube
16. Freeze the eluted protein in liquid nitrogen or continue with the next purification step

3.2.2 Immobilized metal ion affinity chromatography (IMAC)

3.2.2.1 Chemicals

IMAC resin (Ni-NTA or TALON)
Imidazole

1. Take 1–2 mL of IMAC resin slurry (Ni-NTA or TALON) (final bed volume 0.5–1 mL)
2. Resuspend in 10 volumes of lysis buffer supplemented with 10 mM imidazole in a 50 mL Falcon tube
3. Collect resin by centrifugation, 500 × *g* at 4 °C
4. Repeat steps 2 and 3 once
5. Add cell lysate together with 10 mM imidazole to the washed IMAC resin and incubate for 1–2 h rotating in cold room
6. Collect resin by centrifugation, 500 × *g* at 4 °C
7. Resuspend the resin in 50 mL lysis buffer supplemented with 10 mM imidazole and collect by centrifugation, 500 × *g* at 4 °C
8. Resuspend in a small volume of wash buffer (with 10 mM imidazole) and transfer resin to a gravity chromatography column. Wash the Falcon tube with additional buffer to retain all of the resin
9. Wash the resin in the column with an additional 25 mL of wash buffer (with 10 mM imidazole)
10. Stop the flow with a bottom cap and add 2 mL elution buffer (wash buffer with 200 mM imidazole)
11. Incubate 10 min on ice before eluting into a 15 mL Falcon tube
12. Repeat steps 10 and 11 twice
13. Wash the column with 1 mL elution buffer into the same Falcon tube
14. Freeze the eluted protein in liquid nitrogen or continue with the next purification step

3.2.3 Calmodulin

3.2.3.1 Chemicals

Calmodulin resin
$CaCl_2$
$Mg(OAc)_2$
ATP
EDTA
EGTA

1. Take 1–2 mL of calmodulin resin slurry (final bed volume 0.5–1 mL)
2. Resuspend in 10 volumes of lysis buffer (without EDTA) in a 50 mL Falcon tube
3. Collect resin by centrifugation, 500 × *g* at 4 °C
4. Repeat steps 2 and 3 once
5. Add cell lysate to the washed calmodulin resin together with 2 mM $CaCl_2$ and incubate for 1–2 h rotating in cold room

6. Collect resin by centrifugation, 500 × *g* at 4 °C
7. Resuspend the resin in 50 mL lysis buffer with 2 mM $CaCl_2$ and collect by centrifugation, 500 × *g* at 4 °C
8. Resuspend in 25 mL of wash buffer with 2 mM $CaCl_2$ and added $MgCl_2$ (10 mM) and ATP (1 mM) and incubate for 15 min rotating in cold room (ATP wash)
9. Collect resin by centrifugation, 500 × *g* at 4 °C
10. Resuspend in a small volume of wash buffer with 2 mM $CaCl_2$ and transfer resin to a gravity chromatography column. Wash the Falcon tube with additional buffer to retain all of the resin
11. Wash the resin in the column with an additional 25 mL of wash buffer with 2 mM $CaCl_2$
12. For purification of CDK and Csm3/Tof1, continue with elution by TEV protease below. For all other proteins, continue with step 13
13. Stop the flow with a bottom cap and add 2 mL elution buffer (wash buffer with addition of EDTA and EGTA, see specific protein details below)
14. Incubate 10 min on ice before eluting into a 15 mL Falcon tube
15. Repeat steps 12 and 13 twice
16. Wash the column with 1 mL elution buffer into the same Falcon tube
17. Freeze the eluted protein in liquid nitrogen or continue with the next purification step

3.2.4 GST-tag

3.2.4.1 Chemicals

Glutathione sepharose
PreScission protease

1. Take 2 mL of glutathione resin slurry (final bed volume 1 mL)
2. Resuspend in 10 volumes of lysis buffer in a 50 mL Falcon tube
3. Collect resin by centrifugation, 500 × *g* at 4 °C
4. Repeat steps 2 and 3 once
5. Add cell lysate to the washed GST resin and incubate for 1–2 h rotating in cold room
6. Collect resin by centrifugation, 500 × *g* at 4 °C
7. Resuspend the resin in 50 mL lysis buffer and collect by centrifugation, 500 × *g* at 4 °C
8. Resuspend in a small volume of lysis buffer and transfer resin to a gravity chromatography column. Wash the Falcon tube with additional buffer to retain all of the resin
9. Wash the resin in the column with an additional 25 mL of lysis buffer

10. Stop the flow with a bottom cap and resuspend the resin in 1 mL lysis buffer
11. Add 100 μL PreScission protease (stock at 2 mg/mL = 2 Units/μL) and put a cap on top of the column
12. Incubate for 2 h rotating in cold room
13. Collect the eluate and elute more by adding 1 mL wash buffer three times
14. Freeze the eluted protein in liquid nitrogen or continue with the next purification step

3.2.5 IgG (only for Sld3/Sld7)

3.2.5.1 Chemicals

IgG resin
$Mg(OAc)_2$
ATP

1. Take 2 mL of IgG resin slurry (final bed volume 1 mL)
2. Resuspend in 10 volumes of lysis buffer in a 50 mL Falcon tube
3. Collect resin by centrifugation, 500 × *g* at 4 °C
4. Repeat steps 2 and 3 once
5. Add cell lysate to the washed IgG resin and incubate for 45 min rotating in cold room
6. Collect resin by centrifugation, 500 × *g* at 4 °C
7. Resuspend the resin in 50 mL lysis buffer and collect by centrifugation, 500 × *g* at 4 °C
8. Resuspend in 25 mL of wash buffer with added $MgCl_2$ (10 mM) and ATP (1 mM) and incubate for 15 min rotating in cold room (ATP wash).
9. Collect resin by centrifugation, 500 × *g* at 4 °C
10. Resuspend in a small volume of wash buffer and transfer resin to a gravity chromatography column. Wash the Falcon tube with additional buffer to retain all of the resin
11. Wash the resin in the column with an additional 25 mL of wash buffer
12. For Sld3/7 purification follow the steps under elution by TEV protease

3.2.6 Elution by TEV protease

3.2.6.1 Chemicals

TEV protease

1. Stop the flow with a bottom cap and resuspend the resin in 1 mL wash buffer
2. Add 100 μg/mL TEV protease and put a cap on top of the column
3. Incubate for 1.5 h rotating in cold room

4. Collect the eluate and elute more by adding 2.5 mL wash buffer two times
5. See below for removal of the TEV protease by IMAC
6. Freeze the eluted protein in liquid nitrogen or continue with the next purification step

3.2.7 TEV protease removal by IMAC

3.2.7.1 Chemicals

IMAC resin (Ni-NTA or TALON).

Imidazole.

1. Take 0.5 mL of IMAC resin slurry (final bed volume 0.25 mL)
2. Resuspend in 10 volumes of lysis buffer supplemented with 10 mM imidazole in a 50 mL Falcon tube
3. Collect resin by centrifugation, 500 × *g* at 4 °C
4. Repeat steps 2 and 3 once
5. Add eluate with TEV protease (supplemented with 10 mM imidazole) to the washed IMAC resin and incubate for 30 min rotating in cold room
6. Collect the flow-through and wash by adding 1 mL wash buffer (supplemented with 10 mM imidazole) two times. Collect washes together with the flow-through
7. Freeze the eluted protein in liquid nitrogen or continue with the next purification step

3.3 Ion exchange chromatography

The ion exchange chromatography is performed using a chromatography system.

3.3.1 MonoQ and SP FF

1. If necessary, dilute the eluate from the initial affinity chromatography step to get a final salt concentration according to the loading conditions of the individual protein (see details for specific protein below).
2. Load the protein on a column equilibrated in loading buffer (see details for specific protein below)
3. Wash the column with wash buffer and volume specified for the individual protein (see details for specific protein below)
4. Elute the protein with step elution or a salt gradient according to the elution conditions of the individual protein (see details for specific protein below) collecting 0.35 mL fractions
5. Run fractions around the peak on an SDS-PAGE and pick the main peak fractions for continued purification

3.3.2 *Heparin*

Purification using heparin affinity is performed using a chromatography system and Heparin HP 1 mL columns. Follow the same steps as for ion exchange chromatography.

3.3.3 *Hydroxyapatite*

3.3.3.1 Chemicals

Hydroxyapatite (HTP).

1. Mix 2 g dry HTP powder with 12 mL wash buffer in a 15 mL Falcon tube
2. Rotate 5 min at room temperature
3. Allow beads to settle for 2 min and remove buffer and fine particles with aspirator
4. Add 10 mL wash buffer and repeat step 2–4 two more times
5. Resuspend beads as 50% slurry in wash buffer
6. Add 2 mL of beads (4 mL slurry) to new gravity chromatography column
7. Wash two times with 4 mL wash buffer and add a cap to the bottom
8. Transfer the solution containing the protein of interest to the prepared HTP column and put a cap on the top of the column
9. Incubate rotating for 15 min in the cold room
10. Discard flow through from HPT column
11. Wash beads with 2 mL wash buffer
12. Wash beads with 5 mL rinse buffer
13. Elute with 10 mL elution buffer in 1 mL fractions
14. Run all samples on SDS-PAGE to evaluate which fractions to continue with
15. Freeze the eluted protein in liquid nitrogen or continue with the next purification step

3.4 Size exclusion chromatography

3.4.1 *Hiload Superdex 16/600 200 pg and Superdex 200 increase 10/300 GL*

1. Concentrate the protein solution to ~0.5 mL for Superdex 200 Increase 10/300 GL and ~2.5 mL for Hiload Superdex 16/600 200 pg with a centrifugal filter column (see below).
2. Centrifuge at full speed and 4 °C for 5 min in a benchtop centrifuge for Eppendorf tubes to remove any precipitates

3. Load the protein on a column equilibrated in buffer (see specific protein details below) using a 1 mL loop for Superdex 200 Increase 10/300 GL and a 5 mL loop for Hiload Superdex 16/600 200 pg
4. Elute the protein and collect 0.5 mL fractions for Superdex 200 Increase 10/300 GL or 2 mL fractions for Hiload Superdex 16/600 200 pg
5. Run fractions around the peak on an SDS-PAGE and pick the main peak fractions for continued purification

3.5 Dialysis and concentration

3.5.1 Buffer exchange and concentration by dialysis

3.5.1.1 Equipment

D-Tube Dialyzer Midi/Maxi MWCO 3.5 kDa (Millipore).

1. Prepare 2 L of cold dialysis buffer (add DTT just before use).
2. Load protein solution in a D-Tube Dialyzer Midi/Maxi MWCO 3.5 kDa (Millipore).
3. Dialyze for 2 h in 1 L buffer with gentle stirring in cold room changing the buffer after 1 h

3.5.2 Concentration by centrifugal filter column

3.5.2.1 Equipment

Amicon Ultra centrifugal filter units (Millipore) with the correct volume (4 or 15 mL depending on sample size) and MWCO (see details for specific proteins below).

To estimate the final protein concentration in a sample use a Bradford assay. We recommend that the final protein concentration is determined by SDS-PAGE before assembling reactions (see below).

1. Wash the filter unit by adding 2–4 mL of the buffer of the protein of interest and spin for a few minutes at a speed according to the filter unit of choice
2. Discard flow through and any buffer retained in the column
3. Apply the protein sample
4. Spin for 5 min
5. Check volume level and mix gently by pipetting up and down a few times
6. Repeat steps 4 and 5 until the wanted volume or concentration is reached, possibly adjusting spin time

3.5.3 Protein concentration determination by SDS-PAGE

3.5.3.1 Equipment

Gel doc, scanner or equivalent

3.5.3.2 Chemicals

BSA

Coomassie stain (Instant Blue)

1. Run a standard of BSA (50, 100, 200, 400 and 800 ng) together with at least three different sized samples of the protein of interest on an SDS-PAGE. If possible, aim to load between 50 and 800 ng per subunit in each sample using the Bradford protein concentration as guide. Generally, Bradford overestimates the concentration compared to gel quantification
2. Stain the gel with Instant Blue and scan or take a photo using a Gel doc or equivalent
3. Open the image in ImageJ, individually select all lanes and then plot lanes
4. Define peaks using the line tool and measure each peak area using the wand tool
5. Transfer results to a chart and make a standard curve from the five BSA concentrations
6. Calculate the amount (ng) in each peak of the samples of unknown concentration
7. Calculate the concentration (ng/μL) in each peak (peptide)
8. Transform concentration from ng/μL to μM for each peptide
9. If the sample is a multimer of equimolar proportions, calculate the average from the individual subunits as the final concentration

3.5.4 Buffers used in the purification of proteins

3.5.4.1 Chemicals

HEPES
Tris
HCl
KOH
K_2HPO_4
KH_2PO_4
Glycerol
NP-40-S
Triton-X
Tween-20
EDTA
EGTA
DTT

ATP
Sucrose
$Mg(OAc)_2$
NaCl
KCl
KOAc
K-Glutamate

3.6 Protocols for purification of each protein in the replication system

3.6.1 *Orc*

Expression: Yeast (strain ySDOrc) with alpha factor arrest.

Lysis buffer: Buffer $1^{(300)}$ with protease inhibitors.

1. Calmodulin: Supplement lysate in buffer $1^{(300)}$ with 2 mM $CaCl_2$. Wash with buffer $1^{(300)}$ with 2 mM $CaCl_2$. Elute with buffer $1^{(300)}$ supplemented with 1 mM EDTA and 2 mM EGTA
2. Hiload Superdex 16/600 200 pg: Concentrate eluate from calmodulin resin with Amicon Ultra centrifugal filter unit with MWCO 100 kDa to roughly 3 mL prior to loading on column. Equilibrate and elute column with buffer $2^{(200)}$
3. Concentration: Concentrate with Amicon Ultra centrifugal filter unit with MWCO 100 kDa to roughly 0.5 mg/mL as determined by Bradford
4. Freeze 100 µL aliquots

3.6.2 *Cdt1/Mcm2–7*

Expression: Yeast (strain yAM33) with alpha factor arrest.

Lysis buffer: Buffer $3^{(100)}$ with protease inhibitors.

1. Calmodulin: Supplement lysate in buffer $3^{(100)}$ with 2 mM $CaCl_2$. Wash with buffer $3^{(100)}$ with 2 mM $CaCl_2$. Elute with buffer $3^{(300)}$ supplemented with 1 mM EDTA and 2 mM EGTA
2. Hiload Superdex 16/600 200 pg: Concentrate eluate from calmodulin resin with Amicon Ultra centrifugal filter unit with MWCO 100 kDa to roughly 3 mL prior to loading on column. Equilibrate and elute column with buffer $3^{(100)}$.
3. Concentration: Concentrate with Amicon Ultra centrifugal filter unit with MWCO 100 kDa to roughly 1.7 mg/mL as determined by Bradford
4. Freeze 100 µL aliquots

3.6.3 Cdc6

Expression: *E. coli* (BL21, plasmid pAM1) in 1 L. Induce expression with 1 mM IPTG and express for 5 h at 18 °C.

Lysis buffer: Buffer $4^{(150)}$ with protease inhibitors.

1. GST: Wash column with buffer $4^{(150)}$. Elute 1 column volume with buffer $5^{(75)}$ after incubation with PreScission protease. Elute further with 3 × 1 mL of buffer $5^{(75)}$. Dilute eluate with 1 mL of buffer $5^{(0)}$
2. Hydroxyapatite: Column equilibrated in buffer $5^{(75)}$. First wash with 2 mL buffer $5^{(75)}$, then an additional wash with 5 mL of buffer $6^{(150)}$. Elute with buffer $6^{(400)}$
3. Dialysis: Dialysis against Buffer $7^{(100)}$, aim for 0.2 mg/mL
4. Freeze 100 µL aliquots

3.6.4 DDK

Expression: Yeast (strain ySDK8).

Lysis buffer: Buffer $8^{(500)}$ with protease inhibitors.

Note: Extra lambda phosphatase step during calmodulin affinity purification, see below.

1. Calmodulin: Supplement lysate in buffer $8^{(400)}$ with 2 mM $CaCl_2$. Wash with buffer $8^{(400)}$ with 2 mM $CaCl_2$. Resuspend in 1 mL buffer $8^{(400)}$ after wash in step 11 (see calmodulin affinity purification protocol). Add 100 µL (from a 2 mg/mL stock) lambda PPase together with 1 mM $MnCl_2$ and incubate for 1 h rotating in cold room. Discard flow through and wash with 10 mL buffer $8^{(400)}$. Continue with elution (step 13 of calmodulin affinity purification protocol) in buffer $8^{(400)}$ supplemented with 2 mM EDTA and 2 mM EGTA
2. Superdex 200 Increase 10/300 GL: Concentrate eluate from calmodulin resin with Amicon Ultra centrifugal filter unit with MWCO 30 kDa to roughly 0.5 mL prior to loading on column. Equilibrate and elute column with buffer $9^{(200)}$.
3. Concentration: Concentrate with Amicon Ultra centrifugal filter unit with MWCO 30 kDa to roughly 0.3 mg/mL as determined by Bradford
4. Freeze 100 µL aliquots

3.6.5 Sld3/Sld7

Expression: Yeast (strain yTD6) with alpha factor arrest.

Lysis buffer: Buffer $10^{(500)}$ with protease inhibitors.

1. IgG: Equilibrate IgG resin with Buffer $10^{(500)}$. Incubate IgG resin with lysate, wash with Buffer $10^{(500)}$, and elute Sld3/Sld7 by TEV protease cleavage as described above
2. Superdex 200 Increase 10/300 GL: Concentrate eluate from IgG resin with Amicon Ultra centrifugal filter unit with MWCO 30 kDa to roughly 0.5 mL prior to loading on column. Equilibrate and elute column with buffer $10^{(500)}$
3. Concentration: Concentrate with Amicon Ultra centrifugal filter unit with MWCO 30 kDa to roughly 0.2 mg/mL as determined by Bradford
4. Freeze 100 μL aliquots

3.6.6 *Sld2*

Expression: *E. coli* (BL21 codonplus(DE3)-RIL, plasmid pGC441) in 8 L. Induce with 0.2 mM IPTG and express o/n at 16 °C.

Lysis buffer: Buffer $11^{(500)}$ with protease inhibitors.

1. GST: Wash column with buffer $11^{(500)}$. Elute 1 column volume with buffer $11^{(500)}$ after incubation with PreScission protease. Elute further with 3 × 1 mL of buffer $11^{(500)}$
2. SP FF: Dilute eluate from glutathione resin 1:1 with buffer $11^{(0)}$. Load on an SP FF column equilibrated in buffer $11^{(250)}$ and wash with 20 CV buffer $11^{(250)}$. Elute by step elution, 10 CV of buffer $11^{(700)}$
3. Dialysis: Dialysis against buffer 12 with either 350 mM NaCl or 700 mM KOAc (see below regarding the effect that salt has on the replication assay)
4. Freeze 100 μL aliquots with as high protein concentration as possible

3.6.7 *Cdc45*

Expression: Yeast (strain yJY13) with alpha factor arrest.

Lysis buffer: Buffer $13^{(500)}$ with protease inhibitors.

1. FLAG: Equilibrate and wash column with buffer $13^{(500)}$. Elute in buffer $13^{(300)}$
2. Superdex 200 Increase 10/300 GL: Concentrate the eluate from the M2-resin with Amicon Ultra centrifugal filter unit with MWCO 10 kDa to less than 0.5 mL prior to loading on column. Equilibrate and elute column with buffer $13^{(300)}$
3. Concentration: Concentrate with Amicon Ultra centrifugal filter unit with MWCO 10 kDa to roughly 0.5 mg/mL as determined by Bradford
4. Freeze 100 μL aliquots

3.6.8 GINS

Expression: *E. coli* (Rosetta, plasmid pFJD12) in 2L. Induce with 1 mM IPTG and express for 2.5 h at 37 °C.

Lysis buffer: Buffer $14^{(200)}$ with protease inhibitors.

1. IMAC: Equilibrate and wash column with buffer $14^{(200)}$ with 10 mM imidazole. Elute in buffer $14^{(200)}$ with 200 mM imidazole
2. Dialysis: Dialys against Buffer $14^{(50)}$
3. MonoQ: Load on MonoQ column equilibrated in buffer $14^{(50)}$ and wash with 20 column volumes of buffer $14^{(50)}$. Elute with gradient elution over 30 column volumes from buffer $14^{(50)}$ to buffer $14^{(500)}$
4. Superdex 200 Increase 10/300 GL: Concentrate the eluate from the IMAC-resin with Amicon Ultra centrifugal filter unit with MWCO 30 kDa to less than 0.5 mL prior to loading on column. Equilibrate and elute column with buffer $15^{(200)}$
5. Concentration: Concentrate with Amicon Ultra centrifugal filter unit with MWCO 30 kDa to roughly 3–5 mg/mL as determined by Bradford
6. Freeze 100 μL aliquots

3.6.9 Dpb11

Expression: Yeast (strain yVP8) with alpha factor arrest.

Lysis buffer: Buffer $10^{(500)}$ with protease inhibitors.

1. FLAG: Equilibrate and wash column with buffer $10^{(500)}$, followed by a second wash and elution in buffer $10^{(300)}$
2. Superdex 200 Increase 10/300 GL: Concentrate eluate from M2-resin with Amicon Ultra centrifugal filter unit with MWCO 10 kDa to less than 0.5 mL prior to loading on column. Equilibrate and elute column with buffer $10^{(300)}$
3. Concentration: Concentrate with Amicon Ultra centrifugal filter unit with MWCO 10 kDa to roughly 0.4 mg/mL as determined by Bradford
4. Freeze 100 μL aliquots

3.6.10 S-CDK

Expression: Yeast (strain yAE88) with addition of 5 μg/mL nocodazole together with galactose.

Lysis buffer: Buffer $16^{(300)}$ with protease inhibitors.

1. Calmodulin: Supplement lysate in Buffer $16^{(300)}$ with 2 mM $CaCl_2$. Wash with buffer $16^{(300)}$ with 2 mM $CaCl_2$. Elute with TEV protease in with buffer $16^{(300)}$ (see elute with TEV protease under affinity chromatography above). Remove TEV protease by introducing an IMAC step using buffer $16^{(300)}$ with 10 mM imidazole (see TEV protease removal by IMAC under affinity chromatography above)

2. Superdex 200 Increase 10/300 GL: Concentrate eluate from IMAC with Amicon Ultra centrifugal filter unit with MWCO 10 kDa to less than 0.5 mL prior to loading on column. Equilibrate and elute column with buffer $16^{(300)}$
3. Concentration: Concentrate with Amicon Ultra centrifugal filter unit with MWCO 10 kDa to roughly 0.4 mg/mL as determined by Bradford
4. Freeze 100 μL aliquots

3.6.11 Pol ε

Expression: Yeast (strain AJ2) with alpha factor arrest.

Lysis buffer: Buffer $17^{(100)}$ with protease inhibitors.

1. Calmodulin: Supplement lysate in Buffer $17^{(100)}$ with 2 mM $CaCl_2$. Wash with buffer $17^{(100)}$ with 2 mM $CaCl_2$. Elute with buffer $17^{(400)}$ supplemented with 2 mM EDTA and 2 mM EGTA. Immediately proceed to Heparin purification
2. Heparin: Load on HiTrap Heparin column equilibrated in buffer $17^{(200)}$ and wash with 20 column volumes buffer $17^{(450)}$. Elute with gradient elution over 30 column volumes from buffer $17^{(450)}$ to buffer $17^{(1000)}$
3. Superdex 200 Increase 10/300 GL: Concentrate eluate from HiTrap Heparin column with Amicon Ultra centrifugal filter unit with MWCO 100 kDa to less than 0.5 mL prior to loading on column. Equilibrate and elute column with buffer $17^{(500)}$
4. Concentration: Concentrate with Amicon Ultra centrifugal filter unit with MWCO 100 kDa to roughly 1 mg/mL as determined by Bradford
5. Freeze 100 μL aliquots

3.6.12 Pol α

Expression: Yeast (strain yAE95).

Lysis buffer: Buffer $18^{(400)}$ with protease inhibitors.

1. Calmodulin: Supplement lysate in Buffer $18^{(400)}$ with 2 mM $CaCl_2$. Wash with buffer $18^{(400)}$ with 2 mM $CaCl_2$. Elute with buffer $18^{(400)}$ supplemented with 2 mM EDTA and 2 mM EGTA. Immediately proceed to MonoQ purification
2. MonoQ: Dilute eluate from calmodulin with buffer $19^{(0)}$ to equivalent of 150 mM NaCl. Load on a 1 mL MonoQ column equilibrated in buffer $19^{(150)}$ and wash with 10 column volumes buffer $19^{(150)}$. Elute with gradient elution over 20 column volumes from buffer $19^{(150)}$ to buffer $19^{(1000)}$
3. Superdex 200 Increase 10/300 GL: Concentrate eluate from MonoQ column with Amicon Ultra centrifugal filter unit with MWCO 30 kDa to less than 0.5 mL prior to loading on column. Equilibrate and elute column with buffer $19^{(150)}$

4. Concentration: Concentrate with Amicon Ultra centrifugal filter unit with MWCO 30 kDa to roughly 1 mg/mL as determined by Bradford
5. Freeze 100 μL aliquots

3.6.13 RPA

Expression: Yeast (strain yAE31) with alpha factor arrest.

Lysis buffer: Buffer $20^{(500)}$ with protease inhibitors.

1. Calmodulin: Supplement lysate in Buffer $20^{(500)}$ with 2 mM $CaCl_2$. Wash with buffer $20^{(500)}$ with 2 mM $CaCl_2$. Elute with buffer $20^{(500)}$ supplemented with 2 mM EDTA and 2 mM EGTA
2. Heparin: Dilute eluate from calmodulin resin with buffer $21^{(0)}$ to equivalent of 250 mM NaCl. Load on HiTrap Heparin column equilibrated in buffer $21^{(250)}$ and wash with 20 column volumes buffer $21^{(250)}$. Elute with gradient elution over 20 column volumes from buffer $21^{(250)}$ to buffer $21^{(1000)}$
3. Superdex 200 Increase 10/300 GL: Concentrate eluate from HiTrap Heparin column with Amicon Ultra centrifugal filter unit with MWCO 10 kDa to less than 0.5 mL prior to loading on column. Equilibrate and elute column with buffer $21^{(150)}$
4. Concentration: Concentrate with Amicon Ultra centrifugal filter unit with MWCO 10 kDa to roughly 1.5 mg/mL as determined by Bradford
5. Freeze 100 μL aliquots

3.6.14 Mcm10

Expression: *E. coli* (BL21 codonplus(DE3)-RIL, plasmid pMD132) in 2 L. Induce with 1 mM IPTG and express for 3 h at 25 °C.

Lysis buffer: Buffer $22^{(500)}$ with protease inhibitors.

1. FLAG: Equilibrate and wash column with buffer $22^{(500)}$, followed by a second wash and elution in buffer $22^{(300)}$
2. IMAC: Load eluate from M2-resin on a prepared IMAC resin equilibrated in buffer $22^{(500)}$ supplemented with 10 mM imidazole. Wash with buffer $22^{(500)}$ with 10 mM imidazole. Elute in buffer $22^{(500)}$ with 200 mM imidazole
3. Superdex 200 Increase 10/300 GL: Concentrate eluate from IMAC resin with Amicon Ultra centrifugal filter unit with MWCO 30 kDa to less than 0.5 mL prior to loading on column. Equilibrate and elute column with buffer $23^{(200)}$
4. Concentration: Concentrate with Amicon Ultra centrifugal filter unit with MWCO 30 kDa to roughly 0.2 mg/mL as determined by Bradford
5. Freeze 100 μL aliquots

3.6.15 *Ctf4*

Expression: Yeast (strain yAE40) with alpha factor arrest.

Lysis buffer: Buffer $20^{(200)}$ with protease inhibitors.

1. Calmodulin: Supplement lysate in Buffer $20^{(200)}$ with 2mM $CaCl_2$. Wash with buffer $20^{(200)}$ with 2mM $CaCl_2$. Elute with buffer $20^{(200)}$ supplemented with 2mM EDTA and 2mM EGTA
2. MonoQ: Load on MonoQ column equilibrated in buffer $20^{(150)}$ and wash with 20 column volumes buffer $20^{(150)}$. Elute with gradient elution over 30 column volumes from buffer $20^{(150)}$ to buffer $20^{(1000)}$
3. Superdex 200 Increase 10/300 GL: Concentrate eluate from MonoQ column with Amicon Ultra centrifugal filter unit with MWCO 30kDa to less than 0.5mL prior to loading on column. Equilibrate and elute column with buffer $21^{(150)}$
4. Concentration: Concentrate with Amicon Ultra centrifugal filter unit with MWCO 30kDa to roughly 0.3mg/mL as determined by Bradford
5. Freeze 100μL aliquots

3.6.16 *TopoI*

Expression: Yeast (strain yAE42).

Lysis buffer: Buffer $18^{(300)}$ with protease inhibitors.

1. Calmodulin: Supplement lysate in Buffer $18^{(300)}$ with 2mM $CaCl_2$. Wash with buffer $18^{(300)}$ with 2mM $CaCl_2$. Elute with buffer $18^{(300)}$ supplemented with 2mM EDTA and 2mM EGTA
2. Superdex 200 Increase 10/300 GL: Concentrate eluate from calmodulin resin with Amicon Ultra centrifugal filter unit with MWCO 10kDa to less than 0.5mL prior to loading on column. Equilibrate and elute column with buffer $18^{(150)}$
3. Concentration: Concentrate with Amicon Ultra centrifugal filter unit with MWCO 10kDa to roughly 0.3mg/mL as determined by Bradford
4. Freeze 100μL aliquots

3.6.17 *TopoII*

Expression: Yeast (strain yAE46) with alpha factor arrest.

Lysis buffer: Buffer $18^{(300)}$ with protease inhibitors.

1. Calmodulin: Supplement lysate in Buffer $18^{(300)}$ with 2mM $CaCl_2$. Wash with buffer $18^{(300)}$ with 2mM $CaCl_2$. Elute with buffer $18^{(300)}$ supplemented with 2mM EDTA and 2mM EGTA
2. MonoQ: Dilute eluate from calmodulin with buffer $18^{(0)}$ to equivalent of 150mM NaCl. Load on MonoQ column equilibrated in buffer $18^{(150)}$ and wash with 20 column volumes buffer $18^{(150)}$. Elute with gradient elution over 25 column volumes from buffer $18^{(150)}$ to buffer $18^{(800)}$

3. Superdex 200 Increase 10/300 GL: Concentrate eluate from MonoQ with Amicon Ultra centrifugal filter unit with MWCO 30 kDa to less than 0.5 mL prior to loading on column. Equilibrate and elute column with buffer $18^{(150)}$
4. Concentration: Concentrate with Amicon Ultra centrifugal filter unit with MWCO 30 kDa to roughly 0.3 mg/mL as determined by Bradford
5. Freeze 100 μL aliquots

3.6.18 Mrc1

Expression: Yeast (strain yAE71).

Lysis buffer: Buffer $19^{(400)}$ with protease inhibitors.

1. FLAG: Equilibrate and wash column in Buffer $19^{(400)}$, followed by elution in Buffer $19^{(400)}$ with 3xFLAG peptide
2. MonoQ: Dilute eluate from calmodulin with buffer $19^{(0)}$ to equivalent of 150 mM NaCl. Load on MonoQ column equilibrated in buffer $19^{(150)}$ and wash with 20 column volumes buffer $19^{(150)}$. Elute with gradient elution over 10 column volumes from buffer $19^{(150)}$ to buffer $19^{(1000)}$
3. Dialysis: Dialysis against Buffer $24^{(150)}$, aim for 0.2 mg/mL
4. Freeze 100 μL aliquots

3.6.19 Csm3/Tof1

Expression: Yeast (strain yAE48).

Lysis buffer: Buffer $18^{(200)}$ with protease inhibitors.

1. Calmodulin: Supplement lysate in Buffer $18^{(200)}$ with 2 mM $CaCl_2$. Wash with buffer $18^{(200)}$ with 2 mM $CaCl_2$. Elute with TEV protease in buffer $18^{(200)}$ (see elute with TEV protease under affinity chromatography above)
2. MonoQ: Dilute eluate from calmodulin with buffer $19^{(0)}$ to equivalent of 150 mM NaCl. Load on MonoQ column equilibrated in buffer $19^{(150)}$ and wash with 20 column volumes buffer $19^{(150)}$. Elute with gradient elution over 30 column volumes from buffer $19^{(150)}$ to buffer $19^{(1000)}$
3. Superdex 200 Increase 10/300 GL: Concentrate eluate from MonoQ with Amicon Ultra centrifugal filter unit with MWCO 30 kDa to less than 0.5 mL prior to loading on column. Equilibrate and elute column with buffer $19^{(150)}$
4. Concentration: Concentrate with Amicon Ultra centrifugal filter unit with MWCO 30 kDa to at least 0.4 mg/mL as determined by Bradford
5. Freeze 100 μL aliquots

3.6.20 *RFC*

Expression: Yeast (strain yAE41).

Lysis buffer: Buffer $25^{(400)}$ with protease inhibitors.

1. Calmodulin: Supplement lysate in Buffer $25^{(400)}$ with 2 mM $CaCl_2$. Wash with buffer $25^{(400)}$ with 2 mM $CaCl_2$. Elute with buffer $25^{(400)}$ supplemented with 2 mM EDTA and 2 mM EGTA
2. SP FF: Dilute eluate from calmodulin with buffer $25^{(0)}$ to equivalent of 200 mM NaCl. Load on HiTrap SP FF column equilibrated in buffer $25^{(200)}$ and wash with 10 column volumes buffer $25^{(200)}$. Elute with gradient elution over 30 column volumes from buffer $25^{(200)}$ to buffer $25^{(1000)}$
3. Superdex 200 Increase 10/300 GL: Concentrate eluate from MonoQ with Amicon Ultra centrifugal filter unit with MWCO 30 kDa to less than 0.5 mL prior to loading on column. Equilibrate and elute column with buffer $25^{(150)}$
4. Concentration: concentrate with Amicon Ultra centrifugal filter unit with MWCO 30 kDa to about 0.8–1.0 mg/mL as determined by Bradford
5. Freeze 100 μL aliquots

3.6.21 *PCNA*

Expression: *E. coli* (BL21 codonplus(DE3)-RIL, plasmid pET28a-PCNA) in 0.75 L. Induce with 0.5 mM IPTG and express for 2.5 h at 37 °C.

Lysis buffer: Buffer 26 with protease inhibitors.

1. Salt precipitation: Add ammonium sulfate to 150 mM and polymin P to 0.4% to the cell lysate and stir for 10 min on ice. Centrifugate at 48000 × *g* at 4 °C for 20 min. Collect the supernatant and add 0.23 g/mL solid ammonium sulphate slowly while stirring on ice. Stir for 10 min and then centrifugate, 48,000 × *g* at 4 °C for 20 min. Slowly add 0.24 g/mL solid ammonium sulfate to the supernatant while stirring on ice. Stir for 10 min on ice and centrifugate, 48,000 × *g* at 4 °C for 20 min. Resuspended precipitated material in 3 mL of Buffer $21^{(100)}$
2. Heparin/SP FF/DEAE: Adjust sample conductivity to the equivalent of buffer $21^{(150)}$ by dilution in buffer $21^{(0)}$. Connect a HiTrap SP FF, a HiTrap Heparin and a HiTrap DEAE column in tandem in that order (Protein sample flows from SP FF, through Heparin to DEAE last) equilibrated in $21^{(150)}$. Load sample on columns in buffer $21^{(150)}$. Remove the HiTrap Heparin and HiTrap SP FF columns. Wash the HiTrap DEAE column with 5 column volumes buffer $21^{(150)}$. Elute with gradient elution over 25 column volumes from buffer $21^{(150)}$ to buffer $21^{(600)}$

3. MonoQ: Dilute eluate from DEAE column 1:1 with buffer $21^{(0)}$. Load on MonoQ column equilibrated in buffer $21^{(150)}$ and wash with 10 column volumes buffer $21^{(150)}$. Elute with gradient elution over 30 column volumes from buffer $21^{(150)}$ to buffer $21^{(600)}$
4. Superdex 200 Increase 10/300 GL: Concentrate eluate from MonoQ with Amicon Ultra centrifugal filter unit with MWCO 3 kDa to less than 0.5 mL prior to loading on column. Equilibrate and elute column with buffer $21^{(150)}$
5. Concentration (If necessary): Concentrate with Amicon Ultra centrifugal filter unit with MWCO 3 kDa to about 0.65 mg/mL as determined by Bradford
6. Freeze 100 μL aliquots

3.6.22 Pol δ

Expression: Yeast (strain yAE34).

Lysis buffer: Buffer $18^{(400)}$ with protease inhibitors.

1. Calmodulin: Supplement lysate Buffer $18^{(400)}$ with 2 mM $CaCl_2$. Wash with buffer $18^{(400)}$ with 2 mM $CaCl_2$. Elute with buffer $18^{(400)}$ supplemented with 2 mM EDTA and 2 mM EGTA. Immediately proceed to Heparin purification
2. Heparin: Dilute eluate from calmodulin with buffer $19^{(0)}$ to equivalent of 200 mM NaCl. Load on HiTrap Heparin column equilibrated in buffer $19^{(200)}$ and wash with 10 column volumes buffer $19^{(200)}$. Elute with gradient elution over 15 column volumes from buffer $19^{(200)}$ to buffer $19^{(1000)}$
3. Superdex 200 Increase 10/300 GL: Concentrate eluate from Heparin column with Amicon Ultra centrifugal filter unit with MWCO 30 kDa to less than 0.5 mL prior to loading on column. Equilibrate and elute column with buffer $18^{(150)}$
4. Concentration: Concentrate with Amicon Ultra centrifugal filter unit with MWCO 30 kDa to about 0.7–0.9 mg/mL as determined by Bradford
5. Freeze 100 μL aliquots

3.6.23 Ligase 1

Expression: Yeast (strain yJY33).

Lysis buffer: Buffer $27^{(400)}$ with protease inhibitors.

1. FLAG: Equilibrate and wash column in Buffer $27^{(400)}$, followed by elution in Buffer $27^{(400)}$ with 3xFLAG peptide

2. MonoQ: Dilute eluate from calmodulin with buffer $27^{(0)}$ to equivalent of 100 mM NaCl. Load on MonoQ column equilibrated in buffer $27^{(100)}$ and wash with 10 column volumes buffer $27^{(100)}$. Elute with gradient elution over 20 column volumes from buffer $27^{(100)}$ to buffer $27^{(1000)}$
3. Dialysis: Dialysis against Buffer $27^{(100)}$, aim for 0.4 mg/mL
4. Freeze 100 μL aliquots

3.6.24 Fen1

Expression: Yeast (strain yJY31).

Lysis buffer: Buffer $28^{(150)}$ with protease inhibitors.

1. FLAG: Equilibrate and wash column in Buffer $28^{(150)}$, followed by elution in Buffer $28^{(150)}$ with 3xFLAG peptide
2. Heparin: Load on HiTrap Heparin column equilibrated in buffer $28^{(150)}$ and wash with 10 column volumes buffer $28^{(150)}$. Elute with gradient elution over 20 column volumes from buffer $28^{(150)}$ to buffer $28^{(1000)}$
3. Dialysis: Dialysis against Buffer $28^{(100)}$, aim for 0.3 mg/mL
4. Freeze 100 μL aliquots

4. Optimizations of the in vitro replication assay

4.1 Setup of replication assay

Because DDK and S-CDK regulate the initiation of DNA replication, in vitro replication reactions must be assembled in sequential steps following the order of the reaction in vivo. The overall scheme can vary but in general are the following four steps required:

1. Loading of Mcm2–7
2. DDK phosphorylation
3. CMG formation
4. Origin firing

4.2 Tuning of the assay

The assay can be tuned in many ways. For example, if absolute coordination of origin firing is not needed, steps 3 and 4 can be combined, or if Mcm2–7 loading need to be kept strict, CDK alone can be added after the DDK step to inhibit further Mcm loading during CMG formation (Taylor et al., 2018). Timing of the individual steps also affect the outcome of the assay. If reduced origin firing is preferred, the time for Mcm2–7 loading or DDK phosphorylation can be shortened (Fig. 1). Individual concentrations of loading

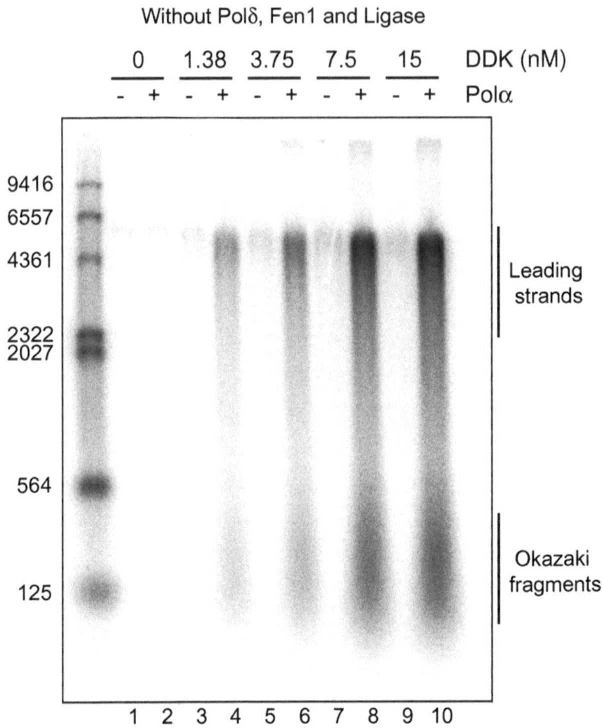

Fig. 1 DDK titration. Replication assay using 5.8 kb ARS1 template. This assay was done with 5 min Mcm2–7 loading, 10 min DDK step, 10 min CMG formation and 20 min replication. The assay was performed without Polδ, Fen1 and Ligase.

reaction proteins or DDK can also be lowered. In general, our experience is that less efficient loading of Mcm2–7 give an increased specificity for origins. Simply by having lower concentrations of the loading proteins and shorter incubation times, good specificity for firing at origins can be obtained even on an ARS1 origin (Fig. 2). Other ways of achieving origin specificity include adding KCl during Mcm loading to reduce promiscuous loading (Taylor et al., 2018).

Depending on what is studied different templates can be used. We routinely use a circular plasmid containing the ARS1 origin with a size of either 5.8 or 10.6 kb (Yeeles et al., 2015, 2017). The template can also be linearized to be able to study specific fork directions (Taylor et al., 2018).

In this protocol we describe a typical in vitro replication assay as used in our lab.

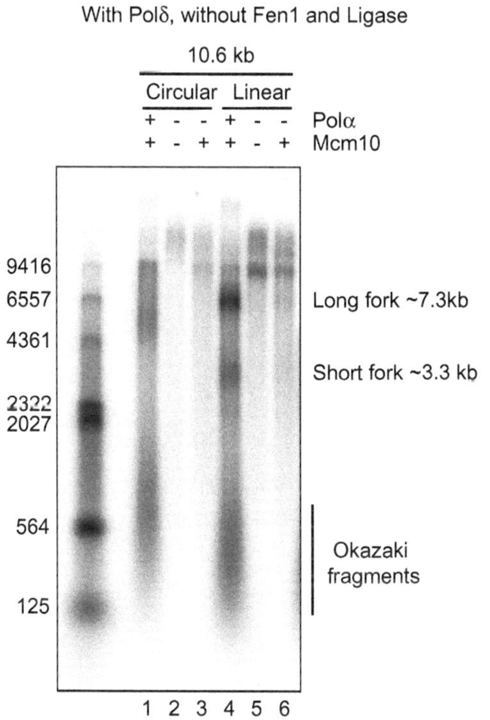

Fig. 2 Origin specificity at lower firing. Replication assay performed as in Fig. 1 with 5 nM DDK and using 10.6 kb ARS1 template. *ScaI* was used to linearize the template meaning that one fork will travel ~7.3 kb to the end and the other 3.3 kb. The assay was performed without Fen1 and Ligase.

4.3 DNA replication assay

4.3.1.1 Equipment

Thermomixer
G-50 columns
Agarose gel equipment
Gel dryer
Phosphoimager or photofilm

4.3.1.2 Chemicals

Replication proteins
Origin-containing plasmid
HEPES
KOH

K-Glutamate
$Mg(OAc)_2$
DTT
NP-40-S
dNTPs
NTPs
[α-^{32}P]dCTP
EDTA
NaOH
Agarose

4.3.1.3 Protocol

All incubations are conducted in a Thermomixer at 30 °C and 1250 rpm. Gently mix and quickly spin down before every incubation. The loading step is performed in 1/2 volume reactions in loading buffer and can be performed in a master mix. The final replication reaction volume is adjusted with replication buffer. A typical final reaction volume is 10–20 μL.

Buffers

1. Prepare a loading buffer adjusted for a final loading reaction concentration of 25 mM HEPES-KOH pH 7.6, 100 mM K-glutamate, 10 mM $Mg(OAc)_2$, 0.01% NP-40-S and 1 mM DTT. The exact composition will differ depending on the stock concentration of proteins and DNA in step 3
2. Prepare a replication buffer adjusted for a final replication reaction concentration of 25 mM HEPES-KOH pH 7.6, 225 mM K-glutamate, 10 mM $Mg(OAc)_2$, 0.01% NP-40-S and 1 mM DTT. The increase in K-Glutamate in the replication step is to minimize the strand displacement activity of Pol δ. If the reactions are performed without Pol δ, the K-Glutamate level can remain 100 mM through the replication step as well. The exact composition of the buffer will differ depending on the stock concentration of proteins added in step 10

Loading of Mcm2–7

3. Make a loading reaction master mix with loading buffer from step 1 containing 5 mM ATP, 10 nM ORC, 15 nM Cdc6, 20 nM Cdt1–Mcm2–7 and 4 nM DNA. Incubate for 20 min

DDK phosphorylation

4. Add DDK to a final concentration of 10 nM to the loading reaction master mix. Incubate for a further 10 min while preparing step 5

CMG formation

5. Prepare a dNTP mix giving a final reaction concentration of 80 μM each. The dNTP mix is added during the CMG step to minimize the nuclease activity from Pol ε. If the CMG formation step and origin firing step is combined, add the dNTPs to the nucleotide/buffer mix in step 8
6. Prepare a CMG formation protein master mix that gives a final replication reaction concentration of 20 nM GINS, 20 nM Cdc45, 10 nM Pol ε, 5 nM CDK, 5 nM Dpb11, 10 nM Sld3/7 and 5 nM Sld2. This mix should ideally be ready exactly when the step 3 incubation is done to minimize the time the proteins are in the protein master mix. If the CMG formation step and origin firing step is combined, add these proteins to the replication protein master mix in step 9 and continue with step 8
7. Transfer the reaction master mix to ice and add the dNTP mix and the CMG formation protein master mix. Incubate 10 min. Start preparing step 8

Origin firing

8. Prepare a nucleotide/buffer mix with replication buffer giving final replication reaction concentrations of 200 μM each of CTP, GTP and UTP and 33 nM [α-^{32}P]dCTP. This mix should also give a final replication reaction concentration of 225 mM K-Glutamate. This mix can be aliquoted in the individual reaction tubes
9. Make a replication protein master mix giving final replication reaction concentrations of 10 nM RFC, 25 nM Ligase 1, 25 nM Fen1, 25 nM PCNA, 0.5 nM Pol δ, 30 nM RPA, 5 nM Ctf4, 5 nM TopoI, 5 nM TopoII, 7.5 nM Csm3-Tof1, 2.5 nM Mcm10, 7.5 nM Mrc1 and 40 nM Pol α. This mix should ideally be ready exactly when the step 7 incubation is done to minimize the time the proteins are in the protein master mix
10. Transfer the reactions to ice. Add the replication protein master mix. Mix and spin down. Aliquot to the prepared tubes containing the replication buffer. Move to 30 °C with shaking and incubate for 20 min. Stop the reaction by addition of 50 mM EDTA

Analysis

11. Remove free [α-^{32}P]dCTP by passing the reactions over a G-50 column
12. Separate DNA products on a 0.9% alkaline agarose gel in 30 mM NaOH, 2 mM EDTA

4.4 Troubleshooting

4.4.1 Salt

Both the salt concentration and type of salt will influence the replication reaction (Fig. 3), but if the reaction contains PCNA/RFC and Mrc1/Csm3/Tof1, both NaCl and Acetate are tolerated at moderate levels. We routinely use glutamate as added salt in our reactions. Acetate is in some cases used to substitute glutamate, but our experience is that it has a negative effect on DNA replication at concentration close to what is used in those assays (Fig. 3). It is important to calculate total salt contributions from all included proteins and perform an experiment titrating in salt to know the salt sensitivity of your assay. One protein that usually contributes with high levels of salt is Sld2. If the chloride concentration of the reaction is higher than wanted, then Sld2 can be dialyzed with KOAc instead of NaCl.

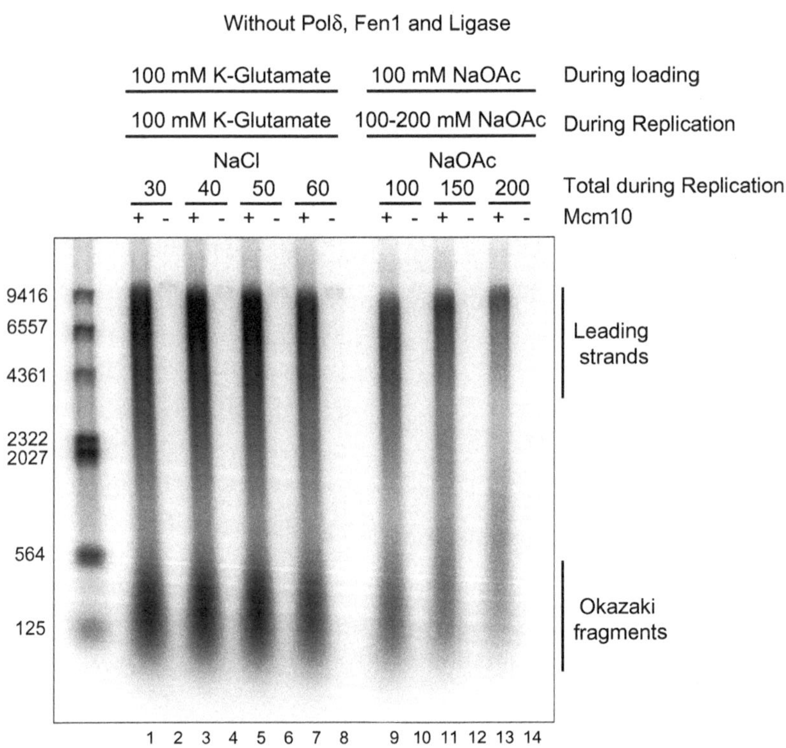

Fig. 3 The effect of chloride and acetate on the replication assay. Replication assay as described in the protocol using the 10.6 kb template. The assay was performed without Polδ, Fen1 and Ligase.

4.4.2 Challenging proteins

There are a few proteins known to cause problems due to variations in activity between preparations. These proteins include Sld2, Sld3/Sld7, Mcm10 and Pol α. These are also known to be less stable while in a protein master mix, hence it is best to add these proteins as late as possible. If there is a problem with the reaction, a titration of these proteins might give valuable insight into what is wrong. Problems caused by Sld2, Sld3/Sld7 or Mcm10 will give an overall lower signal, where Mcm10 has been seen to cause problems both at low and high levels. Problems with Pol α will usually not cause to much problems with the leading strand but rather give fewer and more diffuse Okazaki fragments (Fig. 4).

Another protein where we have experienced variable activity is Pol δ. In many cases Pol δ activity has been assessed by examining Okazaki fragment size. Our experience is that Okazaki fragment ligation is a better way of optimizing the Pol δ concentration in the assay. As seen in Fig. 5 the difference in Okazaki fragment length at no Pol δ, low and high Pol δ levels is quite small, however higher levels seem to inhibit Okazaki fragment ligation.

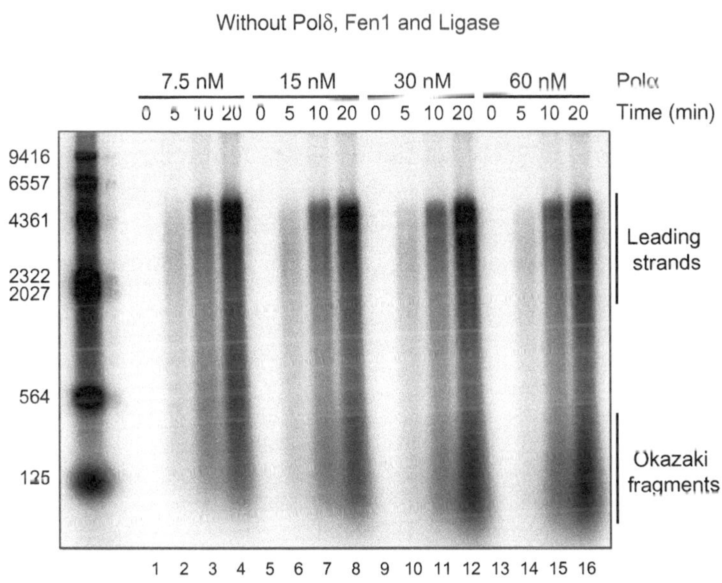

Fig. 4 Titration of Pol α. Replication assay as described in the protocol using the 5.8 kb template. The assay was performed without Polδ, Fen1 and Ligase.

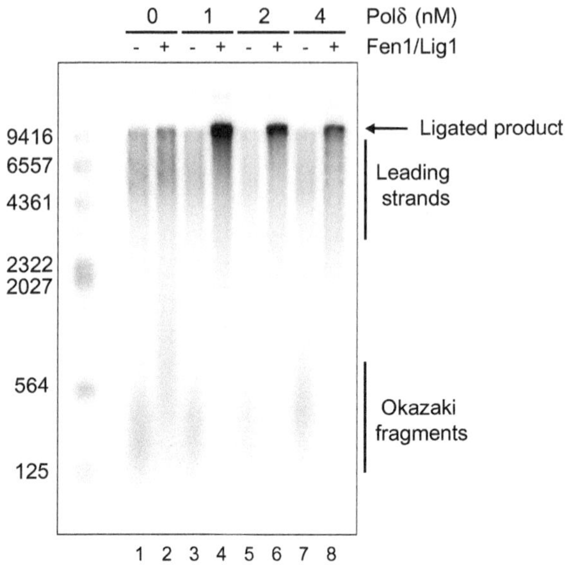

Fig. 5 Titration of Pol δ. Replication assay as described in the protocol using the 10.6 kb template.

5. Summary

The reconstitution of eukaryotic DNA replication has already led to many mechanistic discoveries, both regarding the replisome and its function under unperturbed DNA replication as well as in specific situations such as replication initiation and termination, replication of telomeres and replication of damaged templates. Using this reconstituted origin dependent DNA replication system, we can continue to discover mechanisms at the replisome as well as connected processes.

Acknowledgments

We thank A. Alidoust and N. Patel in the Structural Biology Laboratory for yeast protein expression. Berta Canal and Masashi Minamino, and all current and past members of the Diffley group for stimulating discussions and advice. This work was supported by the Francis Crick Institute, which receives its core funding from Cancer Research UK (FC001066), the UK Medical Research Council (FC001066), and the Wellcome Trust (FC001066). This work was also funded by Stiftelsen Bengt Lundqvists Minne (V.P.), Vetenskapsrådet (2018-00373 to V.P.), Wenner-Gren Foundations (E.J.) along with a Wellcome Trust Senior Investigator Awards (106252/Z/14/Z and 219527/Z/19/Z to J.F.X.D.), and a European Research Council Advanced Grant (669424-CHROMOREP to J.F.X.D.). For the purpose of Open Access, the author has applied a CC BY public copyright license to any Author Accepted Manuscript version arising from this submission.

References

de Godoy, L. M., et al. (2008). Comprehensive mass-spectrometry based proteome quantification of haploid versus diploid yeast. *Nature*, *455*(7217), 1251–1254.

Deegan, T. D., et al. (2019). Pif1-family helicases support fork convergence during DNA replication termination in eukaryotes. *Molecular Cell*, *74*(2), 231–244, (e9).

Devbhandari, S., et al. (2017). Chromatin constrains the initiation and elongation of DNA replication. *Molecular Cell*, *65*(1), 131–141.

Ghaemmaghami, S., et al. (2003). Global analysis of protein expression in yeast. *Nature*, *425*(6959), 737–741.

Kurat, C. F., et al. (2017). Chromatin controls DNA replication origin selection, lagging-strand synthesis, and replication fork rates. *Molecular Cell*, *65*(1), 117–130.

Sharp, P. M., et al. (1987). The codon adaptation index—A measure of directional synonymous codon usage bias, and its potential applications. *Nucleic Acids Research*, *15*(3), 1281–1295.

Taylor, M. R. G., et al. (2018). The initial response of a eukaryotic replisome to DNA damage. *Molecular Cell*, *70*(6), 1067–1080, (e12).

Yeeles, J. T., et al. (2015). Regulated eukaryotic DNA replication origin firing with purified proteins. *Nature*, *519*(7544), 431–435.

Yeeles, J. T. P., et al. (2017). How the eukaryotic replisome achieves rapid and efficient DNA replication. *Molecular Cell*, *65*(1), 105–116.

CHAPTER TWO

Monitoring the replication of structured DNA through heritable epigenetic change

Guillaume Guilbaud and Julian E. Sale*

Division of Protein & Nucleic Acid Chemistry, MRC Laboratory of Molecular Biology, Cambridge, United Kingdom

*Corresponding author: e-mail address: jes@mrc-lmb.cam.ac.uk

Contents

Abstract

DNA can adopt non-B form structures that create significant blocks to DNA synthesis and seeking understanding of the mechanisms cells use to resolve such impediments continues to be a very active area of research. However, the ability to monitor the stalling of DNA synthesis at specific sites in the genome in living cells, of central importance to elucidating these mechanisms, poses a significant technical challenge. Replisome stalling is often transient with only a small fraction of events leading to detectable genetic changes, making traditional reporter assays insensitive to the stalling

Methods in Enzymology, Volume 661
ISSN 0076-6879
https://doi.org/10.1016/bs.mie.2021.08.009

event per se. On the other hand, the imprint stalling leaves on the epigenome can be exploited as a form of biological 'tape recorder' that captures episodes of fork stalling as heritable changes in histone modifications and in transcription. Here we describe a detailed protocol for monitoring DNA structure-dependent epigenetic instability of the *BU-1* locus in the avian cell line DT40, which has proved a sensitive tool for understanding the mechanisms by which structured DNA is replicated in a vertebrate system.

1. Introduction

One of the principal challenges to DNA replication is the behaviour of the DNA template itself. DNA can adopt a wide variety of non-B form secondary structures, promoted by local sequence, DNA melting, topology and the presence of RNA:DNA hybrids. These nucleic acid structures present a significant challenge to DNA replication with current evidence supporting their very frequent occurrence. Indeed, a significant fraction of the DNA secondary structures found in a cell may be generated as a consequence of DNA replication itself (discussed further in Lerner & Sale, 2019). How the replisome, and the large collection of accessory factors associated with it, responds to DNA secondary structures is still poorly understood, particularly in vertebrate cells. Progress in understanding the in vivo interactions between specific DNA sequences and factors involved in removing secondary structure requires methods to monitor how the replisome behaves at specific loci in defined genetic backgrounds. Monitoring the genetic consequences of structure-forming sequences has been successfully used in the budding yeast, *S. cerevisiae* (e.g., Kim et al., 2008; Lopes et al., 2011) and in the worm, *C. elegans* (Cheung, Schertzer, Rose, & Lansdorp, 2002; Kruisselbrink et al., 2008). However, genetic instability at secondary structure-forming sequences, even when precipitated by replication fork arrest, may only pick up a fraction of fork stalling events.

In vertebrate cells, few approaches exist to monitor site specific fork stalling with high sensitivity and spatial and temporal resolution. Two-dimensional gel electrophoresis of genomic restriction fragments (Brewer & Fangman, 1987) has been widely used to study the replication of secondary structure-forming DNA in yeast (Krasilnikova & Mirkin, 2004), but is more challenging in complex genomes unless performed on plasmids (Follonier, Oehler, Herrador, & Lopes, 2013) or following enrichment for replication intermediates (Mesner, Crawford, & Hamlin, 2006). Use of DNA combing coupled to fluorescence in-situ hybridisation

(FISH) can reveal replication fork pausing in the vicinity of structure-forming sequences, e.g., telomeres (Sfeir et al., 2009) or trinucleotide repeat expansions (Gerhardt et al., 2016) and, recently, DNA fibre labelling has been coupled to direct detection of G quadruplex structures in fibres (Jimeno et al., 2018; Stroik et al., 2020). However, as a method, DNA fibre analysis is very time consuming and lacks spatial resolution. A recent live-cell imaging approach that monitors the doubling in fluorescence that occurs upon the replication of genomically integrated *lacO* and *tetO* arrays bound by fluorescent proteins and separated by a sequence containing the structure-forming sequence under investigation (Dahan et al., 2018; Dovrat et al., 2018) shows significant promise. However, it has not yet been ported to higher eukaryotic systems.

The *BU-1* (also known as chb6 or B6.1) locus of the highly genetically tractable chicken cell line DT40 provides a solution to the problem of monitoring site specific replisome pausing at secondary structure forming sequences in vivo. It does this by harnessing the disruption of epigenetic information by a replication fork in which DNA unwinding and DNA synthesis have become uncoupled (Fig. 1). *BU-1* encodes a surface glycoprotein, the levels of which, as detected by flow cytometry, closely and rapidly mirror the mRNA levels present in the cell (Sarkies et al., 2012), allowing expression to act as a surrogate for transcription of the gene.

The six-exon *BU-1* locus (Fig. 1) contains a prominent G4 motif in the second intron, approximately 3.5 kb from the transcription start site (hereafter termed the +3.5 G4). Our previous work on this locus (Sarkies et al., 2012; Schiavone et al., 2014) is consistent with a model in which pausing of leading strand synthesis at this sequence leads to disruption of histone recycling and interruption of the normal mitotic inheritance of the previous epigenetic state of the locus (Sarkies, Reams, Simpson, & Sale, 2010). Thus, stalling of leading strand synthesis at the +3.5 G4 can lead to a change in the expression state of the gene, which manifests as stochastic generation of Bu-1 loss variants, in which expression of the protein on the surface of the cells is reduced relative to normal levels (Papadopoulou et al., 2015; Sarkies et al., 2012; Schiavone et al., 2014). Importantly, the loss of epigenetic information leading to a change in transcription and protein expression is, in this context, irreversible: cells that have switched from high expression to a lower state do not revert and are thus propagated through the population. The switching away from a high expression state therefore acts as a biological recorder of episodes of stalling of DNA synthesis accompanied by uncoupling of the DNA helicase. In practical terms this phenomenon

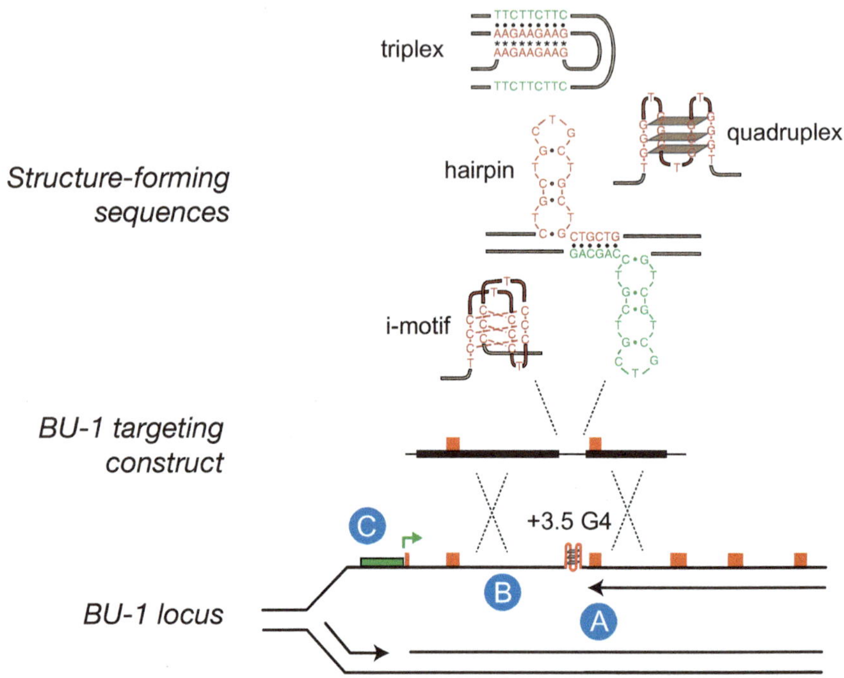

Fig. 1 Principle of the Bu-1 loss variant assay. The six exon *BU-1* locus lies equidistant between two strong replication origins and thus is likely to be replicated bidirectionally with ~50% forks coming from each direction (Schiavone et al., 2014). The leading strand polymerase of forks entering the locus from the 3′ end will encounter the +3.5 G4, or other structure-forming sequence introduced in its place (A). This will result in uncoupling of DNA unwinding by the replicative helicase from DNA synthesis (B) leading to interruption of histone recycling through the 5′ end of the locus (C) (Sarkies et al., 2012). If this involves the promoter, the consequent loss of histone marks results in switching of the expression state of the locus away from the normal Bu-1^{high} state to a lower expression state. The +3.5 G4 can be removed or alternative sequences can be installed in its place using a targeting construct (Papadopoulou, Guilbaud, Schiavone, & Sale, 2015; Schiavone et al., 2014). The first step of targeting introduces a selection cassette, which inactivates expression of the locus, providing a useful control (see Fig. 3A). Following Cre-mediated removal of the selection cassette, Bu-1^{high} expression is restored (Schiavone et al., 2014).

is monitored using the principle of the fluctuation analysis (Luria & Delbrück, 1943) in which multiple small populations of cells are expanded for a fixed number of divisions and the proportion of *BU-1* loss variants analysed. From this, it is possible to recover an approximation of a per division rate of stalling and uncoupling events, using an approach discussed further below (Schiavone et al., 2014).

We have previously used this method to explore the factors that contribute to the replication of DNA sequences that are capable of forming structured DNA, for instance the contribution of specialised translesion polymerases (Schiavone et al., 2014; Wickramasinghe, Arzouk, Frey, Maiter, & Sale, 2015), helicases (Lerner et al., 2020; Sarkies et al., 2012) and the repriming activity of the primase-polymerase PrimPol (Schiavone et al., 2016; Šviković et al., 2019). By replacing the naturally occurring +3.5 G4 with other structure forming sequences, we have begun to understand the mechanisms deployed at different forms of G4. For instance, REV1 appears to play its most significant role at G4s with longer loops and relatively low thermal stability (Schiavone et al., 2014), while PrimPol-dependent repriming appears more important at thermodynamically stable G4s (Schiavone et al., 2016). It has also been possible to explore the factors that drive sequences to form a replication impediment such as the presence of an RNA:DNA hybrid, which is essential for a short GAA repeat to interrupt processive DNA synthesis but is less important for the +3.5 G4 (Šviković et al., 2019). Recently, the *BU-1* locus has been used to test the capability of sequences from heterologous systems, such as *Arabidopsis*, to interrupt DNA synthesis (Baxter, Šviković, Sale, Dean, & Costa, 2021). Finally, the locus can be used to screen for and to test for the ability of small molecules to stabilise sequences in living cells, allowing the discovery of novel G4 stabilising ligands active over a wide range of concentrations without toxicity (Guilbaud et al., 2017).

Here we describe a detailed protocol for monitoring *BU-1* expression instability to follow the replication of the naturally-occurring +3.5 G4, or a defined sequence knocked-in to the locus in its place. We discuss the caveats and controls that are needed for obtaining reliable data and for the interpretation of the observed effects.

2. DT40 lines and their culture

2.1 DT40 culture reagents

Culture medium components are combined and sterile filtered before use:

- RPMI1640 with Glutamax® (Invitrogen)
- Fetal bovine serum (Gibco cat. no. 10270)—7% final
- Chicken serum (Sigma cat. no. C5405)—3% final
- Penicillin/streptomycin Gibco cat. no. 15140122—final diluted 1:100
- 2-Mercaptoethanol 50 μM

Flasks (Corning 75 cm^2 cat no. 353136)
Humidified CO_2 incubator

2.2 DT40 culture conditions

DT40 can be grown at 37 °C in a humidified incubator with 10% CO_2. It is also possible to grow the cells at temperatures between 39 and 41 °C (Saribasak & Arakawa, 2006). At 39.5 °C the wild type doubling time is reduced from ~11 h to ~9 h. The cells, which grow in suspension, should be optimally maintained at a density of between 0.2 and 1×10^6/mL to avoid any stress due to overgrowth.

2.3 Genetic controls

While the wild type *BU-1* locus of DT40 with its naturally occurring +3.5 G4 can be used alone to determine the effects of gene disruptions or exogenous ligands on the replication of this specific G4 motif, to realise the full power of the assay requires additional controls. Key among these controls is a cell line in which the +3.5 G4 motif has been deleted (Papadopoulou et al., 2015; Schiavone et al., 2014). In addition, it is also desirable to include lines in which a mutated G4 has been reintroduced or in which the G4 is inverted (i.e., the G-rich strand is on the Crick rather than Watson strand). The constructs for deleting and manipulating the +3.5 G4 are available on request. In addition, complementation of any genetic mutant studied is essential.

3. Protocol: Bu-1 loss variant fluctuation assay

3.1 Equipment

In addition to that listed for DT40 culture:

96-Well round bottom plates (Corning cat. no. 3799)
24-Well flat bottom plates (Corning cat. no. 3526)
Polystyrene flow cytometry tubes (Sarstedt cat. no. 55.476.305)
Sorting flow cytometer (e.g., Sony Biotechnology Synergy) equipped with a 561 nm laser (for detecting phycoerythrin, PE).
Analytical flow cytometer (e.g., Beckton-Dickinson LSRII) equipped with the following lasers: 407 nm (for Hoechst 33342), 488 nm (for fluorescein isothiocyanate, FITC) and 561 nm (for PE).

3.2 Reagents

In addition to those listed for DT40 culture:

Hoechst 33342 (Biotum cat. no. 40047) 10 mM stock solution in water.

- Anti-*BU-1* antibody (see table)

Specificity	Supplier/cat. no	Clone	Conjugation	Working dilution	Notes
Bu-1a (Does not react with Bu-1b)	Santa Cruz/70447	5K98	None or PE or FITC	1:100	Conjugated versions used in early studies no longer available
Bu-1a	ThermoFisher/ MA5-28702	AV20	PE	1:100	
Bu-1b (Does not react with Bu-1a)	Santa Cruz/70449	5K94	None or PE or FITC	1:100	Conjugated versions used in early studies no longer available
Bu-1b	ThermoFisher/ MA5-28757	5-11G2	FITC	1:100	

PE, phycoerythrin; FITC, fluorescein isothiocyanate.

3.3 Experimental design

The principle of the experiment follows the outline of a classical Luria-Delbrück fluctuation analysis (Luria & Delbrück, 1943): multiple parallel cultures are expanded for a given number of divisions before assaying the prevalence of a stochastic mutational process in each culture to derive a rate for that process. In practical terms, we have found that reliable results can be obtained from seeding four Bu-1 positive cells into round bottomed 96 wells plates and growing for ~20 cell divisions before transferring to 24-well plates for another six cell divisions. For best results, combined cell counting and analysis of Bu-1 expression is carried out from standard flow cytometry tubes after the additional expansion of cells in 24-well plates, but can also be performed straight from the 96-well plate stage using an automated 96-well plate sampler.

3.4 Performing and analysing a Bu-1a fluctuation analysis experiment

3.4.1 Before setting up the plates for fluctuation analysis, check the viability of the cells and the fraction with Bu-1a^{high} expression for each line to be analysed. If the populations are >95% Bu-1a^{high}, the plates can be set up using limiting dilution at 4 cells per well in 200 μL medium. If not, cell sorting of four Bu-1a^{high} cells per well of a 96-well plate containing 200 μL medium per well should be performed.

3.4.2 Culture the plates in a humidified environment for ~20 cell divisions (which for wild type cells equates to ~10 days) aiming for a colony in the bottom of each well of ~2 mm diameter. Avoiding evaporation or significant temperature fluctuations (for instance, due to the incubator being opened and closed frequently) is important for good results.

3.4.3 For each condition, pick a minimum of 24 wells to be transferred to 2 mL medium in a 24-well plate. Incubate for another six cell divisions (further 3 days for wild type cells). Do not allow the cells to overgrow, indicated by the medium turning yellow.

3.4.4 At the end of the culturing period, resuspend the cells thoroughly and transfer 300 μL to a flow cytometry tube.

3.4.5 It is important to include the following flow cytometry controls in addition to the genetic controls discussed above (Fig. 3A):

- Bulk wild type population stained (positive control for staining)
- Bulk wild type population unstained (negative control)
- Ideally, include a line harbouring a genetic disruption of the *BU-1* locus stained with anti-Bu-1 to provide a true negative control for antibody background.

3.4.6 Optional: add Hoechst 33342 at a final concentration of 10 μM. Gently vortex and incubate for 1 h in the cell incubator.

3.4.7 Bu-1 staining. Add 3 μL anti-Bu-1 antibody to give the desired final 1:100 dilution. Gently vortex. Incubate for 5 min at room temperature. The flow cytometry analysis should be performed as soon as possible after this incubation and certainly within 2 h. Longer incubations may lead to cell death (Funk, Tregaskes, Young, & Thompson, 1997).

3.4.8 Set up the cytometer to gate for living cells by forward scatter (FSC) and side scatter (SSC) (Fig. 2A) and, optionally, by Hoechst 33342 staining to further select for cells actively engaged in the cell cycle and against cell doublets (Fig. 2B and C).

3.4.9 Acquire at least 10,000 events in the FSC/SSC gate for each sample.

3.4.10 Gate for Bu-1 positive cells on the wild type control as indicated in Fig. 2D. Apply this gate to all samples and record the number of cells

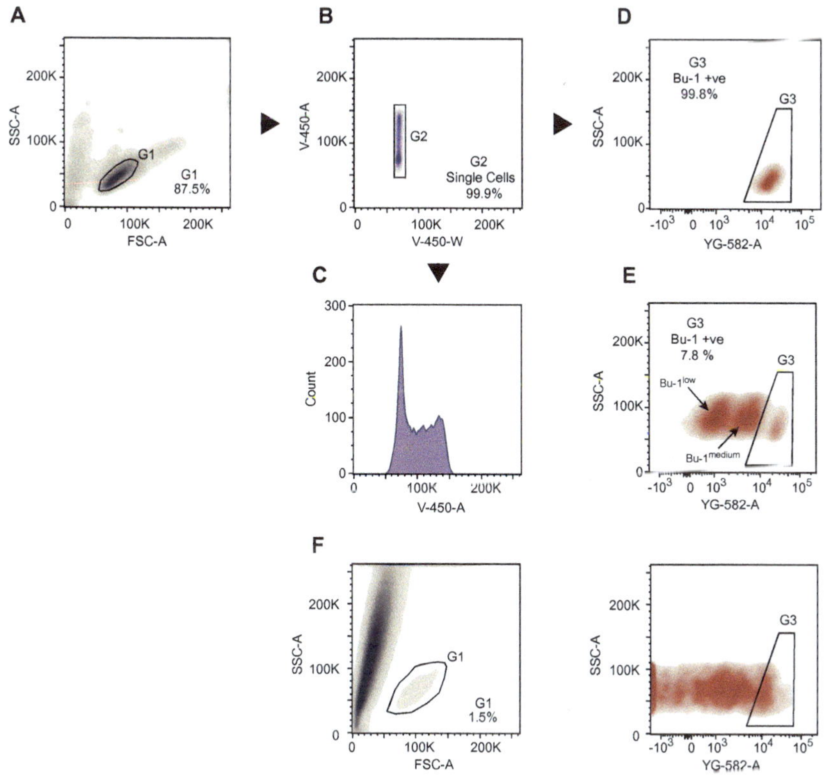

Fig. 2 Gating procedure for measurement of Bu-1 loss variants. The example here is for cells stained with Hoechst 33342 for DNA and anti-Bu-1A-PE. (A) Cells are first gated (gate G1) on forward scatter (FSC) and side scatter (SSC) to select for live cells. (B) G1 is then further gated on the Hoechst 33342 channel pulse area (V-450-A) and pulse width (V-450-W) to select for singlets (gate G2) and to ensure that they are cycling to further confirm viability. (C) If desired, the cell cycle profile can be checked using the violet channel. (D) Gate G2 is used to examine Bu-1A-PE staining, which is gated as shown for Bu-1 positive cells. (E) An example of a clone of PrimPol-deficient cells exhibiting a significant Bu-1 loss population. See text for discussion of the Bu-1^{medium} and Bu-1^{low} populations, which are not relevant for calculating the proportion of Bu-1 loss variants in the population. Percentage Bu-1 loss variants is expressed as (100 − Bu-1^{high}). (F) Example of an unreliable Bu-1 loss population in stressed cells. The cells are mostly dead with only 1.5% within an acceptable FSC/SSC gate. The Bu-1 loss population is smudged across the whole range of expression.

that have lost the Bu-1^{high} expression state (i.e., 100—% positive). See note below concerning states of Bu-1 loss (Fig. 2E). Fig. 2F provides an example of Bu-1 staining in highly stressed cells, discussed further below. Such samples should not be analysed. Example experimental results are shown in Fig. 3.

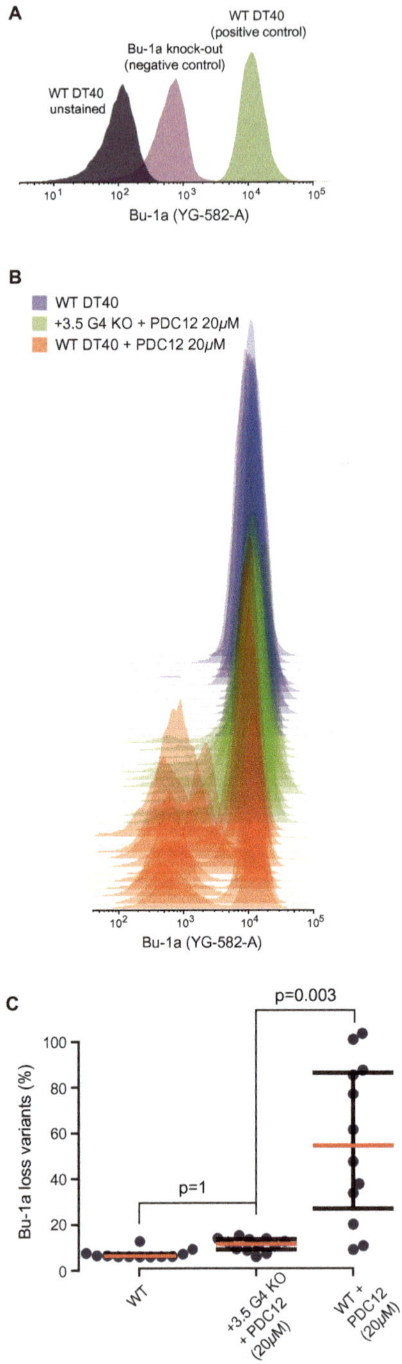

Fig. 3 See figure legend on opposite page.

3.4.11 An R script (www.github.com/Sale-lab/bu1_assay/ along with the example data used to create Fig. 3) can be used to automate the analysis of the data exported from FlowJo to a spreadsheet to provide scatter plots, median and statistics. This analysis can also be performed using other graphing and analysis tools such as Prism.

Preparation of files and running the script.

In this example, the flow cytometry analyses was performed with FlowJo® (Becton Dickinson) and the data exported as an Excel (.xlsx) file. However, only the layout of the data matters in running the script, thus any flow cytometry software can be used for analysis.

The script retrieves the percentage of the gated Bu-1 positive cells by searching any line in the cytometry output file containing the name of this gate ("*Bu1_Pos*"), and extracting the next column value (in this example 98.5%).

	%	#Cells
WT_NT_Tube_001.fcs	NA	20879
WT_NT_Tube_001.fcs/Single_cells	50.4	10519
WT_NT_Tube_001.fcs/Single_cells/Bu1_Pos	98.5	10363

If the name of the gate ("*Bu1_Pos*") has been changed, this must also be amended within the script in the custom parameters section.

For the script to run, four replacements on the original .xlsx must first be performed:

1. All empty cells should be filled with "NA."
2. All spaces between words should be replaced by an underscore.

Fig. 3 Example of Bu-1 fluctuation experiment. (A) Gating controls. In green is Bu-1a staining in wild type cells. In lilac is Bu-1a staining of cells carrying a disrupted *BU-1* locus, i.e., with the selection cassette still in situ. In grey are unstained wild type cells. The difference between the lilac and grey populations gives the background reactivity of the antibody. (B) Example of raw cytometry data from a fluctuation experiment. In blue are wild type populations expanded as described. In red are the wild type populations expanded in the presence of the G quadruplex ligand PDC12 and in green are cells, also expanded in the presence of PDC12, but in which the +3.5 G4 has been deleted (Guilbaud et al., 2017). (C) Scatter plots showing the proportion of Bu-1 loss variants in this experiment with median (red) bar and inter-quartile range (grey bar) as generated by the analysis script (described in 3.4.11 in Section 3.4). *P* value calculated with Fisher's exact test with a bin size of 20%.

3. Column names should be removed.
4. The file should be saved as a .csv file (or tabulated .txt, in which case this must be recorded within the script in the custom parameters section).

The R script should be placed in a folder also containing the Input and Output subfolders.

Input folder: CSV files with data for all conditions to be compared should be put within the same folder inside the ./Input folder. Note: each folder within ./Input is processed independently.

Fisher exact test: The user should define the bin size (in the custom parameters section) for the Fisher exact test. The bin is set to contain 90% of the control sample. Based on our previous experience, well-stained wild type DT40 in optimum condition do not exhibit more than 20% apparent loss variants, most often < 10%.

Output: For each subfolder in ./Input, a subfolder of same name is generated within .//Output. After the script has run, each folder should contain two files:

1. SVG file containing scatter plots of the percentage of Bu-1 loss variants in each sample together with lower and upper quartile (black lines) and median (red line).
2. TXT file containing the P value from the Fisher exact test between conditions.

3.5 A note on statistical tests and data interpretation

A problem with the analysis of data generated by this assay is that the generation of Bu-1 loss variants is both stochastic and prone to some background noise caused by the cytometry-based assay. While the latter can be minimised by careful cell husbandry and analysis, it is usually unwise to draw conclusions about any loss-variant generation up to a median of ~20% as, depending on growth conditions, staining and gating, this is the range for an epigenetically stable wild type population. We have therefore previously adopted an approach of testing for significance using Fisher's hypergeometric distribution following division of the data for each clone into 20% bins. More recently, particularly since switching to the AV20 anti-Bu-1 antibody (see table), we have found that the application of a more straightforward test of variance (ANOVA or Kruskal-Wallis) has yielded satisfactory results.

3.6 Calculation of per-division rate of BU-1 loss variant generation

The rate at which *BU-1* loss variants are generated through epimutations far exceeds any known mutational process. At these very high frequencies, the Luria-Delbruck equations become unreliable and thus we devised a Monte-Carlo approach to simulate the generation of Bu-1 loss variants (Schiavone et al., 2014), which models the process closely based on a limited number of parameters. From this is it possible to estimate the per-division probability of a cell switching from Bu-1^{high} expression to a lower state. In order to apply this model, it is necessary to know the doubling time. This can be derived from the cell number at the end of the experiment which can, in turn, be reasonably accurately derived from the flow cytometry experiments if the flow rate of the instrument is known and the samples are fully resuspended.

4. Optimisation and troubleshooting

Careful cell husbandry is critical to the success of this method. *BU-1* expression becomes unstable in conditions that globally stress replication in DT40 cells, even when the +3.5 G4 is deleted (Papadopoulou et al., 2015). Such conditions can arise, for example, when cells are overgrown, are cooled (Edmunds, Simpson, & Sale, 2008) or have not fully recovered from cryopreservation. Fig. 2F provides an example of Bu-1 staining in a sub-optimal, stressed culture. During culture, care should be taken to avoid temperature fluctuations, overgrowth and evaporation. Hence it is also vitally important to include rigorous controls in each experiment, as discussed above, and to eliminate wells from the analysis that exhibit low viability. If there is uncertainty about whether the observed loss variants result from the epigenetic mechanism described above, which, for instance, may be the case when testing small molecules, isolated variants must be assessed for loss of H3K4me3 around the *BU-1* transcription start site, as previously described (Schiavone et al., 2014).

5. Applications and additional considerations

5.1 Advantages and disadvantages of the method

The *BU-1* locus has provided a sensitive and genetically tightly controllable assay to monitor delays in the replication of specific DNA secondary structure-forming sequences and has allowed insights into the behaviour of the vertebrate replisome in vivo with a sensitivity and refinement that have not been possible with other approaches. However, it is important to also be aware of the limitations of the approach. Principal among these limitations is artefact due to poor culture conditions and lack of controls, as discussed above. It is also important to note that the assay in its current form can only interrogate sequences on the leading strand template as Okazaki repriming will prevent the formation of long post-replicative gaps on the lagging strand. Indeed, inverting a G4 motif so the G-rich strand now falls onto the lagging strand template (Crick strand) can be used as an assay control. Finally, we currently have only extensively validated the *direct* link between unstable expression and secondary structure formation in a limited number of loci in DT40 (Sarkies et al., 2010; Schiavone et al., 2014), although a large number of loci exhibit changes in expression under conditions in which DNA structure may become a more robust replication impediment (Lerner et al., 2020; Papadopoulou et al., 2015; Sarkies et al., 2012).

5.2 Applications

As discussed in the introduction the *BU-1* locus can be adapted to study the capacity of defined sequences to form in vivo replication impediments. Orthogonally, the cells can be genetically modified to explore the factors that relieve replication stalling at these sequences and the factors that drive structure formation, notably RNA. Replacing the endogenous +3.5 G4 with other desired sequences is efficient and significantly facilitated by the ability to screen targeted cells by loss of surface Bu-1 expression, which is due to the presence of the puromycin selection cassette within the targeting construct. Cre recombinase-mediated excision of the cassette restores Bu-1 expression (Schiavone et al., 2014). The method could thus readily be adapted to screen small molecules that interact with a range of secondary structure-forming sequences and for CRISPR screens for factors involved in the replication of specific sequences.

5.3 Discrete expression states of BU-1

Under some circumstances, it is possible to observe that destabilised *BU-1* can adopt three discrete expression states (Fig. 2E) (Guilbaud et al., 2017; Schiavone et al., 2016): the normal Bu-1^{high} state is characterised by its promoter being enriched in H3K4me3 and K9/14ac histone marks.

The Bu-1^{medium} level of expression, is characterised by loss of promoter H3K4me3, but an absence of repressive marks. The Bu-1^{low} state is characterised by loss of H3K4me3 and accompanied by signs of heterochromatinisation (H3K9me3 and DNA methylation). Bu-1^{low} expression is close to that seen in a Bu-1a "knock-out" negative control (Fig. 3A). The first transition, from Bu-1^{high} to Bu-1^{medium} is dependent on the presence of the +3.5 G4, while the transition from Bu-1^{medium} to Bu-1^{low} occurs spontaneously once H3K4me3 has been lost (Guilbaud et al., 2017). For the purposes of the method described here, it is the loss of the Bu-1^{high} state that matters.

5.4 Transvection between the BU-1A and BU-1B alleles

DT40 is derived from an F_1 hybrid chicken (Baba, Giroir, & Humphries, 1985). Consequently, the two alleles of *BU-1, BU-1A* and *BU-1B,* differ and do so sufficiently that they can be recognised by separate antibodies. This ability to monitor the behaviour of the two alleles revealed a curious and poorly understood transvection-like phenomenon, in which instability of one allele is transmitted to the second, even when the second has had its +3.5 G4 deleted. Full stabilisation of both alleles is only observed when the +3.5 G4 is deleted from both (Schiavone et al., 2014). For the assay described here, the practical upshot is that for a negative control for the DNA secondary structure, cells in which the +3.5 G4 has been deleted on both alleles must be used.

6. Summary

The use of the *BU-1* locus of DT40 to study the replication of structured DNA has provided some important insights into the mechanisms operating in vertebrate cells to achieve this feat, as well as revealing that structure formation during replication is a surprisingly common event. When used with care, it is capable of providing detailed information on the factors that promote and counter significant episodes of replisome arrest. While the approach has currently only been extensively validated in DT40 cells, we have observed similar patterns of gene dysregulation in human cells making it likely that loci similar to chicken *BU-1* will be discoverable and that, with the advent of gene editing technology, this will allow exploration of the replication of structured DNA through the lens of the epigenome in mammalian systems as well.

References

Baba, T. W., Giroir, B. P., & Humphries, E. H. (1985). Cell lines derived from avian lymphomas exhibit two distinct phenotypes. *Virology*, *144*(1), 139–151. https://www.ncbi.nlm.nih.gov/pubmed/2998040.

Baxter, C. L., Šviković, S., Sale, J. E., Dean, C., & Costa, S. (2021). The intersection of DNA replication with antisense 3′ RNA processing in Arabidopsis FLC chromatin silencing. *Proceedings of the National Academy of Sciences of the United States of America*, *118*(28). https://doi.org/10.1073/pnas.2107483118. e2107483118.

Brewer, B. J., & Fangman, W. L. (1987). The localization of replication origins on ARS plasmids in *S. cerevisiae*. *Cell*, *51*(3), 463–471. https://doi.org/10.1016/0092-8674(87)90642-8.

Cheung, I., Schertzer, M., Rose, A., & Lansdorp, P. M. (2002). Disruption of dog-1 in Caenorhabditis elegans triggers deletions upstream of guanine-rich DNA. *Nature Genetics*, *31*(4), 405–409. https://doi.org/10.1038/ng928.

Dahan, D., Tsirkas, I., Dovrat, D., Sparks, M. A., Singh, S. P., Galletto, R., et al. (2018). Pif1 is essential for efficient replisome progression through lagging strand G-quadruplex DNA secondary structures. *Nucleic Acids Research*. https://doi.org/10.1093/nar/gky1065.

Dovrat, D., Dahan, D., Sherman, S., Tsirkas, I., Elia, N., & Aharoni, A. (2018). A live-cell imaging approach for measuring DNA replication rates. *Cell Reports*, *24*(1), 252–258. https://doi.org/10.1016/j.celrep.2018.06.018.

Edmunds, C. E., Simpson, L. J., & Sale, J. E. (2008). PCNA ubiquitination and REV1 define temporally distinct mechanisms for controlling translesion synthesis in the avian cell line DT40. *Molecular Cell*, *30*(4), 519–529. https://doi.org/10.1016/j.molcel.2008.03.024.

Follonier, C., Oehler, J., Herrador, R., & Lopes, M. (2013). Friedreich's ataxia-associated GAA repeats induce replication-fork reversal and unusual molecular junctions. *Nature Structural & Molecular Biology*, *20*(4), 486–494. https://doi.org/10.1038/nsmb.2520.

Funk, P. E., Tregaskes, C. A., Young, J. R., & Thompson, C. B. (1997). The avian chB6 (Bu-1) alloantigen can mediate rapid cell death. *Journal of Immunology*, *159*(4), 1695–1702. https://pubmed.ncbi.nlm.nih.gov/9257830.

Gerhardt, J., Bhalla, A. D., Butler, J. S., Puckett, J. W., Dervan, P. B., Rosenwaks, Z., et al. (2016). Stalled DNA replication forks at the endogenous GAA repeats drive repeat expansion in Friedreich's Ataxia cells. *Cell Reports*, *16*(5), 1218–1227. https://doi.org/10.1016/j.celrep.2016.06.075.

Guilbaud, G., Murat, P., Recolin, B., Campbell, B. C., Maiter, A., Sale, J. E., et al. (2017). Local epigenetic reprogramming induced by G-quadruplex ligands. *Nature Chemistry*, *9*(11), 1110–1117. https://doi.org/10.1038/nchem.2828.

Jimeno, S., Camarillo, R., Mejías-Navarro, F., Fernández-Ávila, M. J., Soria-Bretones, I., Prados-Carvajal, R., et al. (2018). The helicase PIF1 facilitates resection over sequences prone to forming G4 structures. *Cell Reports*, *24*(12), 3262–3273.e4. https://doi.org/10.1016/j.celrep.2018.08.047.

Kim, H., Narayanan, V., Mieczkowski, P. A., Petes, T. D., Krasilnikova, M. M., Mirkin, S. M., et al. (2008). Chromosome fragility at GAA tracts in yeast depends on repeat orientation and requires mismatch repair. *The EMBO Journal*, *27*(21), 2896–2906. https://www.embopress.org/doi/pdf/10.1038/emboj.2008.205.

Krasilnikova, M. M., & Mirkin, S. M. (2004). Replication stalling at Friedreich's Ataxia (GAA)n repeats in vivo. *Molecular and Cellular Biology*, *24*(6), 2286–2295. https://doi.org/10.1128/mcb.24.6.2286-2295.2004.

Kruisselbrink, E., Guryev, V., Brouwer, K., Pontier, D. B., Cuppen, E., & Tijsterman, M. (2008). Mutagenic capacity of endogenous G4 DNA underlies genome instability in FANCJ-defective C. elegans. *Current Biology*, *18*(12), 900–905. https://doi.org/10.1016/j.cub.2008.05.013.

Lerner, L. K., Holzer, S., Kilkenny, M. L., Šviković, S., Murat, P., Schiavone, D., et al. (2020). Timeless couples G-quadruplex detection with processing by DDX11 helicase during DNA replication. *The EMBO Journal*, e104185. https://doi.org/10.15252/embj.2019104185.

Lerner, L. K., & Sale, J. E. (2019). Replication of G Quadruplex DNA. *Genes (Basel)*, *10*(2), 95. https://doi.org/10.3390/genes10020095.

Lopes, J., Piazza, A., Bermejo, R., Kriegsman, B., Colosio, A., Teulade-Fichou, M. P., et al. (2011). G-quadruplex-induced instability during leading-strand replication. *The EMBO Journal*, *30*(19), 4033–4046. https://doi.org/10.1038/emboj.2011.316.

Luria, S. E., & Delbrück, M. (1943). Mutations of bacteria from virus sensitivity to virus resistance. *Genetics*, *28*(6), 491–511. https://www.ncbi.nlm.nih.gov/pubmed/17247100.

Mesner, L. D., Crawford, E. L., & Hamlin, J. L. (2006). Isolating apparently pure libraries of replication origins from complex genomes. *Molecular Cell*, *21*(5), 719–726. https://doi.org/10.1016/j.molcel.2006.01.015.

Papadopoulou, C., Guilbaud, G., Schiavone, D., & Sale, J. E. (2015). Nucleotide pool depletion induces G-quadruplex-dependent perturbation of gene expression. *Cell Reports*, *13*(11), 2491–2503. https://doi.org/10.1016/j.celrep.2015.11.039.

Saribasak, H., & Arakawa, H. (2006). Basic cell culture conditions. In J. M. Buerstedde, & S. Takeda (Eds.), *Reviews and protocols in DT40 research* (pp. 345–346).

Sarkies, P., Murat, P., Phillips, L. G., Patel, K. J., Balasubramanian, S., & Sale, J. E. (2012). FANCJ coordinates two pathways that maintain epigenetic stability at G-quadruplex DNA. *Nucleic Acids Research*, *40*(4), 1485–1498. https://doi.org/10.1093/nar/gkr868.

Sarkies, P., Reams, C., Simpson, L. J., & Sale, J. E. (2010). Epigenetic instability due to defective replication of structured DNA. *Molecular Cell*, *40*(5), 703–713. https://doi.org/10.1016/j.molcel.2010.11.009.

Schiavone, D., Guilbaud, G., Murat, P., Papadopoulou, C., Sarkies, P., Prioleau, M.-N., et al. (2014). Determinants of G quadruplex-induced epigenetic instability in REV1-deficient cells. *The EMBO Journal*, *33*(21), 2507–2520. https://doi.org/10.15252/embj.201488398.

Schiavone, D., Jóźwiakowski, S. K., Romanello, M., Guilbaud, G., Guilliam, T. A., Bailey, L. J., et al. (2016). PrimPol is required for replicative tolerance of G quadruplexes in vertebrate cells. *Molecular Cell*, *61*(1), 161–169. https://doi.org/10.1016/j.molcel.2015.10.038.

Sfeir, A., Kosiyatrakul, S. T., Hockemeyer, D., MacRae, S. L., Karlseder, J., Schildkraut, C. L., et al. (2009). Mammalian telomeres resemble fragile sites and require TRF1 for efficient replication. *Cell*, *138*(1), 90–103. https://doi.org/10.1016/j.cell.2009.06.021.

Stroik, S., Kurtz, K., Lin, K., Karachenets, S., Myers, C. L., Bielinsky, A. K., et al. (2020). EXO1 resection at G-quadruplex structures facilitates resolution and replication. *Nucleic Acids Research*, *48*(9), 4960–4975. https://doi.org/10.1093/nar/gkaa199.

Šviković, S., Crisp, A., Tan-Wong, S. M., Guilliam, T. A., Doherty, A. J., Proudfoot, N. J., et al. (2019). R-loop formation during S phase is restricted by PrimPol mediated repriming. *The EMBO Journal*, *38*(3). https://doi.org/10.15252/embj.201899793.

Wickramasinghe, C. M., Arzouk, H., Frey, A., Maiter, A., & Sale, J. E. (2015). Contributions of the specialised DNA polymerases to replication of structured DNA. *DNA Repair (Amst)*, *29*, 83–90. https://doi.org/10.1016/j.dnarep.2015.01.004.

CHAPTER THREE

Visualizing replication fork encounters with DNA interstrand crosslinks

Ryan C. James[a], Marina A. Bellani[b], Jing Zhang[c], Jing Huang[d], Althaf Shaik[b], Durga Pokharel[e], Himabindu Gali[f], Julia Gichimu[b], Arun K. Thazhathveetil[g], and Michael M. Seidman[b,*]

[a]Department of Molecular Biology and Genetics, Cornell University, Ithaca, NY, United States
[b]Laboratory of Molecular Biology and Immunology, National Institute on Aging, National Institutes of Health, Baltimore, MD, United States
[c]Department of Neurosurgery, Shanghai East Hospital, Tongji University School of Medicine, Shanghai, China
[d]Institute of Chemical Biology and Nanomedicine, State Key Laboratory of Chemo/Biosensing and Chemometrics, College of Biology, Hunan University, Changsha, China
[e]Horizon Discovery, Lafayette, CO, United States
[f]Frederick National Laboratory for Cancer Research, Frederick, MD, United States
[g]Nitto Denko Avecia Inc., Cincinnati, OH, United States
*Corresponding author: e-mail address: seidmanm@grc.nia.nih.gov

Contents

Methods in Enzymology, Volume 661
ISSN 0076-6879
https://doi.org/10.1016/bs.mie.2021.08.015

Abstract

Replication forks encounter numerous challenges as they move through eu- and hetero-chromatin during S phase in mammalian cells. These include a variety of impediments to the unwinding of DNA by the replicative helicase such as alternate DNA structures, transcription complexes and R-loops, DNA–protein complexes, and DNA chemical adducts. Much of our knowledge of these events is based on analysis of markers of the replication stress and DNA Damage Response that follow stalling of replisomes. To examine consequences for the replisomes more directly, we developed an approach for imaging collisions of replication forks with the potent block presented by an interstrand crosslink (ICL). The strategy is based on the visualization on DNA fibers of the encounter of replication tracts and an antigen tagged ICL. Our studies revealed an unexpected restart of DNA synthesis past an intact ICL. In addition, and also unexpected, we found two distinct versions of the replisome, one biased toward euchromatin and the other more prominent in heterochromatin. Here, we present details of our experimental procedures that led to these observations.

1. Introduction

1.1 The DNA replication apparatus, the replisome, contains two functional components—a helicase to unwind DNA and polymerases to synthesize it. Although textbooks describe a smooth and unhindered progression the reality is quite the contrary. The genome is filled with impediments that stall the helicase and/or the polymerases. These include adducts introduced by DNA reactive compounds, alternate DNA structures, protein-DNA crosslinks, R-loops, and precursor nucleotide insufficiency. They impose replication stress, which must be resolved by a host of factors that resolve the impediment, repair associated damage, and restore replisome activity. The frequency with which replication stress occurs, and the importance of effective resolution, is underscored by the profound effects of mutations in key genes involved in the stress response. For example, the ATR (Ataxia Telangiectasia and Rad3-related) kinase is an essential gene, and individuals with hypomorphic mutations suffer from severe disorders such as Seckel Syndrome (Alderton et al., 2004; Klingseisen & Jackson, 2011). Mutations in other relevant genes such as BRCA1/2, FANCM, are associated with a high frequency of cancer

1.2 The helicase of the replisome has been defined biochemically. The six MCM (2–7) proteins (M) (Zhai & Tye, 2017) form an open offset ring that encircles duplex DNA in G_1 phase. DNA binding requires loading factors and is referred to as licensing (Abid et al., 2017;

Parker et al., 2019; Yao & O'Donnell, 2019). Licensed sites can become origins of replication. Many more are loaded than used during a "normal" S phase, with the surplus supplying a reservoir of dormant origins that can be activated to rescue stalled or broken forks (Blow, Ge, & Jackson, 2011). Activation in S phase requires association of CDC45 (C) and the GINS (G) complex of four proteins (Douglas, Ali, Costa, & Diffley, 2018; Gambus et al., 2006) which closes a "gate" in the MCM complex (Bochman & Schwacha, 2010). The composition and structure of the active CMG helicase, as defined on model substrates, has been a subject of intense investigation, driven by cryo electron microscopy (Gao et al., 2019; Li, Yao, & O'Donnell, 2020) combined with very accomplished biochemistry. The association with polymerases and additional structural proteins completes the assembly of a functional replisome, again as defined on model substrates. It is important to note that while only about 30 proteins are required to reconstitute a biochemically functional eukaryotic replisome (Bell & Labib, 2016), databases such as BioGrid list hundreds of proteins that interact with MCM proteins

1.3 Studies on replication stress in living cells have been of interest for many years and often involve treatments of cells that result in blocks to either the helicase, or rather more frequently, to the polymerases. Thus, for example, ultraviolet photoproducts do not block the helicase, but do challenge replisome associated polymerases. DNA synthesis restarts past the adducts leaving gaps in the daughter strands (Heller & Marians, 2006; Lehmann & Fuchs, 2006; Rupp & Howard-Flanders, 1968; Yeeles & Marians, 2011). Studies with purified proteins and model templates reveal that the replicative helicase moves past single strand adducts continuing to unwind DNA. Downstream repriming enables the resumption of DNA synthesis (Marians, 2018; Taylor & Yeeles, 2018). For the most part these studies followed replication of templates carrying adducts that could be accommodated within the barrel of the MCM ring. However, recent work from the Walter laboratory argues that the CMG can move past protein-DNA adducts too large to fit within that barrel (Sparks et al., 2019). The absence of CMG dissociation implied a greater flexibility of structure than indicated by static models. Indeed, recent structural studies imply the existence of gates that could enable travel past bulky blocks (Gao et al., 2019; Wasserman, Schauer, O'Donnell, & Liu, 2019; Yang, Seidman, Rupp, & Gao, 2019).

The possibility that the replisome might be able to pass even large single strand adducts is intriguing and has obvious implications for the resilience of replication. However, the structure imposed by compounds that form interstrand crosslinks (ICLs) would appear to defeat even the most robust replisome. Ever since their description they have been considered absolute blocks to replication, more specifically to the replisomal helicase (Marmur & Grossman, 1961; Muniandy, Liu, Majumdar, Liu, & Seidman, 2010). This long held belief led to models in which forks would stall at ICLs until uncoupling of the strands (unhooking) was completed. However, this view has been replaced in recent years by the very elegant work from the Walter group. They showed, in a Xenopus egg extract system, that replication is completed by two forks converging from either side at the ICL before repair can begin (Knipscheer et al., 2009; Raschle et al., 2008; Zhang et al., 2015). This system has proven to be extraordinarily powerful, and their results define our current view of the cellular response to ICLs during replication.

1.4 We were interested in the response to blocks to the replicative helicase in the genome of live mammalian cells. Consequently, the "absolute block" imposed by the ICLs was an attractive model. This presented the experimental challenge of following events at ICLs distributed across the genome, without the site-specific sequence option of the plasmid system so productively exploited by the Walter laboratory. Since we could not examine events at specific sequences, we chose to use the ICL itself as the focus of our inquiry. We took advantage of the properties of the psoralens, the photoactive principal in extracts of medicinal plants used for thousands of years as topical treatments for skin disorders (Benedetto, 1977). These compounds offer several technical advantages over other agents frequently used in experiments with ICLs. Psoralens are nonreactive until exposed to long wave UV light (UVA). They intercalate at 5′ TA/AT sites in DNA, which positions them to react with thymine bases on opposite strands, upon exposure to UVA. Consequently, they form a high frequency of ICLs as a proportion of total reaction products (over 90% of trimethyl psoralen (TMP) adducts, are ICLs) (Lai et al., 2008). Furthermore, psoralens can be covalently linked to other moieties without loss of ICL forming potential. This feature prompted our conjugation of TMP to digoxigenin (Dig), frequently used as an immuno tag

(Huang et al., 2013). The tag allowed us to visualize the ICLs in different contexts: by immunofluorescence in cells, and by an immuno quantum dot in replication tracts in DNA fibers

Analysis of the patterns of replication in the vicinity of ICLs demonstrated both single fork stalling and the dual fork convergence predicted by the Walter model. However, unexpectedly, the most frequent pattern, one that we termed traverse, was consistent with restart of replication on the side of the ICL distal to the replication fork encounter. Although this observation was contrary to decades of conventional wisdom, it has been confirmed recently by another group (Mutreja et al., 2018). The restart patterns were dependent on ATR which, in response to the introduction of ICLs, phosphorylated (among many other substrates) serine 108 on MCM2. The availability of a commercial antibody specific for pMCM2S108 permitted its use as a convenient marker of the "stressed" replisome. Approximately 50% of the restart frequency was dependent on the translocase activity of FANCM. After treatment of cells with TMP, coimmunoprecipitation experiments demonstrated binding of FANCM by replisomes marked by pMCM2S108. This interaction was accompanied by the loss of the GINS proteins from the replisome (Huang et al., 2019).

Although FANCM made a substantial contribution it did not account for all the traverse events. In subsequent experiments, we looked for another factor that was important for the replication restart. We found that the DONSON protein was also involved and that double knockdown of FANCM and DONSON greatly reduced traverse frequencies. This indicated that DONSON and FANCM were not epistatic with one another. We then performed sequential coIP on chromatin proteins with or without bound DNA to ask if DONSON and FANCM were on the same stressed replisomes and which were associated with ICLs. The sequential IP experiments demonstrated that in cells treated with Dig-TMP/UVA there were multiple replisome species. The first IP, against a GINS protein, recovered a fraction containing both DONSON and the GINS complex, but without pMCM2S108. Dig-TMP was absent from the DNA associated with these proteins indicating that these replisomes had not encountered ICLs. The second IP, against the remaining DONSON, captured pMCM2S108, Dig-TMP, but no GINS proteins, and no FANCM. This fraction represented DONSON replisomes that were in the vicinity of ICLs and had been the target of ATR. The remaining supernatant was then challenged with an antibody to FANCM. This IP contained pMCM2S108, Dig-TMP, but no DONSON or GINS proteins. Thus, there were a minimum of three replisomes in cells exposed to the psoralen/UVA: unstressed containing DONSON, a DONSON stressed replisome, and a FANCM stressed replisome (Zhang et al., 2020).

While IP experiments are very informative, they report protein interactions in a homogenate of a population of cells and cannot inform as to conditions in individual cells. To ask if the DONSON and FANCM stressed replisomes were present in the same cells at the same time we took advantage of the proximity ligation assay (PLA) (Koos et al., 2014). We performed PLA between DONSON and pMCM2S108, imaged the cells, and marked the position of the fields on the plates. We then stripped the cells of antibodies and reaction products and repeated the PLA, this time between FANCM and pMCM2S108. We reimaged the same cells and aligned the images of the first and second PLA patterns. This "sequential" PLA yielded some cells with a majority of DONSON: pMCM2S108 signals, while others had greater frequency of FANCM: pMCM2S108 signals. Alignment of the images of the nuclei demonstrated that the DONSON and FANCM stressed replisomes could be in the same cells at the same time, but not in the same place (Zhang et al., 2020).

Our subsequent experiments were directed toward understanding the implications of the different stressed replisomes. We inquired as to the relationship between the two forms and the stage of S phase by collecting cells in very early and very late S phase. We found by PLA that the DONSON stressed replisome (DONSON: pMCM2S108) was predominant in early S phase while the FANCM: pMCM2S108 signals were most frequent in late S phase. DNA sequences in euchromatin replicate early while those in heterochromatin replicate late. Consistent with these distinctions DONSON was preferentially associated with euchromatin marks such as histone H3K4me3, while FANCM was associated with the heterochromatin indicator H3K9me3.

To summarize, our studies of replication fork encounters with the potent block presented by ICLs revealed an unexpected synthesis restart option that enables cells to complete replication. In turn this led to the recognition of different replisomes operating at different times and chromatin compartments. In the following, we describe details of the materials and methods that we developed, and our adjustments to established methods, that were the basis of these discoveries.

2. Experimental procedures

2.1 Synthesis of digoxigenin-trimethyl psoralen (Dig-TMP)

2.1.1 *Chemicals*

Methanol, Fisher Scientific, A452-4
Dichloromethane, Sigma-Aldrich, D65100-42
Chloroform, Acros Organics, 42355-0040
28% Ammonium hydroxide solution, Sigma-Aldrich, 221228-500 mL
Digoxigenin-NHS, Roche, 11333054001
Trisoralen (trimethyl psoralen), TCI chemicals, T2267
1,3,5-Trioxane, Sigma-Aldrich, T81108-100G
Hydrochloric acid, 32%, Sigma-Aldrich, W530574
Toluene. Sigma-Aldrich, 244511-2L
4,7,10-Trioxa-1,13-tridecanediamine, Sigma-Aldrich, 369519-100 g
Chloroform, Sigma-Aldrich, 151823-100G
Preparative TLC UV/Silica G/1000 mm, Analtech, 02013
TLC plates silica gel 60 F254, Merck Millipore, 105715
TLC solvent, (chloroform: Methanol: 28% ammonium hydroxide 8:1:0.1

2.1.2 *Equipment*

Rotary evaporator, Buchi Rotavapor® R-3000 equipped with self-cleaning dry vacuum system
Vacuum pump, Welch 2025
Recirculating chiller, PolyScience
Vacuum concentrator, Savant Speedvac concentrator SVC 100 with refrigerated condensation
trap RT 100 and Welch 1402 N vacuum pump
Nuclear Magnetic Resonance, Bruker Ascend 500 MHZ for ^{1}H NMR
Mass spectrometer, Agilent 6120 Quadrupole - LC/MS
UV spectrophotometer DeNovix DS-11 FX+
Analytical balance Mettler AE160
Column chromatography, Teledyne Combiflash Rf+
RediSep Rf silica gel disposable Flash columns, Teledyne, 692203304
^{1}H NMR spectra were recorded on Ascend Bruker 500 MHz instruments using TMS as internal standard

2.1.3 Synthesis of digoxigenin trimethylpsoralen (Dig-TMP)

Reaction progress was monitored by thin layer chromatography on 60F254 silica plates developed in Chloroform: Methanol: 28% ammonium hydroxide (8:1:0.1) (Scheme 1)

a. Synthesis of 4′-Chloromethyl-4,5′,8-trimethylpsoralen (1)

1. Add 1,3,5-Trioxane (0.71 g, 3.77 mmol) portion wise to concentrated HCl (20 mL) at room temperature under vigorous stirring

2. Add trisoralen (4,5′,8-trimethylpsoralen, TMP) (1 g, 4.38 mmol) portion wise to the initial mixture. Stir the reaction mixture at room temperature for 12 h

3. Dilute the mixture with H_2O (80 mL) and extract with dichloromethane

4. Filter the organic layer, dry with Na_2SO_4 and evaporate under reduced pressure to yield **1** as white solid (0.91 g 75%).

^{1}H NMR ^{1}H NMR (500 MHz, $CDCl_3$): δ 7.52 (s, 1H), 6.19 (s, 1H), 4.67 (s, 2H),2.50–2.45 (m, 9H); LC-MS (ESI) (*m/z*): 277.0 (M + 1).

b. Synthesis of 4′-[N-(13- amino-4,7,10-trioxatrideca)]aminomethyl-4,5′,8-trimethylpsoralen (2)

1. To a solution of 4′-chloromethyl-4,5′,8-trimethylpsoralen (0.1 g, 0.363 mmol) in toluene add 4,7,10-trioxa-1,13-tridecanediamine (0.79 g, 3.623 mmol) and reflux for 12 h

2. Remove the excess of solvent under reduced pressure and subject to flash chromatography

3. Elute the compound in 5% Methanol: DCM and recover (**2**) as a yellow viscous solid (0.146 g, 72%). LC-MS (ESI) (*m/z*): 461.3 (M + 1).

c. Synthesis of Digoxigenin-TMP

1 To a solution of **2** (0.0057 g, 0.012 mmol) in dry DMF add digoxigenin-NHS ester (0.005 g, 0.008 mmol) and then trimethylamine (3.6 μL,

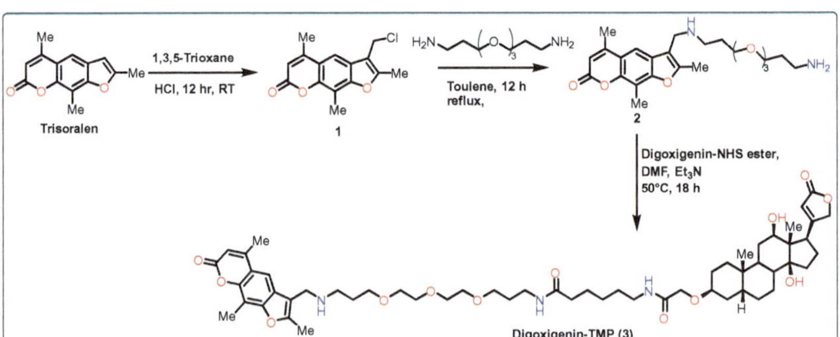

Scheme 1 Synthesis of Digoxigenin-TMP.

0.036 mmol) at room temperature and stir at 50 °C for 18 h. Monitor the progress of the reaction by TLC (60 F254 silica plates) in chloroform: methanol:28% ammonium hydroxide (8:1:0.1).

2 After completion remove the excess solvent by rotary evaporation under reduced pressure
3 Dissolve the resulting residue in a minimum amount of dichloromethane and subject to preparative thin layer chromatography (Analtech Inc., silica gel GF 1000 µm) in Chloroform:Methanol: 28% ammonium hydroxide solution (8:1:0.1).
4 Prepare the preparative TLC glass chamber containing 200 mL of Chloroform:Methanol:28% Ammonium hydroxide (8:1.0.1) mixture
5 Spot the sample across the preparative TLC, 2.5 cm from the lower edge. Place the plate in chamber, seal the top of the lid and allowed to run for ~1 h
6 Remove the plate and visualize the edges (mask the rest of the plate with aluminum foil) under UV 254 to mark the band of interest
7 Gently scrape off the product from plate with a spatula and suspend in Chloroform:Methanol:28% ammonium hydroxide solution (8:1:0.1). Transfer the filtrate to a round bottomed flask and evaporate under reduced pressure to give a viscous yellow pellet (5.8 mg, 66%). (5.8 mg, 66%). LC-MS (ESI) (*m/z*): Exact mass for $C_{56}H_{81}N_3O_{13}$ = 1004.26 found − *m/z* 1004.7 (Fig. 1)

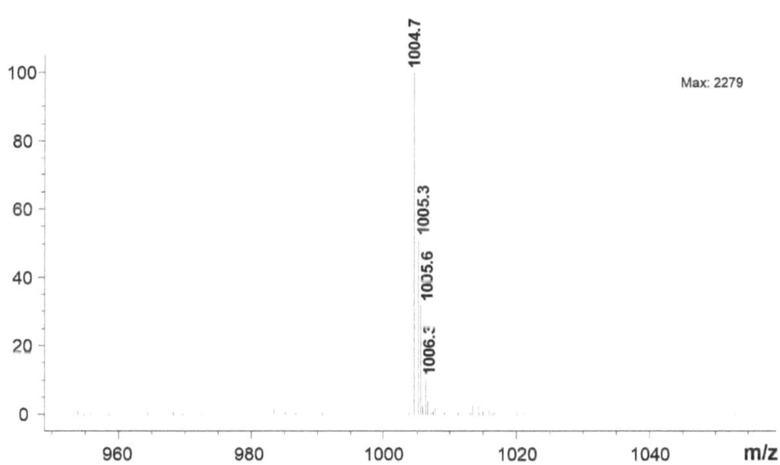

Fig. 1 LC-MS spectral data for Digoxigenin-TMP. Mass spectrometry data were recorded on Agilent quadrupole LC-MS mass spectrometer with ESI ionization.

d. Preparation of Dig-TMP stock solution

1. Dissolve the viscous yellow solid in 1 mL of 50% Ethanol: H_2O and dispense 50 μL to 1.5 mL centrifuge vials (~20 aliquots) and dry in a Speedvac concentrator until solvent is evaporated (medium drying time: ~2 h).
2. Resuspend the pellet in 200 μL of 50% Ethanol: H_2O and dry again. Store the resulting pellet at −20 °C wrapped with aluminum foil
3. To one vial add 25 μL of ethanol and mix gently with a micropipette. Then add an equal volume of water and mix thoroughly with a micropipette. Incubate at room temperature for 10 min. Remove 1 μL and add to 99 μL H_2O
4. Measure the absorbance of the diluted sample at 250 nm
5. Calculate the Dig-TMP stock concentration using following formula: Absorbance $\times 100 \times 10^6/25{,}000 =$ concentration in μM (extinction co-efficient of Dig-TMP = 25,000). Typical stock concentrations are about 4 mM

3. DNA fiber assay

3.1 Cells

HeLa CCL-2 (ATCC)

3.2 Reagents

Dulbecco's Modified Eagle Medium, ThermoFisher # #12430112
CldU, Sigma/Aldrich #C6891
IdU, Sigma/Aldrich #17125
EdU, Sigma/Aldrich ##900584
Rat monoclonal antibody to BrdU, Abcam #ab6326
Mouse monoclonal anti BrdU, BD Biosciences #347580
Rabbit oligoclonal anti Digoxigenin, Invitrogen #710019
Q dot 655 Goat anti rabbit IgG, ThermoFisher #Q-11421MP

The quality of quantum dot: secondary antibody conjugates is variable. Some are prone to aggregation and high background. We screen lots for efficacy and low background.

Goat anti mouse IgG AF-488, ThermoFisher A-11001
Goat anti Rat IgG AF-647, ThermoFisher A-21247
Cell lysis buffer, 0.5% SDS, 200 mM Tris-HCl [pH 7.5], 50 mM EDTA
Slide fixing solution, 3:1 methanol/acetic acid

Depurination solution, 2.5 M HCl
Neutralization buffer, 0.4 M Tris-HCl (pH 7.5)
Prolong gold anti fade reagent, Invitrogen, #P36984

3.3 Equipment

New silane glass slides, NewComer Supply #5070

Not all lots of these slides are effective for fiber assays and before purchase we test lots for DNA binding and low nonspecific binding of antibodies and quantum dots.

UVA chamber, Rayonet
Fluorescence microscope, Zeiss
Heat block, VWR #13259-052
Glass plate
Styrofoam sheet
Centrifuge, Eppendorf centrifuge 5417R
Digital rocking shaker, Thermo scientific #88880019
Coplin jar

3.4 Procedure

3.4.1 Treatment of cells and DNA spreading

1. Rest a metal tray on the heating block and place the assembly in the Rayonet chamber. Before starting the treatment warm the tray to 37 °C
2. Prewarm all media and Styrofoam sheet. When changing medium place plates on the prewarmed Styrofoam sheet to avoid the cold shock associated with placing plates on the steel surfaces of hoods cooled by constant air flow

We have employed this fiber assay in many different cell lines, both adherent and nonadherent, primary, immortalized, and tumor derived. We follow the same treatment procedure for each.

3. Plate HeLa cells in DMEM supplemented with penicillin and streptomycin and 10% calf serum in wells of a 24 well plate
4. When 60–70% confluent add Dig-TMP stock solution (in EtOH: H_2O) to prewarmed medium to a final concentration of 5 μM. The stock should be sufficiently concentrated to keep the final EtOH concentration to about 1% or less. Remove growth medium and replace with 1 mL containing Dig-TMP. Incubate the cells at 37 °C for 30 min
5. Place the plates on the prewarmed metal tray in the Rayonet chamber, remove the lid and cover with a glass sheet and expose to UVA light at

$3 J/cm^2$. Change the medium, maintaining the temperature precautions cited above, to incubate cells in 10 μM CldU for 40 min and then in 100 μM IdU for 40 min

6. Following the incubations with the analogs harvest the cells by trypsinization. Suspend cells in PBS, count, and adjust concentration to 100 cells/μL. Drop approximately 200 cells on a glass slide followed by 10 μL lysis buffer. Mix the buffer and cells with a pipet tip, wait 2 min
7. Tilt the slide 15 degrees. After the liquid has run to the end of the slide air dry the slide and immerse in 3:1 methanol/acetic acid to fix
8. Wash the slides once in 1 × PBS for 5 min and transfer to 2.5 M HCl for 60 min at room temperature, then neutralize the slides with 0.4 M Tris-HCl (pH 7.5) for 10 min, followed by washing once with 1 × PBS for 5 min at room temperature

3.5 Immunostaining

1. Block the slides with freshly prepared blocking solution (5% Bovine serum (BSA) and 10% Goat serum (GS) in PBS-T (Tween-20: 0.1%)
2. Add 4 drops of blocking solution (~100 μL) to each slide and place a piece of parafilm on the top for even distribution of solution
3. Incubate for 1 h at room temperature or at 4 °C for overnight in a humidity chamber
4. Prepare fresh primary and secondary antibody dilution in blocking solution
5. Add 75 μL of rat anti-BrdU (to detect CldU) (1: 200) and cover with piece of parafilm for even distribution
6. Incubate the slide for 1 h 30 min at room temperature in a humidity chamber
7. Rinse the slides three times in 1 × PBS-T for 5 min each (total 15 min).
8. Add 75 μL of Alexa 647 goat anti-rat (1: 100) (as above) and incubate the slide for 1 h 30 min at room temperature in a humidity chamber
9. Rinse the slides three times with 1 × PBS-T for 5 min each (total 15 min) in a Coplin staining jar over digital rocking shaker
10. Prepare Mouse anti-BrdU (IdU), 1:40 + Rabbit anti-Digoxigenin polyclonal antibody 1:200 in blocking solution and add 75 μL mix to the slide
11. Incubate the slide for 1 h 30 min at room temperature in a humidity chamber

12. Rinse the slides three times with 1 × PBS-T for 5 min each (total 15 min) in a Coplin staining jar over digital rocking shaker
13. Add 75 μL of Alexa 488 goat anti mouse (1:100) + Qdot 655 goat anti-rabbit, 1:2500 mix to the slide

(Note: Briefly centrifuge Q dot 655 vial at 5000 × *g* for 1 min and take the supernatant for dilutions/staining)

14. Incubate the slide for 1 h 30 min at room temperature in a humidity chamber
15. Rinse the slides three times with 1 × PBS-T for 5 min each (total 15 min) in a Coplin staining jar over digital rocking shaker
16. Leave the slides flat and allow to air dry for 15–30 min in the dark
17. Mount the slides by placing 3 drops (65 μL) of Prolong Gold anti-fade mounting medium (Invitrogen, #P36984) and cover with glass cover-slip (avoid air bubbles).

3.5.1 Imaging

1. Incubate the slides for curing at 4 °C overnight (in the dark) and image using a fluorescence microscope. We use a Zeiss Axiovert 200 M microscope with the Axio Vision software packages. It is important to view the slides within 24–48 h after mounting as we find that the quantum dots degrade on longer storage in Prolong Gold
2. Locate the DNA fibers on slide using ×10 objective lens. Later change to ×63 oil and focus on the DNA fibers. Images are acquired with ×63 objective lens with oil immersion
3. Detect the quantum dot signal with a Qdot 655 filter
4. Detect the CldU signal with the 647 channel
5. Detect IdU signal with the 488 channel

We find that less than 10% of replication tracts are marked by a quantum dot which indicates an encounter with an ICL. See Fig. 2 for expected Dig+ replication patterns.

4. Sequential proximity ligation assay

GFP-DONSON: pMCM2; FANCM: pMCM2

4.1 Cells

HeLa cells expressing GFP-DONSON

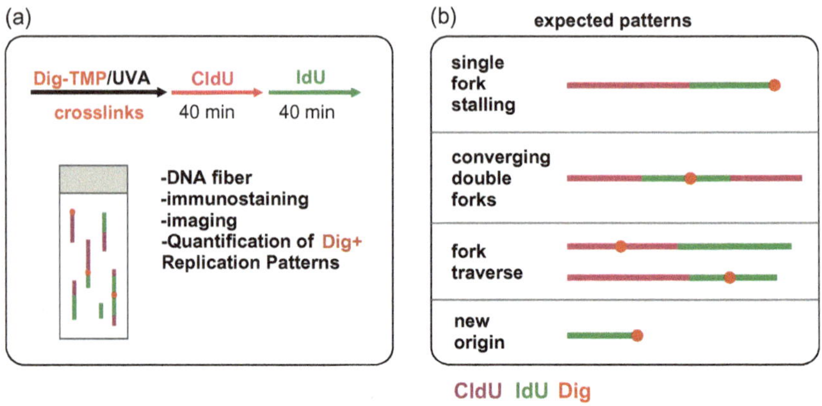

Fig. 2 Quantification of replication patterns in the vicinity of DNA interstrand crosslinks. (A) Experimental Scheme. (B) Expected patterns. Length of replication tracts will vary.

4.2 Reagents

Cell-TAK™, Corning, 354240

rabbit anti-phosphoMCM2S108, Abcam, Cat # ab109271, 1 in 200 in blocking buffer

mouse anti-GFP, Abcam, Cat # ab1218, 1 in 500 in blocking buffer

rabbit anti-FANCM, Bethyl, Cat # 302-637A, 1 in 200 in blocking buffer

Normal goat serum, Jackson Immunoresearch 005-000-121

Duolink® in situ wash buffer A, Sigma Aldrich DUO82046

Duolink® in situ wash buffer B, Sigma Aldrich DUO82048

Duolink® in situ oligonucleotide PLA probe PLUS, Sigma-Aldrich DUO92002

Duolink® in situ oligonucleotide PLA probe MINUS, Sigma-Aldrich DUO92004

Duolink® in situ Detection Reagent Orange Sigma-Aldrich DUO92007

Duolink® in situ Detection Reagent Red, Sigma, DUO92013

Duolink® ligase stock buffer, Sigma-Aldrich

Duolink® ligase, Sigma-Aldrich

Duolink® amplification stock buffer, Sigma-Aldrich

Duolink® DNA polymerase, Sigma-Aldrich

Ultrapure water

Formaldehyde 16%, Fisher Scientific, PI28906

5% Acetic acid

0.1 M $NaHCO_3$, pH 8.0 Filter sterilized
1 N NaOH
RNAse A, Qiagen, 19,101
RNAse A solution: 1 μL of 100 mg/mL RNAse A per 100μL of 5 mM EDTA in 1 ×
Phosphate Buffered Saline (PBS)
0.1% formaldehyde in PBS
4% formaldehyde in PBS
MeOH, cooled to −20 °C
CSK-R buffer: 10 mM PIPES, pH 7.0, 100 mM NaCl, 300 mM sucrose, 3 mM $MgCl_2$, 0.5% Triton X-100, 300 μg/mL RNAse
Triton X permeabilization solution: 1 × PBS, 100 mM glycine, 0.2 mg/mL BSA, 0.5% Triton X EDTA, 1%
Blocking buffer: 5% BSA in 1 × PBS, 0.01% Na Azide. Prior to use add goat serum to 10%.
Stripping buffer: 6 M guanidinium hydrochloride pH 4.0 in 5% sucrose, buffered with 0.05 M citric acid/sodium citrate (0.12 g citric acid/0.17 g trisodium citrate in 100 mL water)
PBS-T: 1 L Phosphate buffered saline, 500 μL Tween 20
Prolong Gold with DAPI, ThermoFisher Scientific, P36931

4.3 Equipment

Glass bottom Plates, MATTEK p35G-1.5-10-C
Diamond tip scriber, Thomas Scientific 1209L21
Confocal microscope such as Nikon TE2000 spinning-disk confocal microscope, with a Plan Fluor ×60/1.25 numerical aperture oil objective

4.4 Software

IMARIS, Bitplane
Volocity, Quantum Technologies

4.5 Procedures

4.5.1 Marking Mattek plates

1. Place a vertical stripe on the side of a Mattek plate with an indelible pen. This will mark the 12:00 o'clock reference position (Fig. 3A).
2. Use the diamond tip pen to etch a cross in the center of the glass bottom well with the vertical in line with the reference mark on the side of the plate. A light touch is recommended to avoid breaking the glass (Fig. 3A).

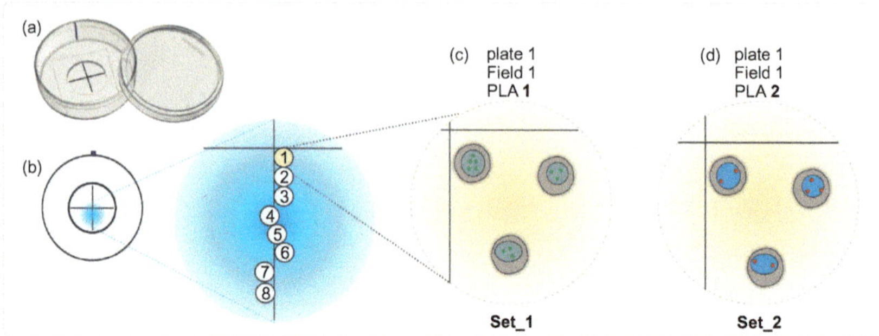

Fig. 3 Imaging the same fields for the first and second PLA. (A) Use permanent marker to generate the 12:00 o'clock mark on the outside of the plate. Use a diamond pen to scratch a cross on the glass coverslip, with one of the lines aligned with the 12:00 mark. (B) Image the plate after PLA1 starting at the cross, following one (or more) line(s), recording a "map" of the fields photographed, so that the same fields can be searched for when imaging PLA2. (C and D) After PLA 2, with the help of the map, photograph the same fields. For each plate and field there will be a Set_1 (stack of images covering the cells in that field, with DAPI and PLA1 channels) and a Set-2 (stack of images with DAPI and PLA2 channels).

3. Sterilize the plate with an EtOH wash and dry in a hood
4. In subsequent manipulations carefully aspirate solutions from the glass bottom well so as to keep the glass intact. It is best to keep plates in a larger plate, such as a 150 mm plastic dish to avoid cracking or breaking the glass that can occur when Mattek plates are placed directly on the bench

4.5.2 Preparation of CellTak coated Mattek plates

Cell-TAK is an adhesive preparation that improves attachment of cells to surfaces. We have found that it greatly improves retention of cells through the extensive washes required by the PLA.

The glass bottomed well in the middle of the Mattek plates can accommodate 100 μL of reagents. To coat 11 plates:

1. Mix 12.7 μL of Cell-TAK (3.02 μg/μL) and 23.9 μL 5% acetic acid, add 1045 μL 0.1 M $NaHCO_3$ (pH 8.0) and 18.3 μL NaOH
2. Immediately apply 100 μL of the mixture to each glass bottomed well. Incubate 20 min at room temperature
3. Aspirate and wash 3–4 times with sterile PBS
4. Seed HeLa cells expressing GFP-DONSON to 35% confluency so that the next day they are 70% confluent

4.5.3 Sequential PLA of DONSON: pMCM2 and FANCM: pMCM2 in cells treated with TMP/UVA

Nonspecific binding of antibodies is a major source of artifactual results with proximity ligation assays. We find that treatments of cells with detergents, RNAse A, and MeOH reduce the high background which can be mistaken for genuine signals.

1. Incubate cells in the glass bottomed well with 100 μL 1.5 μM TMP in medium for 30 min. Expose to UVA at 3 J/cm^2 in the pre warmed Rayonet chamber as above. Replace the medium with fresh warm medium, again as above
2. Incubate at 37 °C for 60 min
3. Remove medium, wash with PBS
4. Incubate in 0.1% formaldehyde in PBS for 10 min at room temperature. This helps retain cells on the plate
5. Wash with PBS
6. Incubate with CSK-R buffer for 3 min, remove the buffer and repeat with CSK-R for another 3 min
7. Incubate with 4% formaldehyde for 10 min
8. Wash with PBS
9. Incubate in precooled MeOH at −20 °C for 20 min
10. Wash with PBS three times
11. Incubate with Triton X permeabilization solution
12. Wash with PBS three times
13. Incubate with RNAse A solution for 30 min at 37 °C
14. Wash with PBS three times
15. Incubate cells with blocking buffer for 2 h at room temperature or overnight at 4 °C
16. Remove blocking buffer. Then add 20 μL each primary antibody to the glass bottom well in the plate (anti GFP, 1/500; anti pMCM2, 1/200, in blocking buffer). Incubate 30 min at 37 °C. It is important to avoid drying the cells in the well. Place the covered plates in a humid chamber during the incubation
17. After the incubation with the primary antibodies wash the plate with PBS-T for 5 min three times
18. During the washes dilute PLA plus and minus probes 1:5 in blocking buffer during the incubation with primary antibodies. For one well add 8 μL plus and 8 μL minus probe to 24 μL blocking buffer. Vortex solution

19. Remove the last PBS-T wash and add the 40 μL probe solution to the well. Incubate 1 h at 37 °C
20. Wash with Buffer A (from manufacturer) three times 10 min on shaker
21. During the last wash prepare the ligation solution. Mix 31 μL ultrapure water and 8 μL 5 × ligation stock
22. Remove the last wash of Buffer A, add 1 μL ligase to the 1 × ligation buffer, vortex, and add immediately to the well. Incubate 30 min at 37 °C in a humid chamber
23. Wash three times for 2 min each with Buffer A
24. Mix 8 μL in situ detection reagent Orange 5 × stock buffer with 31.5 μL ultrapure water
25. Remove the last wash of Buffer A from the well. Add 0.5 μL of DNA polymerase (kept at −20 °C) to the amplification buffer. Vortex and add immediately to the well
26. Incubate for 100 min at 37 °C in a humid chamber
27. Remove amplification solution and wash five times for 3 min each with Buffer B
28. Wash once with PBS
29. Add Prolong Gold mounting medium in the well. Care should be taken to avoid trapping air bubbles under the coverslip
30. Place the plate on the stage of the confocal microscope with the mark on the side positioned at 12:00 o'clock
31. Focus the 60 × oil objective on the intersection of the cross scratched on the glass
32. Take a bright field image of the cells at the intersection and record the location of the field as field #1 (Fig. 3B and C).
33. Take 16 stacks covering 3.2 μm of the field (Volocity) and export the stack of images as .OMETIFF (so that they can be imported in Imaris) using a file name that will identify the PLA, the plate and the field number. It is important to take enough stacks to cover the whole cell nucleus
34. Depending on the distribution of cells choose a scratch line progressing out from the intersection, e.g., horizontal to the right. Record the choice of line
35. Move the plate along the scratch line such that the next field is immediately adjacent to field. #1. Take a bright field image of these cells and 16 stacks of the fluorescent PLA signals in this field as above. Record as field #2

36. Continue to move along the chosen line taking images and recording location along the line as above until a few hundred cells have been examined (Fig. 3B).
37. Remove the coverslip from the plate and wash the well with PBS-T to remove the mounting medium
38. To remove antibodies and reaction products add 100 µL 6 M guanidine: HCl in 5% of sucrose to the well and incubate at 40 °C for 10 min. Then wash with PBST
39. Examine the fields for residual fluorescence, there should be none
40. Treat with blocking buffer and repeat the PLA between FANCM and pMCM2S108 as above (steps 14–29), except use the Duolink Detection Reagent Red
41. Position the mark on the side of the plate at 12:00 o'clock as before and locate the intersection of the scratched cross. Referencing the bright field photo of field #1 locate that field and take 16 stacks of the fluorescent PLA and DAPI signals as above (Fig. 3B–D).
42. Move along the scratch line as before locating each field and capturing 16 stacks of the PLA signal as above, and export as .OMETIFF using a file name that will identify the PLA, the plate and the field number
43. After the two successive PLA experiments, for each cell imaged there will be

- Set_1: one stack of images with 2 channels: PLA1 (green) and DAPI from PLA1 (pseudo-colored light blue) (Fig. 4A)
- Set_2: one stack of images with 2 channels: PLA2 (red) and DAPI from PLA2 (pseudocolored darker blue or a fourth color) (Fig. 4B)

The next step involves merging the two sets to compare the pattern obtained for PLA1 and PLA2. Before doing this, the two sets need to be aligned in x, y and z using their DAPI channel (or another nuclear compartment or nuclear membrane marker). 3D imaging software such as IMARIS, allows the user to register in x, y, z, a cell that has been imaged at different timepoints. The software can be tricked to import Set_2 as a second timepoint for Set_1, generating a time-series containing both sets (Fig. 4C). Next, create a surface of each nucleus using the DAPI channels and perform correction of rotational and translational drift between the two sets. Having at least 3 cells in the field will increase the accuracy of drift correction (Fig. 4D).

44. Verify they are correctly aligned using the DAPI channels from both sets, crop to the cell of interest and save the correctly registered and cropped Set_1 and Set_2 (Fig. 4E).

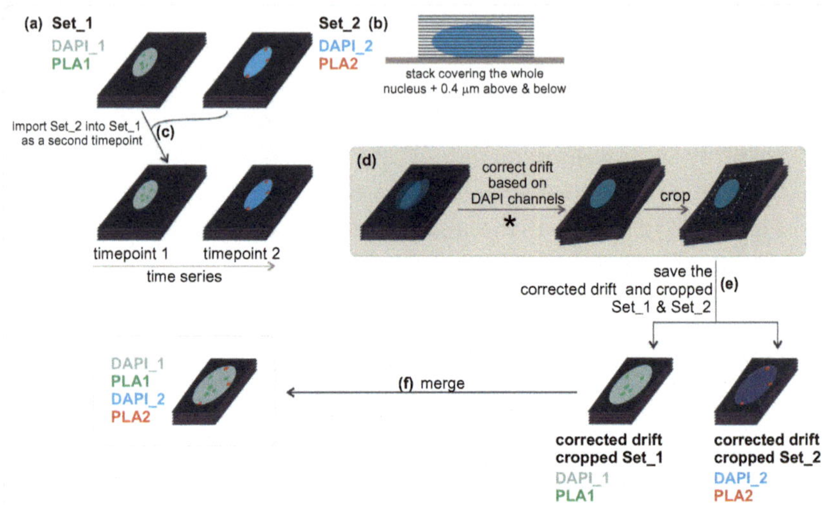

Fig. 4 Aligning both PLAs in xyz coordinates. (A and B) For each plate and field locate Set_1 and Set_2 (stack of images covering the whole nucleus). (C) Open Set-1 in IMARIS and import Set-2 as a second time point. (D) Generate a surface from the DAPI channel and use it to correct translational and rotational drift between the two timepoints. * The scheme shows only one cell in the field for simplicity but note that the field should contain at least 3 cells for the software to accurately align in 3D space the two "timepoints." Crop to the selected cell. Make a copy. (E) For copy #1: delete the second time point and save as corrected_drift_cropped_Set_1. For Copy#2: delete the first time point and save as corrected_drift_cropped_Set_2. (F) Merge to generate a composite containing both DAPI channels, PLA1 and PLA2.

45. Merge the correctly registered cropped sets to visualize both PLAs in the same image (Fig. 4F). A representative image is shown in Supplementary Video 1 in the online version at https://doi.org/10.1016/bs.mie.2021.08.015

Acknowledgments

This research was supported by the Intramural Research Program of the NIH, National Institute on Aging, United States (Z01-AG000746-08).

Contributions

R.C.J., M.A.B., J.Z., J.H., H.G., and J.G. developed assays and performed experiments to validate the assays. A.K.T., D.P., and A.S. developed synthetic strategies and synthesized compounds. R.C.J., J.H., J.Z., and M.A.B. captured images and performed image analysis. R.C.J., M.A.B., A.S., J.Z., and M.M.S. prepared the manuscript.

References

Abid, A. F., Douglas, M. E., Locke, J., Pye, V. E., Nans, A., Diffley, J. F. X., et al. (2017). Cryo-EM structure of a licensed DNA replication origin. *Nature Communications*, *8*(1), 2241. https://doi.org/10.1038/s41467-017-02389-0 (pii).

Alderton, G. K., Joenje, H., Varon, R., Borglum, A. D., Jeggo, P. A., & O'Driscoll, M. (2004). Seckel syndrome exhibits cellular features demonstrating defects in the ATR-signalling pathway. *Human Molecular Genetics*, *13*(24), 3127–3138. (Retrieved from http://www.ncbi.nlm.nih.gov/pubmed/15496423).

Bell, S. P., & Labib, K. (2016). Chromosome duplication in Saccharomyces cerevisiae. *Genetics*, *203*(3), 1027–1067. doi:203/3/1027 [pii]; https://doi.org/10.1534/genetics.115.186452 [doi].

Benedetto, A. V. (1977). The psoralens. An historical perspective. *Cutis*, *20*(4), 469–471 (Retrieved from http://www.ncbi.nlm.nih.gov/pubmed/338258).

Blow, J. J., Ge, X. Q., & Jackson, D. A. (2011). How dormant origins promote complete genome replication. *Trends in Biochemical Sciences*, *36*(8), 405–414 (Retrieved from http://www.ncbi.nlm.nih.gov/pubmed/21641805).

Bochman, M. L., & Schwacha, A. (2010). The Saccharomyces cerevisiae Mcm6/2 and Mcm5/3 ATPase active sites contribute to the function of the putative Mcm2-7 'gate'. *Nucleic Acids Research*, *38*(18), 6078–6088. https://doi.org/10.1093/nar/gkq422.

Douglas, M. E., Ali, F. A., Costa, A., & Diffley, J. F. X. (2018). The mechanism of eukaryotic CMG helicase activation. *Nature*, *555*(7695), 265–268. doi:nature25787 [pii]; https://doi.org/10.1038/nature25787 [doi].

Gambus, A., Jones, R. C., Sanchez-Diaz, A., Kanemaki, M., van Deursen, F., Edmondson, R. D., et al. (2006). GINS maintains association of Cdc45 with MCM in replisome progression complexes at eukaryotic DNA replication forks. *Nature Cell Biology*, *8*(4), 358–366. doi:ncb1382 [pii]; https://doi.org/10.1038/ncb1382 [doi].

Gao, Y., Cui, Y., Fox, T., Lin, S., Wang, H., de Val, N., et al. (2019). Structures and operating principles of the replisome. *Science*, *363*(6429). https://doi.org/10.1126/science.aav7003.

Heller, R. C., & Marians, K. J. (2006). Replication fork reactivation downstream of a blocked nascent leading strand. *Nature*, *439*(7076), 557–562 (Retrieved from http://www.ncbi.nlm.nih.gov/pubmed/16452972).

Huang, J., Liu, S., Bellani, M. A., Thazhathveetil, A. K., Ling, C., de Winter, J. P., et al. (2013). The DNA translocase FANCM/MHF promotes replication traverse of DNA interstrand crosslinks. *Molecular Cell*, *52*(3), 434–446. doi:S1097-2765(13)00713-2 [pii]; https://doi.org/10.1016/j.molcel.2013.09.021 [doi].

Huang, J., Zhang, J., Bellani, M. A., Pokharel, D., Gichimu, J., James, R. C., et al. (2019) Remodeling of Interstrand crosslink proximal replisomes is dependent on ATR, FANCM, and FANCD2. *Cell Reports*, *27*(6), 1794–1808. doi:S2211-1247(19)30497-8 [pii]; https://doi.org/10.1016/j.celrep.2019.04.032 [doi].

Klingseisen, A., & Jackson, A. P. (2011). Mechanisms and pathways of growth failure in primordial dwarfism. *Genes & Development*, *25*(19), 2011–2024. doi:25/19/2011 [pii]; https://doi.org/10.1101/gad.169037 [doi].

Knipscheer, P., Raschle, M., Smogorzewska, A., Enoiu, M., Ho, T. V., Scharer, O. D., et al. (2009). The Fanconi anemia pathway promotes replication-dependent DNA interstrand cross-link repair. *Science*, *326*(5960), 1698–1701 (Retrieved from http://www.ncbi.nlm.nih.gov/pubmed/19965384).

Koos, B., Andersson, L., Clausson, C. M., Grannas, K., Klaesson, A., Cane, G., et al. (2014). Analysis of protein interactions in situ by proximity ligation assays. *Current Topics in Microbiology and Immunology*, *377*, 111–126. https://doi.org/10.1007/82_2013_334 [doi].

Lai, C., Cao, H., Hearst, J. E., Corash, L., Luo, H., & Wang, Y. (2008). Quantitative analysis of DNA interstrand cross-links and monoadducts formed in human cells induced by psoralens and UVA irradiation. *Analytical Chemistry*, *80*(22), 8790–8798 (Retrieved from http://www.ncbi.nlm.nih.gov/pubmed/18947205).

Lehmann, A. R., & Fuchs, R. P. (2006). Gaps and forks in DNA replication: Rediscovering old models. *DNA Repair (Amst)*, *5*(12), 1495–1498 (Retrieved from http://www.ncbi.nlm.nih.gov/pubmed/16956796).

Li, H., Yao, N. Y., & O'Donnell, M. E. (2020). Anatomy of a twin DNA replication factory. *Biochemical Society Transactions*, *48*(6), 2769–2778. https://doi.org/10.1042/bst20200640.

Marians, K. J. (2018). Lesion bypass and the reactivation of stalled replication forks. *Annual Review of Biochemistry*. https://doi.org/10.1146/annurev-biochem-062917-011921.

Marmur, J., & Grossman, L. (1961). Ultraviolet light induced linking of deoxyribonucleic acid strands and its reversal by photoreactivating enzyme. *Proceedings of the National Academy of Sciences of the United States of America*, *47*, 778–787 (Retrieved from http://www.ncbi.nlm.nih.gov/pubmed/13767019).

Muniandy, P. A., Liu, J., Majumdar, A., Liu, S. T., & Seidman, M. M. (2010). DNA interstrand crosslink repair in mammalian cells: Step by step. *Critical Reviews in Biochemistry and Molecular Biology*, *45*(1), 23–49 (Retrieved from http://www.ncbi.nlm.nih.gov/pubmed/20039786).

Mutreja, K., Krietsch, J., Hess, J., Ursich, S., Berti, M., Roessler, F. K., et al. (2018). ATR-mediated global fork slowing and reversal assist fork traverse and prevent chromosomal breakage at DNA Interstrand cross-links. *Cell Reports*, *24*(10), 2629–2642. doi: S2211-1247(18)31273-7 [pii]; https://doi.org/10.1016/j.celrep.2018.08.019 [doi].

Parker, M. W., Bell, M., Mir, M., Kao, J. A., Darzacq, X., Botchan, M. R., et al. (2019). A new class of disordered elements controls DNA replication through initiator self-assembly. *eLife*, *8*. https://doi.org/10.7554/eLife.48562 (doi];48562 [pii).

Raschle, M., Knipsheer, P., Enoiu, M., Angelov, T., Sun, J., Griffith, J. D., et al. (2008). Mechanism of replication-coupled DNA interstrand crosslink repair. *Cell*, *134*(6), 969–980 (Retrieved from http://www.ncbi.nlm.nih.gov/pubmed/18805090).

Rupp, W. D., & Howard-Flanders, P. (1968). Discontinuities in the DNA synthesized in an excision-defective strain of Escherichia coli following ultraviolet irradiation. *Journal of Molecular Biology*, *31*(2), 291–304 (doi:0022-2836(68)90445-2 [pii).

Sparks, J. L., Chistol, G., Gao, A. O., Raschle, M., Larsen, N. B., Mann, M., et al. (2019). The CMG helicase bypasses DNA-protein cross-links to facilitate their repair. *Cell*, *176*(1–2), 167–181. doi:S0092-8674(18)31452-1 [pii]; https://doi.org/10.1016/j.cell.2018.10.053 [doi].

Taylor, M. R. G., & Yeeles, J. T. P. (2018). The initial response of a eukaryotic replisome to DNA damage. *Molecular Cell*, *70*(6), 1067–1080. doi:S1097-2765(18)30344-7 [pii]; https://doi.org/10.1016/j.molcel.2018.04.022 [doi].

Wasserman, M. R., Schauer, G. D., O'Donnell, M. E., & Liu, S. (2019). Replication fork activation is enabled by a single-stranded DNA gate in CMG helicase. *Cell*, *178*(3), 600–611. doi:S0092-8674(19)30735-4 [pii]; https://doi.org/10.1016/j.cell.2019.06.032 [doi].

Yang, W., Seidman, M. M., Rupp, W. D., & Gao, Y. (2019). Replisome structure suggests mechanism for continuous fork progression and post-replication repair. *DNA Repair (Amst)*, *81*, 102658. doi:S1568-7864(19)30211-3 [pii]; https://doi.org/10.1016/j.dnarep.2019.102658 [doi].

Yao, N. Y., & O'Donnell, M. E. (2019). Getting ready for DNA duplication. *eLife*, *8*. https://doi.org/10.7554/eLife.51291 (doi];51291 [pii).

Yeeles, J. T., & Marians, K. J. (2011). The Escherichia coli replisome is inherently DNA damage tolerant. *Science*, *334*(6053), 235–238 (Retrieved from http://www.ncbi.nlm.nih.gov/pubmed/21998391).

Zhai, Y., & Tye, B. K. (2017). Structure of the MCM2-7 double hexamer and its implications for the mechanistic functions of the Mcm2-7 complex. *Advances in Experimental Medicine and Biology*, *1042*, 189–205. https://doi.org/10.1007/978-981-10-6955-0_9.

Zhang, J., Bellani, M. A., James, R. C., Pokharel, D., Zhang, Y., Reynolds, J. J., et al. (2020). DONSON and FANCM associate with different replisomes distinguished by replication timing and chromatin domain. *Nature Communications*, *11*(1), 3951. https://doi.org/10.1038/s41467-020-17449-1.

Zhang, J., Dewar, J. M., Budzowska, M., Motnenko, A., Cohn, M. A., & Walter, J. C. (2015). DNA interstrand cross-link repair requires replication-fork convergence. *Nature Structural & Molecular Biology*, *22*(3), 242–247. doi:nsmb.2956 [pii]; https://doi.org/10.1038/nsmb.2956 [doi].

CHAPTER FOUR

Single-molecule imaging of replication fork conflicts at genomic DNA G4 structures in human cells

Wei Ting C. Lee*,†, Dipika Gupta*,†, and Eli Rothenberg*
Department of Biochemistry and Molecular Pharmacology, New York University School of Medicine, New York, NY, United States
*Corresponding authors: e-mail address: weitingchelsea.lee@nyulangone.org; dipika.gupta@nyulangone.org; eli.rothenberg@nyulangone.org

Contents

Abstract

DNA G-quadruplexes (G4s) are stable, non-canonical DNA secondary structures formed within guanine(G)-rich sequences. While extensively studied *in vitro*, evidence of the occurrence of G4s *in vivo* has only recently emerged. The formation of G4 structures may pose an obstacle for diverse DNA transactions including replication, which is linked

† These authors contributed equally.

Methods in Enzymology, Volume 661
ISSN 0076-6879
https://doi.org/10.1016/bs.mie.2021.08.008

to mutagenesis and genomic instability. A fundamental question in the field has been whether and how the formation of G4s is coupled to the progression of replication forks. This process has remained undefined largely due to the lack of experimental approaches capable of monitoring the presence of G4s and their association with the replication machinery in cells. Here, we describe a detailed multicolor single-molecule localization microscopy (SMLM) protocol for detecting nanoscale spatial-association of DNA G4s with the cellular replisome complex. This method offers a unique platform for visualizing the mechanisms of G4 formation at the molecular level, as well as addressing key biological questions as to the functional roles of these structures in the maintenance of genome integrity.

1. Introduction

DNA G4s are non-canonical secondary structures formed within guanine(G)-rich repeats. *In vitro* biochemical and structural studies suggest that stable G4 structures can form under physiological conditions (Lane, Chaires, Gray, & Trent, 2008). These are enriched within regions of genomic regulation such as telomeres and gene promoters, with computational algorithms predicting the presence of >370,000 putative G4 sites within the human genome (Huppert & Balasubramanian, 2005; Todd, Johnston, & Neidle, 2005).

DNA G4 levels are elevated in S-phase (Biffi, Tannahill, McCafferty, & Balasubramanian, 2013) and transient chromatin unpackaging and DNA unwinding during replication is suggested to be conducive to G4 assembly. However, if and how the formation of G4 structures affects DNA replication, and the mechanisms of their resolution remain unclear. The deregulated formation of G4s could result in genomic instability, epigenetic alterations and ultimately carcinogenesis (Lerner & Sale, 2019; Valton & Prioleau, 2016). Conversely, strategies enhancing G4 formation as impediments to DNA replication, such as G4 stabilizing ligands or inhibitors targeting G4 destabilizing helicases, are being explored for cancer therapeutics (Bidzinska, Cimino-Reale, Zaffaroni, & Folini, 2013; Islam, Jackson, Rahman, & Thurston, 2016; Neidle, 2016). Thus, understanding the physiological regulation and the pathological consequences of G4 formation is crucial to harnessing its potential as an effective treatment modality.

Despite their importance, the direct visualization of these dynamic processes within complex cellular environments has been challenging. The recent development of G4-recognizing antibodies and chemical probes to visualize G4 formation *in vivo* via immunofluorescence have begun to provide crucial insights into G4 biology (Biffi et al., 2013; Henderson et al., 2014). However, the diffraction limit of light restricts the resolution of conventional microscopy to approximately 250 nm (Huang, Babcock, & Zhuang, 2010), hindering our ability to effectively localize G4 structures and probe their association with various protein factors within the dense nuclear environment. Herein, we describe an experimental approach that utilizes multicolor single molecule localization microscopy (SMLM) (Heilemann et al., 2008; Rust, Bates, & Zhuang, 2006) to resolve and monitor the spatial association of DNA G4s with the replisome at the molecular scale. This protocol can be easily adapted to study various aspects of replication fork progression and repair in the context of G4 structures and other lesions (Lee et al., 2021; Reid et al., 2015; Whelan et al., 2018, 2020; Whelan & Rothenberg, 2021).

2. Single molecule localization microscopy (SMLM)

A major limitation of conventional optical microscopy is the inability to resolve the simultaneous emissions from two adjacent fluorophores within a diffraction-limited area and therefore, molecular features below 250 nm cannot be resolved. Over the past two decades, several new optical microscopy approaches, collectively termed super-resolution microscopy, were developed to circumvent this limitation (Huang et al., 2010). The different modalities of various super-resolution microscopy approaches provide improved resolution ranging from 150 to 80 nm for structured illumination microscopy (SIM) (Gustafsson, 2000) and stimulated emission microscopy (STED) (Hell & Wichmann, 1994; Klar & Hell, 1999) respectively, with the highest resolution (5–20 nm) achieved by methods that utilize SMLM, such as photo-activated localization microscopy (PALM) and stochastic optical reconstruction microscopy (STORM) (Betzig et al., 2006; Heilemann et al., 2008; Hess, Girirajan, & Mason, 2006; Kaksonen & Drubin, 2006; Rust et al., 2006).

SMLM utilizes sequential and stochastic switching of a subset of fluorophores between a bright ("on") and a dark ("off") state. The centroid positions of the sparsely distributed active fluorophores can then be determined by mathematical models with nanometer precision (Pertsinidis, Zhang, & Chu, 2010; Shtengel et al., 2009; Thompson, Larson, & Webb, 2002; Yildiz et al., 2003). Through serial imaging and localization of different activated fluorophores subsets, SMLM resolves fluorophore localization within complex and diffraction-limited biological structures to render a super-resolved image with structural information at the nanometer (5–20 nm) scale (Fig. 1). Importantly, multicolor SMLM has advanced our understanding of DNA transactions such as DNA replication and transcription, as well as DNA damage response pathways that otherwise cannot be achieved by conventional microscopy and ensemble *in vitro* assays (Cong et al., 2021; D'Alessandro et al., 2018; Lee et al., 2021; Reid et al., 2015; Tonzi, Yin, Lee, Rothenberg, & Huang, 2018; Whelan et al., 2018, 2020; Whelan & Rothenberg, 2021).

The general workflow of SMLM includes (a) immunolabeling; (b) image acquisition; (c) single-molecule localization and image reconstruction; (d) image analysis. Briefly, important considerations for a SMLM experiment are outlined below. For a detailed description of the theoretical basis of SMLM, instrumentation and data analysis we refer the reader to several excellent articles (Lambert & Waters, 2017; Xu, Ma, & Liu, 2017).

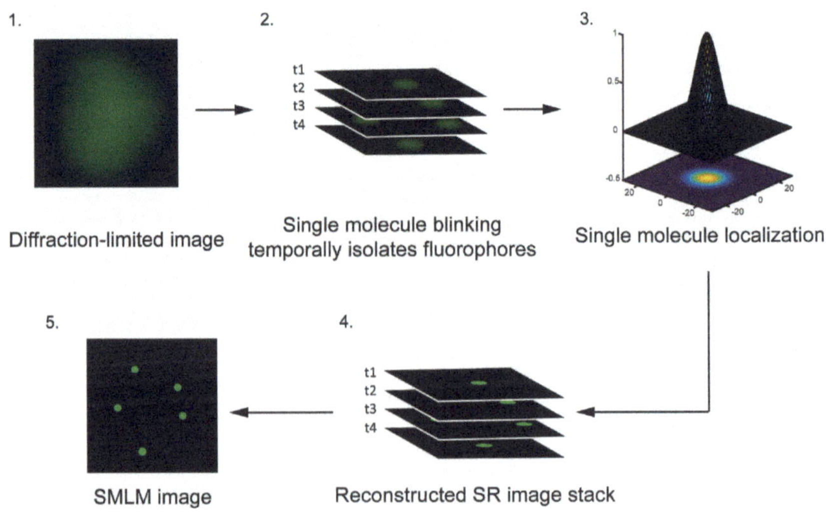

Fig. 1 Single molecule localization microscopy. 1. When all fluorophores simultaneously stay in an "on" state, a diffraction-limited image is obtained. 2. Imaging stochastically photo-switching fluorophores over time across several frames gives rise to a sequence of sparsely populated images, where 3. each fluorophore can be isolated and localized with high precision by fitting the point spread function (PSF) to each spot. 4–5. Stacking these reconstructed images gives rise to a super-resolved image.

2.1 Protein labeling with photoswitchable fluorescent probes

When available, use commercially available fluorophore-conjugated or unlabeled primary antibodies previously used for SMLM to detect targets of interest. Antibodies validated for confocal immunofluorescence also work well for SMLM, but optimal antibody dilutions must be determined for each target and cell line to achieve optimal labeling densities for satisfying the Nyquist sampling theorem (Nyquist, 1928; Shannon, 1949) and good single-molecule localization (Whelan & Bell, 2015). Also, indirect immunofluorescence-based labeling increases the size of the labeled target complex (>30 nm), reducing the labeling density and structural resolution of the sample (Demmerle, Wegel, Schermelleh, & Dobbie, 2015). Therefore, fluorophore-conjugated primary antibodies are advised to be used to reduce the overall label size. Alternatively, increased labeling specificity for SMLM can be achieved by using fluorophore-conjugated ligands to non-fluorescent HaloTag, SNAP-tag, or CLIP-tag fusion proteins of interest (Dellagiacoma et al., 2010; Liss, Barlag, Nietschke, & Hensel, 2015).

2.2 Stochastic photoactivation of labeled fluorophores

Optimal stochastic photoswitching of fluorophores is critical for the resolution enhancement afforded by SMLM. Photophysical properties of different organic dyes have been extensively characterized and we refer the reader to Dempsey et al. (2009), Fernandez-Suarez and Ting (2008), Klein, Proppert, and Sauer (2014). Fluorophores with enhanced emission photon count, low on/off duty cycles and a greater number of switching cycles are ideal for SMLM. Importantly, differences in imaging buffer conditions needed for optimal photoblinking of different fluorophores can complicate multicolor SMLM. Therefore, fluorophore combinations and photoswitching buffer compositions must be empirically tested for optimal single molecule blinking and minimal spectral crosstalk amongst dyes used.

2.3 Localization precision of individual fluorophores

Upon image acquisition, single fluorescent emitters within each frame are identified and localized to generate single molecule coordinate lists using SMLM fitting algorithms (Huang et al., 2013; Small & Stahlheber, 2014; Xu et al., 2017). The precision of single-molecule localization is affected by dye properties, background fluorescence and/or detector noise (Thompson et al., 2002). An imaging system with powerful laser illumination sources, well-aligned excitation and emission paths, and a sensitive camera is therefore crucial for collecting high-quality SMLM movies for precise localization. Details on appropriate SMLM setup and different localization algorithms are extensively discussed in Bates, Jones, and Zhuang (2013), Morten, Reid, and Rothenberg (2014), Small and Stahlheber (2014), Thompson et al. (2002) and Xu et al. (2017).

2.4 Analysis of SMLM data

SMLM enables visualization of dynamic, transient, or rare molecular distributions and interactions that are difficult to resolve using brightness-directed, diffraction-limited imaging techniques (Whelan et al., 2016). Complex analytical tools are however required to derive statistically meaningful data from the large data sets SMLM affords (Bermudez-Hernandez et al., 2017; Yin, Lee, & Rothenberg, 2019; Yin & Rothenberg, 2016). Analysis methods aimed at utilizing coordinate lists rather than rendered images have been developed to attain maximal molecular association information. Therefore, quantitative measures of molecules-of-interest, such as its count, density, spatial distributions, and relationship (colocalization) within distinct structures, can be obtained at near-single-molecule precision (Coltharp, Yang, & Xiao, 2014; Nicovich, Owen, & Gaus, 2017; Wu, Tschanz, Krupnik, & Ries, 2020).

3. Protocol

This protocol describes conditions optimized for visualization of DNA G4 structures at replication forks in the human osteosarcoma U2OS cell line (ATCC® HTB-96™). However, this protocol has been successfully adapted for use with other cell lines such as HeLa (ATCC® CCL-2™), SW 1271 (ATCC® CRL-2177™), etc. Important considerations for optimization in other lines are provided in NOTES.

3.1 Reagents and equipment

Reagents and resources

Reagent or resource	**Source**	**Identifier**
Antibodies		
DNA G4 (1H6)—Mouse (1:200)	Millipore	MABE1126
DNA G4 (1H6)—Mouse (1:200)	Absolute Antibody	Ab00389-1.1
DNA G4 (BG4)—FLAG-tag (1:200)	Millipore	MABE917

Reagents and resources—cont'd

Reagent or resource	**Source**	**Identifier**
DNA G4 (1H6)—Rabbit (1:200)	Absolute Antibody	Ab00389-23.0
MCM6—Rabbit, AF568 (1:1000)	Abcam	ab211916
PCNA—Mouse (1:1000)	SCBT	sc56
Rabbit IgG—AF 750 (1:5000)	Invitrogen	A21039
Mouse IgG—AF 750 (1:5000)	Invitrogen	A21037
Rabbit IgG—AF 647 (1:5000)	Invitrogen	A21246
Mouse IgG—AF 647 (1:5000)	Invitrogen	A21235
Rabbit IgG—AF 568 (1:5000)	Invitrogen	A11036
Mouse IgG—AF 568 (1:5000)	Invitrogen	A11031
Mouse IgG—AF 488 (1:5000)	Invitrogen	A11029
Chemicals, peptides, and recombinant proteins		
Dimethyl sulfoxide (DMSO)	Sigma	D8418
Pyridostatin (PDS)	Selleckchem	S7444
Aphidicolin (APH)	Abcam	142,400
5-Ethynyl-2′-deoxyuridine (EdU)	Invitrogen	E10187
Paraformaldehyde (PFA)	Electron Microscopy Sciences	15714 S
PIPES	Sigma-Aldrich	P1851
Sucrose	ChemCruz™ Biochemicals	SC204311
NaCl	Acros Organics	327300025
$MgCl_2$	Sigma-Aldrich	M1028
KOH	Sigma-Aldrich	06005
Triton X-100	Fisher Scientific	BP151-100
Glycine	Sigma-Aldrich	G8898
Bovine serum albumin (BSA)	Gemini	700-100P
Gelatin	Fisher Scientific	G8-500
NH_4Cl	Fisher Scientific	A649-500

Continued

Reagents and resources—cont'd

Reagent or resource	Source	Identifier
Glucose	Sigma	G8270
Mercaptoethylamine (MEA)	Fisher Scientific	BP2664100
Glucose oxidase	Sigma	G2133
Catalase	Sigma	C3155
Critical commercial assays		
Click-iT™ Plus EdU Cell Proliferation Kit for Imaging with Alexa Fluor® 488	ThermoFisher	C10637
Click-iT™ Plus EdU Cell Proliferation Kit for Imaging with Alexa Fluor® 647	ThermoFisher	C10640
Other		
6-Well sterile cell culture plates	CELLSTAR®	657160
0.2 μm, polyethersulfone membrane filter unit	Nalgene®	595-4520
0.45 μm, polyethersulfone membrane filter unit	Nalgene®	295-4545
No. 1 Fisherbrand premium cover glass	Fisherbrand	12548B
Premium microscope slides plain	Fisherfinest	12-544-4
0.75 mm miniature Plated Diamond Flat-Tip drill bit	CR Laurence	MD346
TetraSpeck™ Microspheres, 0.1 μm, fluorescent blue/green/orange/dark red	ThermoFisher Scientific	T7279
Variable speed hand-held Rotary tool	DREMEL	DREMEL 4300
5-min epoxy	Devcon	14250
Scotch® Permanent Double-Sided Tape	Scotch®	
Scalpel		

3.2 Reagent setup

1. Culture medium: Dulbecco's modified Eagle's medium (DMEM) supplemented with 10% fetal bovine serum and 1% penicillin-streptomycin
2. Starvation medium: DMEM with 1% penicillin-streptomycin

3. Pyridostatin (PDS): Prepare PDS stock solution in DMSO at a concentration of 20 mM
4. Aphidicolin (APH) (< 1 μM): Prepare APH stock solution in DMSO at a concentration of 1 mM
5. 5-Ethynyl-2′-deoxyuridine (EdU) (Also supplied with Click-iT® Plus EdU Cell Proliferation Kit for Imaging): Prepare EdU stock solution in DMSO at a concentration of 10 mM. Aliquot to minimize freeze thaw cycles and store at −20 °C protected from light
6. Click-iT® Plus EdU Cell Proliferation Kit for Imaging with Alexa Fluor® 488 or Alexa Fluor® 647 conjugated picolyl azide. For SMLM detection of EdU incorporation into nascent DNA fluorescent azides purchased from other vendors (e.g., Fluoroprobes, Biotium, click chemistry tools etc.) have also been successfully used
7. 4% paraformaldehyde (PFA): Dilute 32% paraformaldehyde (electron microscopy grade) in PBS to a final concentration of 4%. Prepare fresh and keep at room temperature in the dark until use
8. Cytoskeleton (CSK) buffer: 10 mM PIPES, 300 mM Sucrose, 100 mM NaCl, 3 mM $MgCl_2$ in ultrapure water. Adjust to pH 7.4 using 4 M KOH. Filter using a 0.2 μm, polyethersulfone membrane filter unit, and store at 4 °C for up to 6 months
9. Pre-extraction buffer: Prepare fresh by addition of Triton X-100 to CSK buffer to a final concentration of 0.5% using a cut tip. Vortex until all Triton X-100 is in solution and store at 4 °C until use
10. Blocking buffer: 2% glycine, 2% bovine serum albumin (BSA), 0.2% gelatin, 50 mM NH_4Cl in PBS. Place on a magnetic stirrer overnight at 4 °C. The gelatin will not completely go into solution. Filter using a 0.45 μm, polyethersulfone membrane filter unit, aliquot and store at 4 °C for up to 2 months
11. Primary and secondary antibodies with recommended starting dilutions. See Lee et al. (2021) for additional information on G4 antibody validation
12. SMLM Imaging Buffer: 10% glucose, 100 mM mercaptoethylamine (MEA), 1 × Gloxy in PBS
    a. Stock solutions of 40% glucose and MEA can be prepared ahead of time and stored at 4 °C. Fresh MEA should be made every 2 weeks for optimal blinking efficiency
    b. 100 × Gloxy: 100 mg/mL glucose oxidase and 2 mg/mL catalase in PBS. Once made, the stock gloxy should be aliquoted and stored at −20 °C for up to 2 weeks

c. Imaging buffer (without Gloxy) should be prepared fresh before each imaging session. 1 × Gloxy should only be added to the imaging buffer and injected into the imaging chamber immediately before image acquisition to ensure pH maintenance

3.3 Cell seeding, drug treatments and EdU incorporation

1. Place one cover glass in each well of a 6 well plate. Seed 1.5 mL of U2OS cell suspension per well. Gently press down on the cover glass with a 1 mL pipette tip to remove air bubbles between the cover glass and the cell culture dish.
2. Place cells in the incubator for 60–72 h. Ideally, cells must be at 60%–70% confluence at the time of fixation.

   NOTE: To enrich for cells in S phase, aspirate media and replace with 1.5 mL of serum-free growth media 24 h after seeding cells onto cover glass. Incubate for 48–72 h to obtain a synchronized cell population arrested at the G0/G1 cell cycle phase. U2OS cells enter early S-phase approximately 12–14 h after release into serum-containing media. As serum starvation is not a 100% efficient, some cell growth is observed while in serum free media. Therefore, staining for S phase specific markers such as PCNA or thymidine analog incorporation into nascent DNA (e.g., EdU) is recommended while imaging. Additionally, these markers can be used to identify early S phase cells based on staining patterns resulting from the spatiotemporal DNA replication program.
3. To induce G4 stabilization, treat cells with 20 μM G4-stabilizing ligand, PDS, for 1 h prior to harvest. Other available G4-stabilizing ligands can also be tested for use. Alternatively, to promote G4 formation at the replication fork treat cells with low concentrations of APH, (< 1 μM) for 1 h prior to harvest, to induce ssDNA formation.
4. To label newly synthesized DNA at replication forks, add EdU to the cell culture medium to a final concentration of 10 μM, 15 min prior to harvest.

3.4 Pre-extraction and fixation

1. Gently wash cells twice with 1 × PBS, 1 mL/well. Be quick, as cells can detach in PBS without Ca^{2+} or Mg^{2+}.
2. Aspirate PBS and add 1 mL of cold freshly prepared pre-extraction buffer into each well. Incubate at room temperature for 10 min and minimize disturbance to the plate.

NOTE: The pre-extraction step removes much of the cytosolic and soluble nuclear components to detect only chromatin bound proteins. However, this pre-extraction step can cause cells to detach from the cover glass particularly with less adherent cell types or at high cell confluency. Cells can be observed using a bright-field microscope after fixation to determine optimal extraction time with minimal cell loss. Using other cell types, the pre-extraction time has been reduced to a minimum of 3 min without significant changes in G4 staining specificity.

3. Aspirate pre-extraction buffer. Gently wash with 1 mL of 1 × PBS thrice. Be gentle as cells are prone to detach at this step.
4. Fix cells with 1 mL 4% PFA in PBS per well for 15 min at room temperature in the dark.
5. Aspirate PFA solution and wash cells with 1 mL of 1 × PBS twice.
6. Wash cells with 1 mL of blocking buffer for 5 min, twice at room temperature. At this point, cells can be stored in blocking buffer at 4 °C for up to 1 week.

3.5 EdU detection

Only 100 μL of Click-iT® reaction cocktail or antibody solution is needed for all labeling reactions on cover glass. If performing labeling in 6-well plates, make sure that all liquid within the well is aspirated and that the cover glass is centered with no corners touching the sides before careful application of 100 μL of reagent directly onto the cover glass. Alternatively, 100 μL of Click-iT® reaction cocktails or antibody solution can be applied to a cover glass placed on parafilm, cell side up. For long incubations, we recommend the use of a humidified chamber to prevent the cover glass from drying out.

1. Prepare Click-iT® reaction cocktail using the components from the Click-iT™ Plus EdU Cell Proliferation Kit for Imaging as per manufacturer's instructions.

   NOTE: The concentration of fluorescent picolyl azide or fluorescent azide must be titrated to achieve optimal labeling densities for good single molecule localization. For Alexa Fluor® 647 picolyl azide provided with the Thermofisher Click-iT® Plus EdU Cell Proliferation Kit, we recommend a starting dilution range of 0.12–0.24 μL of Alexa Fluor®AF647 picolyl azide/100 μL of Click-iT® reaction cocktail. When using other fluorescent picolyl azides, concentrations could require titration to get optimal single-molecule resolution.

2. Aspirate blocking buffer and gently pipette 100 μL of Click-iT® reaction cocktail onto cover glass and incubate at room temperature in the dark for 30 min.
3. After incubation, aspirate Click-iT® reaction cocktail and wash cover glass with 1 mL of blocking buffer for 5 min, thrice.
4. Store cells in blocking buffer overnight at 4 °C in the dark.

   NOTE: While a blocking step can be performed for 1 h at room temperature before proceeding to incubate with primary antibody solutions, it results in increased nonspecific antibody binding. Therefore, an overnight blocking step is recommended at this step. Once Click-iT® reaction is performed, cells can be stored at 4 °C in blocking buffer in the dark for up to 3 days.

3.6 Immunostaining

1. Aspirate blocking buffer, and gently pipette 100 μL of primary antibody diluted in blocking buffer (see Section 3.2) onto each cover glass. Incubate in the dark for 1 h at room temperature.

   NOTE: In our experience, staining for PCNA using anti-PCNA antibody from Santa Cruz Biotechnology (sc56) in PFA fixed samples requires that the Click-iT® reaction be performed. If not detecting EdU, perform incubation with 100 μL of Click-iT® reaction cocktail without fluorescent azide as described in Section 3.5.
2. Wash cover glass with 1 mL of blocking buffer for 5 min, thrice.
3. Aspirate blocking buffer, and gently pipette 100 μL of secondary antibody diluted in blocking buffer (see Section 3.2) onto each cover glass. Incubate in the dark for 30 min at room temperature.
4. Wash cover glass with 1 mL of blocking buffer for 5 min, thrice.

3.7 Assembly of imaging chamber

Fluorophore blinking during SMLM is achieved through redox reactions carried out by the reductant (MEA) in the imaging buffer (Vogelsang et al., 2008). Therefore, the assembly of a microfluidic imaging chamber (Fig. 2) is necessary to replace and maintain samples in the SMLM imaging buffer during image acquisition.

1. Place glass slide on a wet acrylic surface. Wet top surface of the slide with water. The presence of water minimizes heating of the drill bit as you drill two 0.75 mm diameter holes on the glass slide. The holes form the inlet and outlet of the chamber, and should be vertically separated

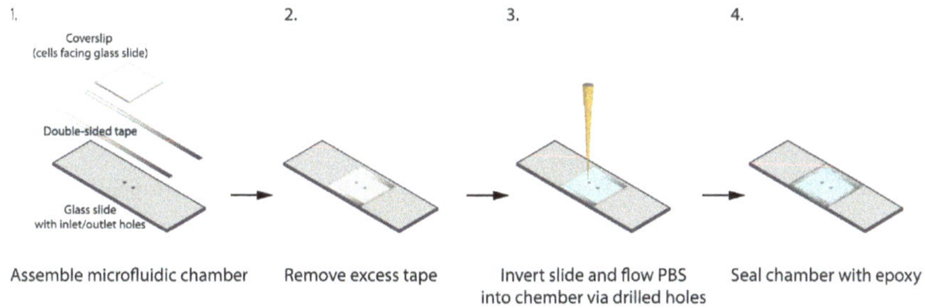

Fig. 2 Schematic illustration of sample chamber assembly (Section 3.7).

by ~1 cm, which is narrower than the cover glass but wide enough to ensure adequate sample imaging area. Rinse the slides with deionized MilliQ H_2O and wipe dry.

2. Attach two pieces of double-sided tape along the long edges of the cleaned glass slide as a spacer.
3. Remove the cover glass from the 6-well plate and carefully wipe the non-cell surface dry. Wipe dry ~ 2 mm off from two opposing edges of the cell side of the cover glass. Carefully place the cover glass over the glass slide, such that the two holes are in the center of the cover glass, the dry edges of the cover glass align with the double-sided tape and the cells face the glass slide. Gently push down on the edges of the cover glass along the tape using a micropipette tip to get a good seal. Through one of the drilled holes on the slide, carefully pipette enough PBS (~50 μL) into the chamber to just cover the entire cover glass surface and keep the sample wet.
4. Cut and remove excess double-sided tape around the cover glass using a scalpel. Seal the chamber by applying 5-min epoxy along the edges of the cover glass and allow epoxy to set. The seal must be tight such that any liquid injected into the chamber through one hole would only escape through the second hole (Fig. 2).

3.8 SMLM

Multicolor SMLM enables visualization of different targets of interest at sub-cellular resolution. However, chromatic aberration can commonly cause a shift in apparent positions of different targets imaged in different channels, leading to incorrect localization coordinates. Therefore, immediately before or after an imaging experiment, images of dispersed TetraSpeck™ Microspheres in each imaging channel should be acquired to calibrate the SMLM setup for any localization shifts.

1. Immediately prior to imaging, replace the PBS in the imaging chamber by injecting 100 μL of freshly prepared SMLM Imaging buffer using a 100 μL pipette tip and place the slide onto an appropriate SMLM setup.
2. At low illumination power, identify S-phase nuclei using markers such as EdU or PCNA. Adjust depth of field to bring the center plane of the nuclei of interest into focus.

   NOTE: If possible, use highly inclined and laminated optical sheet (HILO) illumination. In this setup, the angled excitation beam escapes out the cover glass-sample interface, affording an imaging depth of up to 10 μm, allowing tissues, nuclei, or other subcellular compartments sitting beyond the cell membrane to be imaged. Importantly, HILO yields an angled thin beam of illumination, thus significantly reducing out-of-focus background signal while maintaining a high excitation intensity (Tokunaga, Imamoto, & Sakata-Sogawa, 2008).
3. Increase the excitation laser power to achieve single molecule blinking and acquire 2000–20,000 frames at 20–100 Hz.
4. Switch dichroic and emission filters to sequentially image the same field of view in all fluorophore channels.

   NOTE: The order in which targets are imaged should be optimized because spectral crosstalk can cause different lasers to bleach and/or photoswitch fluorophores in other channels. Typically, channels should be imaged in order of decreasing wavelength, but this may differ based on the target of interest and fluorophore combinations.
5. Repeat steps 2–4 for other fields of view until sufficient data are collected.
6. Process raw multi-frame image stacks of single-molecule data using any one of the commercially or openly available software or code packages. The molecular coordinates generated from these programs can then be rendered into a SMLM-image of DNA G4 and replisome complexes (Fig. 3).

3.9 SMLM analysis

We have successfully used two independent SMLM analyses to tease out the association of DNA G4s with the replication machinery (Lee et al., 2021). First, we coupled DBSCAN clustering and Nearest Neighbor Distance analysis (Nicovich et al., 2017) to calculate the frequency of replication sites, as marked by PCNA foci, that colocalize with G4. Furthermore, to resolve the specific molecular arrangement of G4s at replication forks, we applied

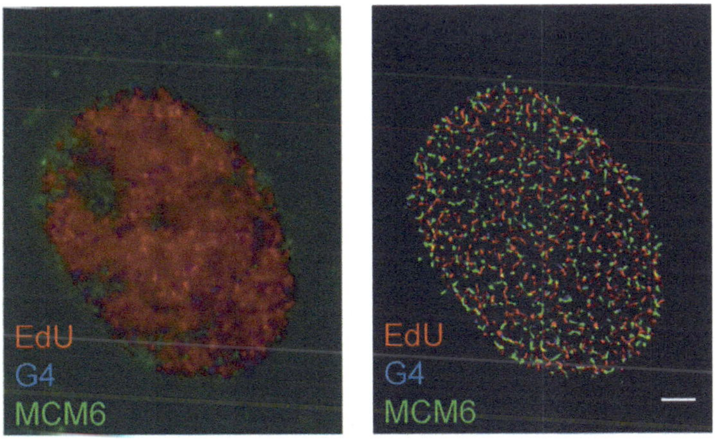

Fig. 3 Representative three-color diffraction-limited (left) and SMLM (right) images of nascent DNA (EdU, red), G4 (blue) and MCM6 (green) in a U2OS nucleus. Scale bar 2 μm. *Adapted from Lee, W. T. C., Yin, Y., Morten, M. J., Tonzi, P., Gwo, P. P., Odermatt, D. C., et al. (2021). Single-molecule imaging reveals replication fork coupled formation of G-quadruplex structures hinders local replication stress signaling.* Nature Communications, 12*(1), 2525. doi:10.1038/s41467-021-22830-9.*

Triple-Correlation (TC) analysis (Yin et al., 2019), an unbiased automated pattern recognition algorithm that identifies triplet molecular patterns of G4 and the replisome components (EdU and MCM) within a single nucleus. Both methods revealed a significant association between G4 and replisomes.

To conclude, this SMLM protocol enables nanoscale spatiotemporal mapping of G4 and DNA replication in cells, offering a new approach to decipher the biological role of G4 formation and its impact on genomic integrity.

References

Bates, M., Jones, S. A., & Zhuang, X. (2013). Stochastic optical reconstruction microscopy (STORM): A method for superresolution fluorescence imaging. *Cold Spring Harbor Protocols*, *2013*(6), 498–520. https://doi.org/10.1101/pdb.top075143.

Bermudez-Hernandez, K., Keegan, S., Whelan, D. R., Reid, D. A., Zagelbaum, J., Yin, Y., et al. (2017). A method for quantifying molecular interactions using stochastic modelling and super-resolution microscopy. *Scientific Reports*, 7(1), 14882. https://doi.org/10.1038/s41598-017-14922-8.

Betzig, E., Patterson, G. H., Sougrat, R., Lindwasser, O. W., Olenych, S., Bonifacino, J. S., et al. (2006). Imaging intracellular fluorescent proteins at nanometer resolution. *Science*, *313*(5793), 1642–1645. https://doi.org/10.1126/science.1127344.

Bidzinska, J., Cimino-Reale, G., Zaffaroni, N., & Folini, M. (2013). G-quadruplex structures in the human genome as novel therapeutic targets. *Molecules*, *18*(10), 12368–12395. https://doi.org/10.3390/molecules181012368.

Biffi, G., Tannahill, D., McCafferty, J., & Balasubramanian, S. (2013). Quantitative visualization of DNA G-quadruplex structures in human cells. *Nature Chemistry*, *5*(3), 182–186. https://doi.org/10.1038/nchem.1548.

Coltharp, C., Yang, X., & Xiao, J. (2014). Quantitative analysis of single-molecule superresolution images. *Current Opinion in Structural Biology*, *28*, 112–121. https://doi.org/10.1016/j.sbi.2014.08.008.

Cong, K., Peng, M., Kousholt, A. N., Lee, W. T. C., Lee, S., Nayak, S., et al. (2021). Replication gaps are a key determinant of PARP inhibitor synthetic lethality with BRCA deficiency. *Molecular Cell*. https://doi.org/10.1016/j.molcel.2021.06.011.

D'Alessandro, G., Whelan, D. R., Howard, S. M., Vitelli, V., Renaudin, X., Adamowicz, M., et al. (2018). BRCA2 controls DNA:RNA hybrid level at DSBs by mediating RNase H2 recruitment. *Nature Communications*, *9*(1), 5376. https://doi.org/10.1038/s41467-018-07799-2.

Dellagiacoma, C., Lukinavicius, G., Bocchio, N., Banala, S., Geissbuhler, S., Marki, I., et al. (2010). Targeted photoswitchable probe for nanoscopy of biological structures. *Chembiochem*, *11*(10), 1361–1363. https://doi.org/10.1002/cbic.201000189.

Demmerle, J., Wegel, E., Schermelleh, L., & Dobbie, I. M. (2015). Assessing resolution in super-resolution imaging. *Methods*, *88*, 3–10. https://doi.org/10.1016/j.ymeth.2015.07.001.

Dempsey, G. T., Bates, M., Kowtoniuk, W. E., Liu, D. R., Tsien, R. Y., & Zhuang, X. (2009). Photoswitching mechanism of cyanine dyes. *Journal of the American Chemical Society*, *131*(51), 18192–18193. https://doi.org/10.1021/ja904588g.

Fernandez-Suarez, M., & Ting, A. Y. (2008). Fluorescent probes for super-resolution imaging in living cells. *Nature Reviews. Molecular Cell Biology*, *9*(12), 929–943. https://doi.org/10.1038/nrm2531.

Gustafsson, M. G. (2000). Surpassing the lateral resolution limit by a factor of two using structured illumination microscopy. *Journal of Microscopy*, *198*(Pt. 2), 82–87. https://doi.org/10.1046/j.1365-2818.2000.00710.x.

Heilemann, M., van de Linde, S., Schuttpelz, M., Kasper, R., Seefeldt, B., Mukherjee, A., et al. (2008). Subdiffraction-resolution fluorescence imaging with conventional fluorescent probes. *Angewandte Chemie (International Ed. in English)*, *47*(33), 6172–6176. https://doi.org/10.1002/anie.200802376.

Hell, S. W., & Wichmann, J. (1994). Breaking the diffraction resolution limit by stimulated emission: Stimulated-emission-depletion fluorescence microscopy. *Optics Letters*, *19*(11), 780–782. https://doi.org/10.1364/ol.19.000780.

Henderson, A., Wu, Y., Huang, Y. C., Chavez, E. A., Platt, J., Johnson, F. B., et al. (2014). Detection of G-quadruplex DNA in mammalian cells. *Nucleic Acids Research*, *42*(2), 860–869. https://doi.org/10.1093/nar/gkt957.

Hess, S. T., Girirajan, T. P., & Mason, M. D. (2006). Ultra-high resolution imaging by fluorescence photoactivation localization microscopy. *Biophysical Journal*, *91*(11), 4258–4272. https://doi.org/10.1529/biophysj.106.091116.

Huang, B., Babcock, H., & Zhuang, X. (2010). Breaking the diffraction barrier: Super-resolution imaging of cells. *Cell*, *143*(7), 1047–1058. https://doi.org/10.1016/j.cell.2010.12.002.

Huang, F., Hartwich, T. M., Rivera-Molina, F. E., Lin, Y., Duim, W. C., Long, J. J., et al. (2013). Video-rate nanoscopy using sCMOS camera-specific single-molecule localization algorithms. *Nature Methods*, *10*(7), 653–658. https://doi.org/10.1038/nmeth.2488.

Huppert, J. L., & Balasubramanian, S. (2005). Prevalence of quadruplexes in the human genome. *Nucleic Acids Research*, *33*(9), 2908–2916. https://doi.org/10.1093/nar/gki609.

Islam, M. K., Jackson, P. J., Rahman, K. M., & Thurston, D. E. (2016). Recent advances in targeting the telomeric G-quadruplex DNA sequence with small molecules as a strategy for anticancer therapies. *Future Medicinal Chemistry*, *8*(11), 1259–1290. https://doi.org/10.4155/fmc-2015-0017.

Kaksonen, M., & Drubin, D. G. (2006). PALM reading: Seeing the future of cell biology at higher resolution. *Developmental Cell*, *11*(4), 438–439. https://doi.org/10.1016/j.devcel.2006.09.008.

Klar, T. A., & Hell, S. W. (1999). Subdiffraction resolution in far-field fluorescence microscopy. *Optics Letters*, *24*(14), 954–956. https://doi.org/10.1364/ol.24.000954.

Klein, T., Proppert, S., & Sauer, M. (2014). Eight years of single-molecule localization microscopy. *Histochemistry and Cell Biology*, *141*(6), 561–575. https://doi.org/10.1007/s00418-014-1184-3.

Lambert, T. J., & Waters, J. C. (2017). Navigating challenges in the application of superresolution microscopy. *The Journal of Cell Biology*, *216*(1), 53–63. https://doi.org/10.1083/jcb.201610011.

Lane, A. N., Chaires, J. B., Gray, R. D., & Trent, J. O. (2008). Stability and kinetics of G-quadruplex structures. *Nucleic Acids Research*, *36*(17), 5482–5515. https://doi.org/10.1093/nar/gkn517.

Lee, W. T. C., Yin, Y., Morten, M. J., Tonzi, P., Gwo, P. P., Odermatt, D. C., et al. (2021). Single-molecule imaging reveals replication fork coupled formation of G-quadruplex structures hinders local replication stress signaling. *Nature Communications*, *12*(1), 2525. https://doi.org/10.1038/s41467-021-22830-9.

Lerner, L. K., & Sale, J. E. (2019). Replication of G Quadruplex DNA. *Genes (Basel)*, *10*(2). https://doi.org/10.3390/genes10020095.

Liss, V., Barlag, B., Nietschke, M., & Hensel, M. (2015). Self-labelling enzymes as universal tags for fluorescence microscopy, super-resolution microscopy and electron microscopy. *Scientific Reports*, *5*, 17740. https://doi.org/10.1038/srep17740.

Morten, M. J., Reid, D. A., & Rothenberg, E. (2014). Single-molecule fluorescence imaging techniques. In *Encyclopedia of analytical chemistry* (pp. 1–20).

Neidle, S. (2016). Quadruplex nucleic acids as novel therapeutic targets. *Journal of Medicinal Chemistry*, *59*(13), 5987–6011. https://doi.org/10.1021/acs.jmedchem.5b01835.

Nicovich, P. R., Owen, D. M., & Gaus, K. (2017). Turning single-molecule localization microscopy into a quantitative bioanalytical tool. *Nature Protocols*, *12*(3), 453–460. https://doi.org/10.1038/nprot.2016.166.

Nyquist, H. (1928). Certain topics in telegraph transmission theory. *Transactions of the American Institute of Electrical Engineers*, *47*(2), 617–644. https://doi.org/10.1109/T-AIEE.1928.5055024.

Pertsinidis, A., Zhang, Y., & Chu, S. (2010). Subnanometre single-molecule localization, registration and distance measurements. *Nature*, *466*(7306), 647–651. https://doi.org/10.1038/nature09163.

Reid, D. A., Keegan, S., Leo-Macias, A., Watanabe, G., Strande, N. T., Chang, H. H., et al. (2015). Organization and dynamics of the nonhomologous end-joining machinery during DNA double-strand break repair. *Proceedings of the National Academy of Sciences of the United States of America*, *112*(20), E2575–E2584. https://doi.org/10.1073/pnas.1420115112.

Rust, M. J., Bates, M., & Zhuang, X. (2006). Sub-diffraction-limit imaging by stochastic optical reconstruction microscopy (STORM). *Nature Methods*, *3*(10), 793 795. https://doi.org/10.1038/nmeth929.

Shannon, C. E. (1949). Communication in the presence of noise. *Proceedings of the IRE*, *37*(1), 10–21. https://doi.org/10.1109/JRPROC.1949.232969.

Shtengel, G., Galbraith, J. A., Galbraith, C. G., Lippincott-Schwartz, J., Gillette, J. M., Manley, S., et al. (2009). Interferometric fluorescent super resolution microscopy resolves 3D cellular ultrastructure. *Proceedings of the National Academy of Sciences of the United States of America*, *106*(9), 3125–3130. https://doi.org/10.1073/pnas.0813131106.

Small, A., & Stahlheber, S. (2014). Fluorophore localization algorithms for super-resolution microscopy. *Nature Methods*, *11*(3), 267–279. https://doi.org/10.1038/nmeth.2844.

Thompson, R. E., Larson, D. R., & Webb, W. W. (2002). Precise nanometer localization analysis for individual fluorescent probes. *Biophysical Journal, 82*(5), 2775–2783. https://doi.org/10.1016/S0006-3495(02)75618-X.

Todd, A. K., Johnston, M., & Neidle, S. (2005). Highly prevalent putative quadruplex sequence motifs in human DNA. *Nucleic Acids Research, 33*(9), 2901–2907. https://doi.org/10.1093/nar/gki553.

Tokunaga, M., Imamoto, N., & Sakata-Sogawa, K. (2008). Highly inclined thin illumination enables clear single-molecule imaging in cells. *Nature Methods, 5*(2), 159–161. https://doi.org/10.1038/nmeth1171.

Tonzi, P., Yin, Y., Lee, C. W. T., Rothenberg, E., & Huang, T. T. (2018). Translesion polymerase kappa-dependent DNA synthesis underlies replication fork recovery. *eLife, 7*. https://doi.org/10.7554/eLife.41426.

Valton, A. L., & Prioleau, M. N. (2016). G-Quadruplexes in DNA replication: A problem or a necessity? *Trends in Genetics, 32*(11), 697–706. https://doi.org/10.1016/j.tig.2016.09.004.

Vogelsang, J., Kasper, R., Steinhauer, C., Person, B., Heilemann, M., Sauer, M., et al. (2008). A reducing and oxidizing system minimizes photobleaching and blinking of fluorescent dyes. *Angewandte Chemie (International Ed. in English), 47*(29), 5465–5469. https://doi.org/10.1002/anie.200801518.

Whelan, D. R., & Bell, T. D. (2015). Image artifacts in single molecule localization microscopy: Why optimization of sample preparation protocols matters. *Scientific Reports, 5*, 7924. https://doi.org/10.1038/srep07924.

Whelan, D. R., Lee, W. T. C., Marks, F., Kong, Y. T., Yin, Y., & Rothenberg, E. (2020). Super-resolution visualization of distinct stalled and broken replication fork structures. *PLoS Genetics, 16*(12), e1009256. https://doi.org/10.1371/journal.pgen.1009256.

Whelan, D. R., Lee, W. T. C., Yin, Y., Ofri, D. M., Bermudez-Hernandez, K., Keegan, S., et al. (2018). Spatiotemporal dynamics of homologous recombination repair at single collapsed replication forks. *Nature Communications, 9*(1), 3882. https://doi.org/10.1038/s41467-018-06435-3.

Whelan, D. R., & Rothenberg, E. (2021). Super-resolution mapping of cellular double-strand break resection complexes during homologous recombination. *Proceedings of the National Academy of Sciences of the United States of America, 118*(11). https://doi.org/10.1073/pnas.2021963118.

Whelan, D. R., Yin, Y., Bermudez-Hernandez, K., Keegan, S., Fenyo, D., & Rothenberg, E. (2016). Single molecule localization microscopy of DNA damage response pathways in Cancer. *Microscopy and Microanalysis, 22*(Suppl. 3), 1016–1017. https://doi.org/10.1017/S1431927616005924.

Wu, Y. L., Tschanz, A., Krupnik, L., & Ries, J. (2020). Quantitative data analysis in single-molecule localization microscopy. *Trends in Cell Biology, 30*(11), 837–851. https://doi.org/10.1016/j.tcb.2020.07.005.

Xu, J., Ma, H., & Liu, Y. (2017). Stochastic optical reconstruction microscopy (STORM). *Current Protocols in Cytometry, 81*. https://doi.org/10.1002/cpcy.23. 12 46 11-12 46 27.

Yildiz, A., Forkey, J. N., McKinney, S. A., Ha, T., Goldman, Y. E., & Selvin, P. R. (2003). Myosin V walks hand-over-hand: Single fluorophore imaging with 1.5-nm localization. *Science, 300*(5628), 2061–2065. https://doi.org/10.1126/science.1084398.

Yin, Y., Lee, W. T. C., & Rothenberg, E. (2019). Ultrafast data mining of molecular assemblies in multiplexed high-density super-resolution images. *Nature Communications, 10*(1), 119. https://doi.org/10.1038/s41467-018-08048-2.

Yin, Y., & Rothenberg, E. (2016). Probing the spatial organization of molecular complexes using triple-pair-correlation. *Scientific Reports, 6*, 30819. https://doi.org/10.1038/srep30819.

CHAPTER FIVE

Studying the DNA damage response in embryonic systems

Elena Lo Furno[†], **Bénédicte Recolin**[‡,*], **Siem van der Laan, Antoine Aze, and Domenico Maiorano***
Institut de Génétique Humaine, Université de Montpellier, Genome Surveillance and Stability, CNRS-UMR9002, Montpellier, France
*Corresponding authors: e-mail address: benedicterecolin@hotmail.com; domenico.maiorano@igh.cnrs.fr

Contents

[†] Present address: San Raffaele Telethon Institute for Gene Therapy (SR-TIGET), IRCCS San Raffaele Scientific Institute, Milan, Italy.
[‡] Present address: AstraZeneca, Cambridge, United Kingdom.

Methods in Enzymology, Volume 661
ISSN 0076-6879
https://doi.org/10.1016/bs.mie.2021.08.016

Abstract

Maintenance and surveillance of genome integrity is crucial during the very early steps of embryonic development, since *de novo* mutations generated during this stage can be propagated in differentiated adult cells and may lead to predisposition to diseases including cancer. Surprisingly, early embryos are characterized by a relaxed control of genome integrity, reminiscent of that observed in cancer cells. How embryos manage to produce healthy adult individuals in such conditions remains still unclear. Here, we describe protocols and methods to study and analyze the DNA damage response and genome integrity in two embryonic experimental systems, early *Xenopus laevis* embryos and mouse embryonic stem cells. We describe methods to study gene functions in the DNA damage response by mRNA microinjection in *Xenopus* embryos generated by *in vitro* fertilization, mutagenesis and developmental regulation of the DNA damage response. We also describe methods to analyze the DNA damage response in mESCs, including synchronization experiments that allow studying the DNA damage response at different cell cycle stages. Analysis of genome integrity in these systems may also help to shed light on the molecular mechanisms that preserve genome integrity and become dysregulated in cancer cells.

1. Introduction

Compared to somatic cells, embryonic cells are characterized by a substantial difference in the length and structure of the cell cycle, as well as by a relaxed DNA damage response. In particular, gap phases (G1 and G2) can be extremely short or non-existing, and S-phase very contracted, depending on the organism (see Kermi, Lo Furno, and Maiorano (2017), for review). For example, *Drosophila* pre-blastoderm and cleavage stage *Xenopus* embryos have virtually no gap phases, while mouse embryonic stem cells (mESCs) have a highly contracted G1 phase. The duration of S-phase can be as short as 4 min in *Drosophila*, and 20 min in *Xenopus*. In addition, feedback mechanisms controlling genome integrity are largely suppressed. Understanding how embryos can manage to preserve genome integrity under these circumstances is a fascinating question that remains still unresolved. In somatic cells, DNA damage is first detected by the apical PI3KK kinases ATM, ATR and DNA-Pk. Activation of these kinases leads to phosphorylation of a plethora of downstream substrates to slow down the cell cycle and activate DNA repair mechanisms (Ciccia & Elledge, 2010). Among these are the tumor suppressor p53, the histone variant H2AX (also known as γH2AX) and the Chk1 and Chk2 protein kinases. Albeit some redundancy, PI3KKs respond to specific types of DNA damage. While ATR is mainly implicated in sensing progression of replication forks (Saldivar, Cortez, & Cimprich, 2017), ATM sense DNA double-strand breaks (DSBs) as well as changes in chromatin structure (Bakkenist & Kastan, 2003). DNA-PK is partially redundant with both ATM and ATR (Blackford & Jackson, 2017).

Here, we present methods and protocols that we have developed in our laboratory to study the DNA damage response in two embryonic systems, *Xenopus laevis* and mESCs. The clawed frog *Xenopus laevis* is an excellent experimental system to study vertebrate development. A main advantage of *Xenopus* is that development occurs externally, it is rapid, it can be easily followed in real time under a stereomicroscope and embryos can be easily manipulated. Further, *cell-free* protein extracts derived from fertilized eggs have been extensively used to study many aspects of the cell cycle and DNA replication *in vitro* (Blow & Laskey, 1986; Lohka & Masui, 1983; Mechali & Harland, 1982; Walter, Sun, & Newport, 1998). Previous (Costanzo & Gautier, 2004; Sannino, Pezzimenti, Bertora, & Costanzo, 2017) and more recent protocols (Aparicio Casado & Gautier, 2021; Lin, Ha, & Yan, 2019; Willis, DeStephanis, Patel, Gowda, & Yan, 2012) are available describing the use of these *in vitro* systems to study the DNA damage response. *Xenopus* embryos have virtually absent G1 and G2 phases and suppress checkpoints that respond to DNA damage (Dasso & Newport, 1990; Newport & Dasso, 1989; Newport & Kirschner, 1982a, 1982b). ESCs have the capacity to be differentiated into any cell type of the body (pluripotency), and therefore constitute an attractive system for medical application to cure genetic diseases. However, these cells are characterized by genetic instability that limits their use in the clinic due to risk of tumor formation (Werbowetski-Ogilvie et al., 2009). mESCs have been, and are currently extensively used to study molecular mechanisms governing pluripotency and differentiation. These cells display a highly contracted G1 phase and a relaxed G1/S checkpoint and therefore cannot delay the cell cycle if DNA damage is generated in mitosis or in G1.

2. Studying the DNA damage response in the early *Xenopus* embryo

Xenopus embryos can be easily generated in the laboratory by *in vitro* fertilization (Sive, Grainer, & Harland, 2000). Upon fertilization, the zygote initiates a series of synchronous and rapid cleavages leading to formation of the early embryo. During the first 12 cleavages (the pre-midblastula transition) transcription is undetectable and its onset coincides with a cell cycle slow down at cycle 13 (mid-blastula transition, MBT) (Newport & Kirschner, 1982a, 1982b). Here, we will describe methods to analyze the DNA damage response in the pre-MBT embryos. Embryos can be exposed to DNA damaging agents followed by phenotypic analysis that can be easily monitored under a stereomicroscope equipped with a camera. Embryos can be also injected with damaged plasmid DNA that can be recovered in *E. coli* to study mutagenesis. Plasmid DNA can be prepared using commercially available kits. We use the Qiagen™ system. When using plasmid DNA to study mutagenesis it is important to obtain almost intact supercoiled DNA in order to minimize background mutations generated by DNA repair upon microinjection in the embryo. A method to isolate nearly 100% supercoiled DNA has been published (Carbone, Fioretti, Fucci, Ausió, & Piscopo, 2012). Injection of mRNAs can be also achieved to study gene function in a dominant negative fashion. We have previously used these methods to identify a suppressor of the UV-induced DNA damage checkpoint in *Xenopus* embryos (Kermi et al., 2015).

2.1 UV-C irradiation of *Xenopus* embryos

2.1.1 Equipment

- UV-C (Bio-Link or Stratalinker)

2.1.2 Buffers and reagents

- MBS 1 ×: 88 mM NaCl, 1 mM $CaCl_2$, 1 mM $MgSO_4$, 5 mM HEPES pH 7.8, 2.5 mM $NaHCO_3$
- 2% cysteine hydrochloride pH 7.95

2.1.3 Procedure

1. Fertilize *Xenopus* eggs laid within 1 h as described (Sive, Grainer, & Harland, 2000).
2. Thirty-five minute post-fertilization remove most of the MBS 0.1 ×, while keeping the eggs floating
3. Poor some 2% cysteine solution in the dish and swirl for 2–4 min (do not exceed 10 min) to remove jelly coat from fertilized eggs
4. Remove cysteine solution and wash the eggs extensively with MBS 1 ×
5. Transfer eggs to a new Petri dish containing fresh MBS 1 × and leave them at 23 °C until first mitotic division begins and two blastomeres are formed (90 min post-fertilization).

6. Remove as much MBS as possible
7. Place embryos in a Petri dish in the UV-irradiator
8. Irradiate at the desired UV-C dose (100–500 J/m^2, see notes).
9. Immediately add fresh MBS
10. Leave embryos at 23 °C and collect them at the desired developmental stage following Nieuwkoop and Faber normal tables (Nieuwkoop & Faber, 1994) (digitalized images can be found here: https://www.xenbase.org/anatomy/alldev.do).
11. Embryos can be either processed immediately or snap frozen in liquid nitrogen and stored at −80 °C

2.2 Analysis of DNA damage in isolated nuclei

Damaged nuclei from early embryos can be also analyzed by indirect immunofluorescence with specific antibodies. A protocol for purification of embryonic nuclei has been previously described (Menut, Lemaitre, Hair, & Méchali, 1999).

2.2.1 Equipment

24-well plate (Falcon®, Ref.: 353047, Corning)
Bench centrifuge
Coverslips
Microscope slides
Orbital shaker
Nail polish
Ice

2.2.2 Buffers and reagents

- Paraformaldheyde (16%, Interchim, IP7070)
- Xb buffer: 100 mM KCl, 0.1 mM $CaCl_2$, 1 mM $MgCl_2$, 50 mM sucrose, 10 mM HEPES pH 7.7; 5 μg/mL proteases inhibitors (Sigma, Leupeptin L2884, Pepstatin A P4265 and Aprotinin A1153)
- Triton X-100 (Sigma, T9284)
- Sucrose
- Absolute ethanol
- PBS
- Bovine serum albumin (BSA,Sigma, A2153)
- Tween-20 (Sigma, P1379)
- Prolong Gold antifade reagent with DAPI (Invitrogen, P36931)

2.2.3 Procedure

1. Prepare nuclei from embryos as described (Menut et al., 1999).
2. Dissolve the final pellet in 100 μL of Xb buffer containing 0.3% Triton X-100
3. Incubate 5 min on ice
4. Add an equal volume of Xb buffer/8% paraformaldehyde
5. Incubate 15 min on ice
6. Prepare a 24-well plates with coverslips at the bottom of each well
7. Add 2 mL of sucrose 0.3 M in Xb buffer
8. Overlay the sucrose cushion with the samples
9. Spin in a bench centrifuge on a swing out bucket rotor at 1500 rpm for 10 min
10. Remove the supernatant
11. Post-fix with 1 mL of cold absolute ethanol (−20 °C) for 15 min at room temperature
12. Remove supernatant and rehydrate coverslips with 1 mL of PBS, 1% BSA for 15 min at room temperature
13. Block coverslips with 1 mL PBS/1% BSA for 1 h at room temperature
14. Incubate coverslips overnight at 4 °C with primary antibodies diluted in PBS/1% BSA/0.1% Tween (100–200 μL) by placing the plate in a humidified chamber or at 4 °C
15. Wash coverslips 5 times for 20 min with 2 mL of 0.1% Tween-20 in PBS each wash on an orbital shaker at moderate speed
16. Incubate with secondary antibody (200 μL) for 1 h at room temperature
17. Wash coverslips as in step 13
18. Spot a drop (5 μL) of Prolong Gold with DAPI (to stain nuclei) on glass slides
19. Put coverslip on top of the Prolong Gold with DAPI
20. Dry coverslips at room temperature in the dark
21. Seal with nail polish
22. Store at 4 °C
23. View with fluorescence microscope. A typical result is shown below (Fig. 1)

2.3 Notes

1. Delivering doses of UV-C over 500 J/m^2 on embryos is not recommended, since this can lead to permanent protein-to-DNA crosslinks and strong inhibition of chromatin decondensation (see Fig. 2).

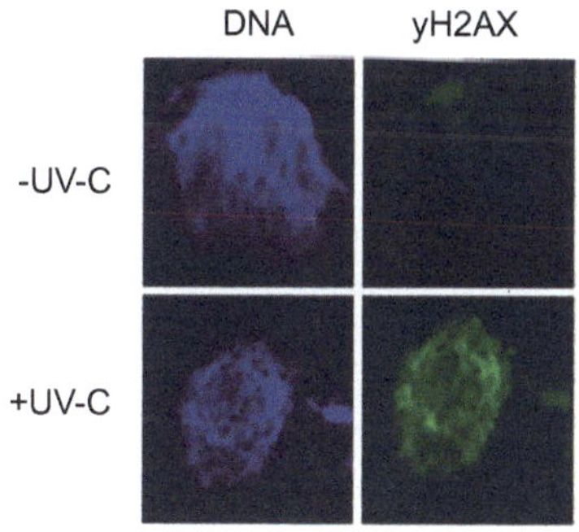

Fig. 1 Observation of UV-C irradiated (300 J/m^2; +UV-C) or not (−UV-C) *Xenopus* nuclei by indirect immunofluorescence using a Zeiss inverted microscope. DAPI stains DNA, γH2AX (Cell Signaling, 2577) stains damaged chromatin.

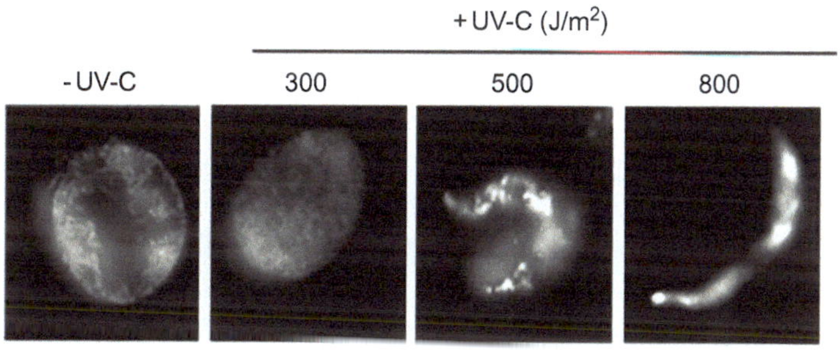

Fig. 2 Morphology of *Xenopus* embryos nuclei either non-irradiated (−UV-C) or irradiated with the indicated doses of UV-C. Nuclei were stained with DAPI to visualize DNA by fluorescence microscopy.

2. Coating coverslips with polylysine is not required and should be avoided since it increases fluorescence background. Paraformaldehyde-fixed nuclei can be stored at 4°C overnight

2.4 UV-C irradiation of plasmid DNA

2.4.1 Equipment

- UV-C (Bio-Link or Stratalinker)
- Parafilm
- 0.5 mL Eppendorf tubes (0030121023)
- Ice

2.4.2 Buffers and reagents

- Plasmid DNA

2.4.3 Procedure

1. Spread between 5 and 10 μL of DNA onto a parafilm square with a P20 micropipette kinked at 45 °C angle, making sure that the liquid is well spread on the parafilm surface
2. Place the parafilm with the DNA in the center of the UV-C irradiator
3. Irradiate the DNA at the desired UV-C intensity ($\lambda = 254$ nm; between 300 and 1000 J/m^2, see notes).
4. Recover the DNA with the micropipette by kinking it at 45 °C angle into a new ice-cold Eppendorf tube and place it on ice
5. Repeat the procedure for the mock treatment, except that the DNA is not irradiated
6. Recover the DNA as in step 4

2.5 Notes

1. Stocks of damaged DNA can be prepared before microinjection into *Xenopus* embryos and stored at −80 °C for a long time
2. We recommend verifying the dose of UV-C delivered by the irradiator by the use of a radiometer (UV - VLX-3 W - Vilber Lourmat), since the delivered dose can vary depending on the age of the UV lamps of the irradiator
3. We recommend testing the efficiency of DNA damage before use in microinjection experiments by performing replication of damaged DNA into Low Speed Supernatant (LSS) *Xenopus* egg extracts (see below). UV-C irradiation results in a DNA synthesis slow down mediated by the ATR-dependent checkpoint (Byun, Pacek, Yee, Walter, & Cimprich, 2005). Protocols for preparing LSS extracts have been previously described (Aparicio Casado & Gautier, 2021; Menut et al., 1999; Murray, 1991; Sannino et al., 2017). We use Murray's protocol for preparing LSS extracts

2.6 Damaging plasmid DNA with MMS

2.6.1 Buffers and reagents

- MMS (Sigma, M4016)
- Alkylation buffer: 10 mM Hepes pH 7.4; 80 mM KCl; 5 mM $MgCl_2$; 1 mM EDTA; 200 mM sucrose; 3% BSA; 1 mM DTT; 5 μg/mL proteases inhibitors (Leupeptin Sigma L2884, Pepstatin A Sigma P4265 and Aprotinin Sigma A1153)

2.6.2 Procedure

1. Dilute the required amount of plasmid DNA with 1 mL of alkylation buffer in 2 mL Eppendorf tubes
2. Add MMS from stock solution to reach the desired final concentration of 80 mM, or water (mock reaction).
3. Incubate on a rotating wheel for 30 min at room temperature
4. Add one volume of phenol/chloroform/Isoamyl alcohol (24/24/1).
5. Vortex and centrifuge 5 min at 13,000 rpm
6. Transfer the aqueous phase to a new tube and precipitated DNA by mixing with 1/10 sodium acetate 3 M and 2 volumes of ice cold absolute Ethanol
7. Centrifuge 10 min at maximum speed in a microcentrifuge
8. Dry the pellet and dissolve it in sterile water
9. Quantify DNA with Nanodrop
10. Test the efficiency of alkylation by replication in *in vitro Xenopus* egg extracts (see below).

2.7 Testing the efficiency of induced DNA damage in *Xenopus* egg extracts

2.7.1 Equipment

- 0.5 mL Eppendorf tubes (0030121023)
- Thermomixer (optional)
- Gloves
- Plexiglass screen
- Water bath or oven

2.7.2 Buffers and reagents

- *Xenopus* egg extract (Low Speed Supernatant)
- Energy mix (20 ×): 200 mg/mL Creatine Kinase; 200 mM Creatine Phosphate; 20 mM ATP; 20 mM $MgCl_2$. Store in aliquots at −20 °C
- Creatin Phosphate (disodium salt, Roche, 10621722001) 1 M in water. Store in aliquots at −20 °C
- Creatin Kinase (Sigma, 10127566001) 10 mg/mL, in Hepes 10 mM pH 7.6, 20% glycerol. Store in aliquots at −20 °C
- ATP/Mg^{2+}: 30 mM ATP; 50 mM $MgCl_2$; pH 7.5. Check pH of the solution on a pH paper. Store in aliquots at −20 °C
- Cycloheximide (Sigma, C7698; 10 mg/mL in water).
- α-[^{32}P]dCTP (or α-[^{32}P]dATP) 3000 Ci/mmol (Perkin Elmer).
- Template plasmid DNA (5–10 kb).

- STOP mix: 40 mM EDTA; 1% SDS
- Proteinase K (10 mg/mL).

2.7.3 Procedure

1. Prepare the required 0.5 mL Eppendorf tubes labeled containing 45 µL of STOP mix + Proteinase K (0.2 mg/mL final concentration).
2. Thaw 50 µL egg extract from −80 °C rapidly in your hands and place it on ice
3. Immediately add 250 µg/mL of cycloheximide (40 × final dilution from stock), mix gently by pipetting
4. Add energy mix (20 × final dilution from stock), mix gently by pipetting
5. Add 5 ng of plasmid DNA/µL egg extract, mix gently by pipetting
6. From this step onwards, reactions and samples must be placed behind a plexiglass screen
7. Add 1 µL α-[^{32}P]dCTP behind a plexiglass screen, mix gently by pipetting
8. Incubate the reaction either at room temperature or in a thermomixer (between 21 and 23 °C).
9. Withdraw 5 µL from the tube every 30 min and mix it in tubes containing STOP mix with Proteinase K by vortexing. Leave at room temperature until the end of the experiment
10. At the end of the experiment incubate samples (and not the remaining master reaction which should stay at room temperature) at 52 °C for 1 h, or at 37 °C overnight, in an oven or a water bath. Make sure to properly shield the reactions with a plexiglass screen to protect the environment from radiation

2.8 DNA precipitation onto glass fiber filters

This procedure allows to separate the free radioactive nucleotide from the nucleotide which has been incorporated into the DNA during replication by means of cold trichloroacetic acid (TCA) precipitation onto glass fibers filters (GF/C). This procedure can be equally applied to monitor incorporation of a radioactive nucleotide into sperm chromatin (Menut et al., 1999).

2.8.1 Equipment

- Glass fiber filters (GF/C, Whatmann, 25 × 100 mm; 3613)
- Plastic Tweezers
- Whatman paper

- 5 mL plastic vials with caps
- Vial racks
- Liquid scintillation counter
- Indian ink
- 500 mL-1 L beaker
- Ice

2.8.2 Buffers and reagents

- 5% TCA/2% pyrophosphate (PPi)
- 5% TCA
- Low grade (96%) EtOH
- Ink pen

2.8.3 Procedure

1. Spot the samples corresponding to each time point onto GF/C filters pre-labeled with Indian ink behind a plexiglass screen using a micropipette
2. Soak the filters into an ice-cold solution of 5% TCA/2% PPi in a beaker on an ice bucket (20 mL TCA per filter) behind a plexiglass screen
3. Leave standing for 30 min (can be left overnight). Stirring is not required but it can be gently applied from time to time
4. Pour the liquid into a disposable flask behind a plexiglass screen and replace with the same volume of cold 5% TCA
5. Leave standing for 15 min
6. Repeat steps 3 and 4 three times
7. Saturate filters in 100 mL of low grade 96% EtOH on ice for 10 min
8. Take the filters out from the beaker using the tweezers
9. Air dry the filters on Whatman paper behind a plexiglass screen
10. Prepare the required number of labeled scintillation rack with vials containing 4 mL of scintillation liquid
11. Put the dried filters into the vials with the tweezers
12. Lock the vials with caps
13. Count 1 μL of the master reaction diluted 10-fold in STOP mix
14. Calculate the amount of α-[^{32}P]dCTP incorporated into DNA at each time point as previously described (Menut et al., 1999)
15. Plot the data in an excel sheet. Typical results are shown below (Fig. 3)

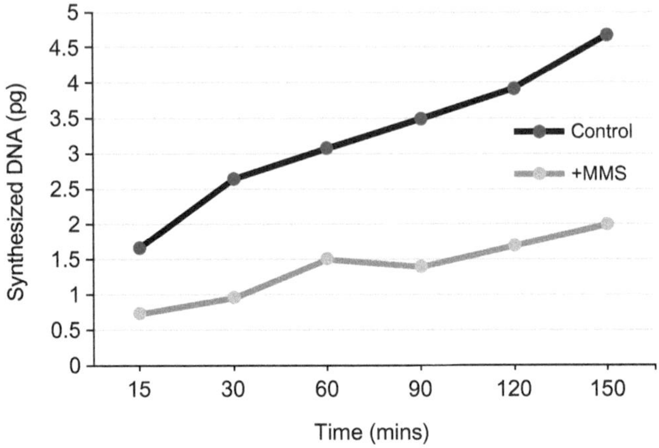

Fig. 3 Replication of plasmid DNA in *in vitro Xenopus* egg extracts exposed (+MMS) or not (Control) to 80 mM of MMS.

2.9 Notes

1. *Xenopus* egg extracts must be handled with care, avoiding strong pipetting. It is highly recommended not to vortex extracts to avoid proteins denaturation which will irreversibly compromise the ability of the extract to replicate the template DNA
2. Replication of plasmid DNA templates in LSS extracts is inefficient (between 5% and 10% of template DNA replicated depending on the extract and DNA length) compared to sperm chromatin (100%). Absolute values of DNA replication (ng or pg of DNA replicated or percentage of template DNA replicated) must be calculated
3. Protocols to analyze protein binding to either damaged chromatin or DNA have been previously described (Aparicio Casado & Gautier, 2021; Sannino et al., 2017).

2.10 Microinjection of damaged plasmid DNA into *Xenopus* eggs

2.10.1 Equipment

- Stereoscope
- Oocyte injector (we use Nanoject II, Drummond Scientific)
- 1.2 mm glass capillary (3.5″3-000-203G/X) Drummond tubes
- Stage micrometer
- P-30 micropipette needle puller (Sutter instruments, 20130418)
- Petri dishes (60 mm × 15 mm)

- Pasteur pipette cut with rounded edges
- Plastic or nylon grid
- Razor blade
- Parafilm
- 1.5 mL Eppendorf tubes (0030120086)
- Nieuwkoop and Faber tables (Nieuwkoop & Faber, 1994) (digitalized images can be found here: https://www.xenbase.org/anatomy/alldev.do).

2.10.2 Buffers and reagents

- Ficoll 400 (SIGMA F9378)
- HS-MBS: 7 mL 0.1 M $CaCl_2$, 100 mL 10 × MBS salts, 4 mL 5 M NaCl, 888 mL water
- MBS 1 ×: 88 mM NaCl, 1 mM $CaCl_2$, 1 mM $MgSO_4$, 5 mM HEPES pH 7.8, 2.5 mM $NaHCO_3$
- 2% cysteine hydrochloride pH 7.95

2.10.3 Procedure(s)

2.10.3.1 Needles preparation and calibration

1. Pull 1.2 mm glass capillary (3.5″3-000-203G/X) Drummond tubes with P-30 micropipette needle puller (Sutter instruments, 20130418; Pulling settings: Heat 1770, Heat 2880, Pull 450).
2. Cut the tip of the needle with the help of a razor blade under a stereoscope to reach a diameter of approximately 20 μM tip diameter. This cut should generate a 45° angle. To facilitate this step, a stabilizing pipette or a broken needle can be employed (Briano, 1983). In this case, the needle to be broken is inserted longitudinally into the tip of the broken needle and perpendicular to the point where the break should take place
3. Confirm the diameter with the use of a stage micrometer and insert the needle into the microinjector
4. Set the microinjector on aspiration mode to calibrate the injection needle or microinject
5. Place a small strip of Parafilm on a 100-mm diameter Petri dish, and dispense a 5 μL drop of the solution to be injected on it
6. Make dot marks on the needle every 0.5 mm corresponding to a volume of 50 nL. The dot marks on the needle are used to monitor the amount of substrate that has been injected

2.10.3.2 Microinjection of damaged plasmid DNA into 2-cell stage *Xenopus* embryos

1. Generate embryos as described in Section 2.1.3 up to step 4
2. Move 2-cell stage embryos with a rounded edges cut Pasteur pipette (this can be done using a Bunsen) to a small Petri dish containing a plastic grid in order to maintain the embryos immobilized. About 50 embryos can be placed on the grid
3. Poor in the Petri dish a solution of 3% Ficoll 400 in HS-MBS
4. Gently introduce the tip of the needle into the vegetal pole of the embryo in proximity of the animal hemisphere, at an approximately 45° angle and inject the appropriate volume of plasmid (2 injections of 9 nL in each blastomer).
5. After injection, gently withdraw the needle and move on to the next embryo
6. Transfer the injected embryos to another Petri dish containing fresh MBS 1 × and leave the embryos to divide until desired stage of collection according to Nieuwkoop and Faber tables in 1.5 mL Eppendorf tubes
7. Remove as much as MBS 1 × buffer as possible
8. Either snap freeze embryos in liquid nitrogen and store at −80 °C or process immediately (Section 2.13) (Fig. 4)

2.11 Microinjection of *in vitro* synthesized mRNA into *Xenopus* embryos

Analysis of gene function in the embryo can be achieved by microinjection of mRNAs. Although embryos contain mRNAs for virtually all genes, *in vitro* translated mRNA can be introduced in excess into *Xenopus* embryos by microinjection to act as a dominant negative using specific mutants compared to injection of the wild-type version of the same mRNA.

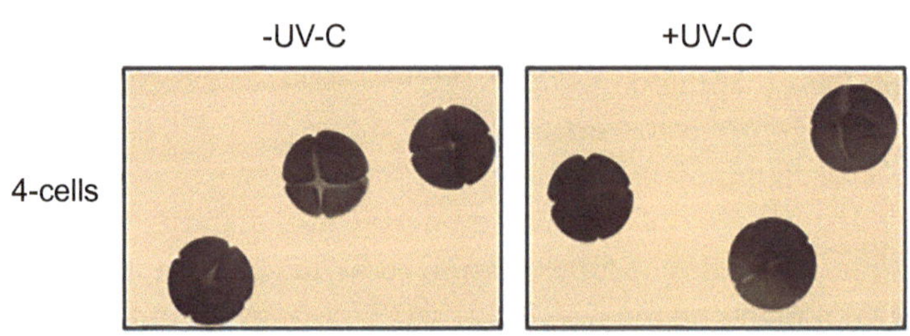

Fig. 4 Stereomicroscope view of *in vitro* fertilized 4-cells *Xenopus* embryos injected at the 2-cell stage with plasmid DNA irradiated (+UV-C) or not (-UV-C) with UV-C (300 J/m^2). Because of inefficient DNA damage response, UV-irradiated embryos continue to cleave and are morphologically indistinguishable from non-irradiated embryos.

2.11.1 *In vitro transcription*

2.11.1.1 Equipment

- Water bath
- Refrigerated microcentrifuge
- Horizontal electrophoresis tank
- Power pack

2.11.1.2 Buffers and reagents

- mMESSAGE mMACHINE kit® (SP6, AM1340; T7, AM1344; Thermofisher)
- Plasmid or linear DNA containing the gene of interest under control of either SP6 or T7 promoter
- RNase-free water
- Phenol/chloroform
- Chloroform
- Isopropanol
- Formaldehyde
- Agarose
- MOPS 1×: 200 mM MOPS; 50 mM sodium acetate; 10 mM EGTA and 10 mM EDTA
- Sodium acetate
- EGTA
- EDTA
- RNA ladders (NEB ssRNA ladder, N0362S)

2.11.1.3 Procedure

1. Set up the *in vitro* transcription reaction following the supplier's instructions
2. Recover mRNAs by phenol-chloroform extraction and isopropanol-mediated precipitation as suggested in the manual provided by the manufacturer (method 4)
3. Dissolve pellet in 20 μL of RNase-free water
4. Denature samples in RNA loading buffer 2× (provided by the *In vitro* transcription kit) by warming at 70 °C for 10 min
5. Check both quality and size of recovered mRNAs by formaldehyde gel electrophoresis (1% Agarose, 2.2% Formaldehyde, 1× MOPS)

2.12 *In vitro* translation

Before microinjection, the ability of synthesized mRNAs to be translated must be verified by *in vitro* translation using the reticulocyte translation system.

2.12.1 *Equipment*

- Water bath
- Vertical electrophoresis apparatus or pre-cast SDS-PAGE gels
- Power pack

2.12.2 *Buffers and reagents*

- Rabbit Reticulocyte Lysate, Nuclease-Treated (Promega, L4960)
- *In vitro* synthesized mRNAs (Section 2.10.1)

2.12.3 *Procedure*

1. Translate synthesized mRNAs using the rabbit reticulocytes lysate system following manufacturer's instructions
2. At the end of the reaction, check the efficiency of translation of the mRNA by loading a sample of the reaction on an SDS-PAGE followed by western blot using antibody that recognizes the translated protein

2.13 mRNA microinjection into 2-cell stage *Xenopus* embryos

2.13.1 *Procedure*

The procedure is identical to that described in Section 2.10, except that 5 ng of *in vitro* synthesized mRNA is injected in each blastomere, with or without plasmid DNA. Water microinjection is used as a negative control.

2.14 DNA isolation from embryos

2.14.1 *Equipment*

- Water bath
- Microcentrifuge
- QIAquick Gel Extraction Kit (#28704)
- Thermomixer or water bath

2.14.2 *Buffers and reagents*

- Xb buffer (see Section 3.4.2)
- *Dpn*I restriction enzyme (New England BioLabs)
- NEB buffer 2 (NEB B7202)
- Proteinase K (Sigma, P6556)
- Phenol Chloroform/Isoamyl alcohol (Thermofisher 15,593,031)
- Glycogen (Sigma G1767-1VL)
- Sodium Acetate pH 5.2
- Absolute Ethanol
- STOP MIX (10 mM EDTA, 0.5% SDS)

2.14.3 Procedure

1. Crush frozen embryos in STOP MIX (20 μL per embryo) supplemented with fresh Proteinase K (0.6 mg/mL).
2. Homogenize embryos in STOP MIX while they are thawing by means of a tip (see notes).
3. Incubate for 1 h at 37 °C in a thermomixer or water bath
4. Immediately after protein digestion at 37 °C, purify DNA by phenol-chloroform extraction and ethanol precipitation overnight at −20 °C
5. Dissolve pellet in 50 μL of nuclease free water and incubate 1 h at 37 °C in a thermomixer or water bath
6. Digest DNA with 100 U of *Dpn*I restriction enzyme to destroy unreplicated plasmids (see notes) by incubation at 37 °C in a water bath for 2 h
7. Add 50 μL of STOP MIX (without Proteinase K) and purify DNA with QIAGEN® gel extraction kit following the manufacturer instructions
8. Elute the DNA with 30 μL of nuclease-free water by standing at room temperature for 4 min followed by centrifugation for 1 min at maximum speed
9. Repeat step 7 to increase the yield of recovered DNA
10. Aliquot the elution in 4 μL in 0.5 mL Eppendorf tubes and store at −20 °C
11. Recover DNA by *E. coli* transformation

2.15 Notes

1. Eggs should be homogenized in STOP MIX supplemented with Proteinase K while they are thawing, by means of either a small pestle or a tip, and incubate the mush 1 h at 37 °C in a thermomixer
2. *Dpn*I digestion is required to clear plasmid DNA that has not been replicated. *Dpn*I cleaves all GATC/CTAG sites having methylated "A" residues on both strands. *Xenopus* embryos lack the ability to remethylate adenosine residues, because, as mammalian cells, they only contain a CpG methylation system. Therefore, plasmid molecules become demethylated when they replicate during subsequent cell cycles and are as such, they become resistant to *Dpn*I digestion

2.16 *Xenopus* embryos protein extracts

2.16.1 Equipment

- Refrigerated microcentrifuge
- 1.5 mL Eppendorf tubes (0030120086)
- Heat block or thermomixer

2.16.2 Buffers and reagents

- Xb buffer (see Section 3.4.2)
- Laemmli buffer 2X
- PhosSTOP (Roche, 4906845001)
- Cytochalasin (10 mg/mL in DMSO, Sigma, C6762)
- Proteases inhibitors (Sigma, 5 μg/mL Leupeptin, L2884; Pepstatin A, P4265; Aprotinin, A1153)

2.16.3 Procedure

1. Transfer embryos to an ice-cold 1.5 mL Eppendorf tube containing Xb buffer supplemented with 10 μg/mL cytochalasin, phosphatases and proteases inhibitors (5 μL of buffer per embryo), or while embryos are thawing if previously snap frozen in liquid nitrogen.
2. Centrifuge 10 min at maximum speed in a microcentrifuge at 4 °C
3. Recover the interphase (middle grayish phase) which contains the protein extract by passing a P1000 pipette through the upper lipid layer
4. Transfer interphase to a 1.5 mL Eppendorf tube containing an equal volume of Laemmli buffer 2 ×
5. Heat at 95 °C for 5 min
6. Analyze protein content by SDS-PAGE followed by western blotting. Typical results are shown below (Fig. 5)

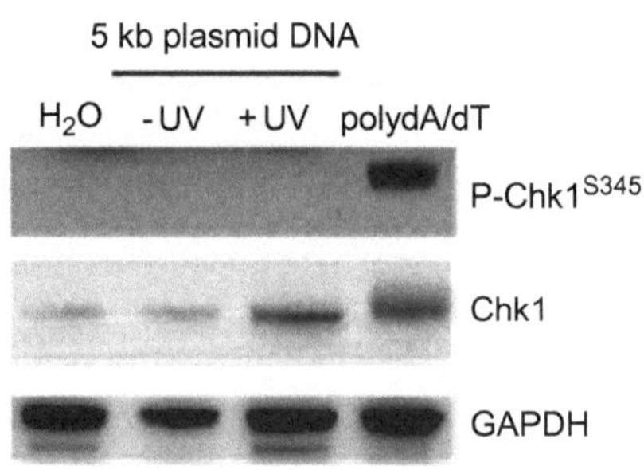

Fig. 5 Western blot of protein extracts obtained from 32-cell stage embryos (pre-MBT) obtained by *in vitro* fertilization of *Xenopus* eggs injected at the 2-cell stage with UV-C irradiated (250 J/m^2; +UV-C) or not (−UV-C) plasmid DNA. No checkpoint activation is observed when injecting UV-C-irradiated plasmid DNA, as monitored by phosphorylation of the Chk1 kinase, while this is observed at the post-MBT stage (Conn, Lewellyn, & Maller, 2004; Kermi et al., 2015). A positive control for Chk1 phosphorylation is made by incubation of polydA/dT DNA in *Xenopus* egg extracts as previously described (Kumagai & Dunphy, 2000).

3. Studying the DNA damage response in mESCs

3.1 Introduction

mESCs in culture can in theory be propagated almost indefinitely in a state that closely resembles that of pluripotent cells of the epiblast of the day 4 embryo (the so-called naïve state). Self-renewal in the naïve state is stabilized by activation of the LIF/GP130/JAK signaling pathway by culturing mESCs in LIF and calf fetal serum (LIF/CFS). mESCs can be also maintained in the presence of inhibitors of both MEK and GSK3b kinases, or 2i (Ying et al., 2008) that will reinforce the pluripotent states (Hackett & Surani, 2014). Here, we will describe protocols to study DNA damage in mESCs (CGR8 cell line) maintained in LIF/CFS. In such conditions, mESCs show an inefficient G1/S checkpoint and fail to arrest in the G1 phase in the presence of DNA damage, notwithstanding normal p53 activation. We have previously shown that this is due to high abundance of the downstream phosphatase CDC25A, stabilized by the Dub3 ubiquitin hydrolase, that restrains the inhibition of the Cdk2/Cyclin E kinase (van der Laan, Tsanov, Crozet, & Maiorano, 2013), similar to what observed in some cancer cells resistant to therapeutic treatments (Falck, Mailand, Syljuasen, Bartek, & Lukas, 2001).

3.2 Damaging mESCs with UV-C

3.2.1 Equipment

- UV-C irradiator (Stratalinker)

3.2.2 Buffers and reagents

- Complete ESC medium: Glutamax Dulbecco's Modified Eagle Medium (Gibco 10567014)
  - 15% FBS (Eurobio, decomplemented at 55 °C for 30 min before use)
  - 1 × Non-Essential Amino Acids (Gibco 11140035)
  - 1 × Penicillin-Streptomycin (Gibco 15140122)
  - 0.1 mM 2-Mercaptoethanol (Gibco 31350010)
  - LIF: 10 μg mL-1 (Eurobio). Filter and keep at 4 °C up to 2 weeks
- Trypsin (TrypLE Express Enzyme Gibco, 12604013)
- 2% gelatin solution (Sigma G1393)
- Mitomycin C (Sigma M4287)
- DPBS (Sigma D8537)
- Trypsin-EDTA (Gibco 25300062)

- Mouse embryonic fibroblasts (feeder cells) inactivated with 10 mg/mL of mitomycin C (Sigma, M4287)
- PBS sterile (cell culture grade)

3.2.3 Procedure

1. Grow mESCs in ESC medium in a humidified atmosphere of 5% CO_2 at 37 °C on feeder cells (amplified in Glutamax DMEM, 10% FBS, 1 × Penicillin-Streptomycin) for at least two passages before performing experiments
2. Separate ESCs from feeder cells by trypsinization
3. Grow on gelatinized tissue culture dishes (0.1% Gelatin)
4. Remove as much as possible medium
5. Wash cells once with sterile PBS
6. Place dishes in the center of the UV-C irradiator
7. Irradiate at the desired UV-C dose (0–6 J/m^2).
8. Remove dish from irradiator
9. Add ESC medium pre-warmed at 37 °C
10. Move dishes to 37 °C incubator for the desired time
11. Cell pellets can be snap frozen in liquid nitrogen and stored at −80 °C or processed immediately. A typical result is shown below (Fig. 6)

3.2.4 Notes

1. We recommend verifying the dose of UV-C delivered by the irradiator by the use of a radiometer (UV - VLX-3 W - Vilber Lourmat), since this can change depending on the age of the UV lamps of the irradiator
2. mESCs are very sensitive to UV-C irradiation. It is recommended to irradiate them with a low dose of UV-C (no more than 6 J/m^2) to avoid apoptosis. This dose is largely sufficient to activate the DNA damage

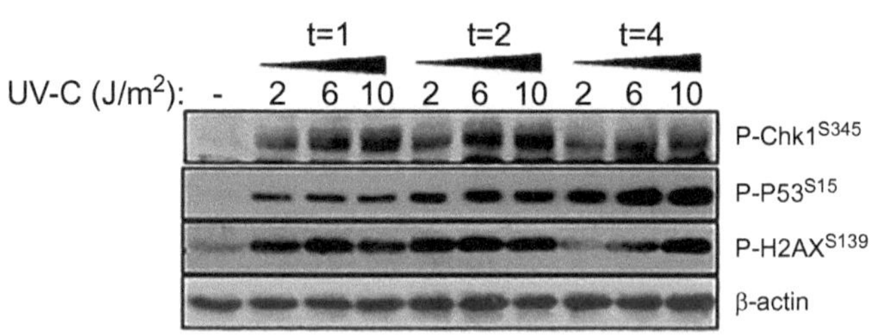

Fig. 6 Western blot of total protein extracts obtained from mESCs (CGR8 cell line) exposed to the indicated doses of UV-C. P-Chk1^{S345}, 2341; γH2AX (p-H2AX^{S139}) 9718; P-p53^{S15} 9284, Cell Signaling Technology; β-actin, A1978, Sigma.

response as illustrated by phosphorylation of the Chk1 kinase and the histone variant H2AX (γH2AX) reaching a maximum level at 2 h post irradiation (Fig. 6).

3.3 Damaging mESCs with MMS

3.3.1 Equipment

- 75 cm^2 flasks

3.3.2 Buffers and reagents

- ESC medium
- MMS (Sigma, M4016)
- PBS sterile (cell culture grade)

3.3.3 Procedure

1. Grow mESCs in ESC medium to 30% confluence in 75 cm^2 flasks
2. Prepare 50 mL of ESC medium with 10 μL of MMS stock solution (0.02% final concentration).
3. Incubate for 30 min at 37 °C
4. Remove MMS by washing cells 3 times with sterile PBS
5. Add ESC medium
6. Sample cells at different time points (every hour up to 6 h after MMS removal)
7. Cells pellets can be snap frozen in liquid nitrogen and stored at −80 °C or processed immediately. A typical result is shown below (Fig. 7)

3.4 Damaging mESCs with bleomycin

3.4.1 Equipment

- Nunc 6-wells plate (Thermofisher 140675)

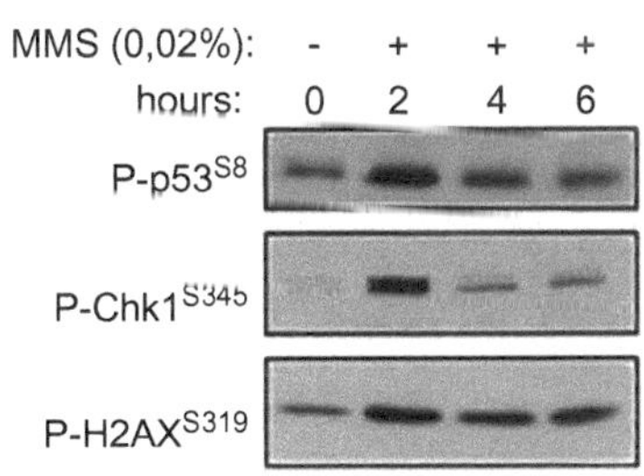

Fig. 7 Western blot of total cell lysates obtained from mESCs (CGR8 cell line) exposed to MMS. Antibodies were as in Fig. 5.

3.4.2 Buffers and reagents

- ESC medium
- Bleomycin (Sigma B8416)
- PBS sterile (cell culture grade)

3.4.3 Procedure

1. Prepare a Bleomycin working stock at 15 U/mL in water
2. Using a gelatin pre-coated 6-wells plate, seed mouse embryonic stem cells (mES) at 30% confluence
3. Next day, carefully wash off the growing medium and add increasing dose of Bleomycin (ranging from 30 milliU/mL (500 × dilution) to 1.875 milliU/mL (8000 × dilution).
4. Incubate wells plate for 6 h at 37 °C
5. Carefully remove Bleomycin-containing medium
6. Wash once with PBS at room temperature
7. Either snap freeze cells in liquid nitrogen or process immediately

3.5 Synchronization of mESCs with Nocodozale

Synchronization of mESCs is useful to study cell cycle-dependent DNA damage response. Due to inefficient checkpoint activation, mESCs are difficult to synchronize with methods currently used in somatic cells. We have found that synchronization with the microtubules inhibitors nocodazole is effective for these cells (van der Laan et al., 2013; van der Laan, Golfetto, Vanacker, & Maiorano, 2014). The Fucci reporter system to sort cells relative to the cell cycle phases has also been recently used (Gonnot, Langer, Bourillot, Doerflinger, & Savatier, 2019).

3.5.1 Equipment

- 75 cm^2 flasks
- Standard inverted-microscope
- Hemocytometer

3.5.2 Buffers and reagents

- Nocodazole (10 mg/mL stock solution in DMSO, Sigma, M1404)

3.5.3 Procedure

1. Grow mESCs in ESC medium to 80–90% confluence in 75 cm^2 flasks
2. Add 1 μg/mL of Nocodazole
3. Incubate for 5 h at 37 °C

4. Check the presence of round cells (should be 40–50%) under an inverted microscope (see notes).
5. Shake off cells by tapping the flask on the bench
6. Recover floating cells in ESC medium
7. Wash cells once with 5 mL of ESC medium
8. Count cells
9. Seed about 7×10^5 cells in a 100 mm flask containing pre-warmed ESC medium
10. Sample cells at time $t = 0$ and every hour, up to 6 h after Nocodazole release
11. Cells pellets can be snap frozen in liquid nitrogen and stored at −80 °C or processed immediately. Check cells synchrony by FACS analysis and western blot. A typical result is shown below (Fig. 8)

3.6 Notes

1. We recommend testing a Nocodazole dose-response, since different cell lines can respond differently
2. The primary parameter to check is visual rounding of the cells before shake-off. Although the process is not efficient in term of number of collected cells, the cells that are collected exit the Nocodazole block efficiently
3. Do not leave cells in Nocodazole overnight, since they may be permanently arrested in mitosis and do not exit from the block

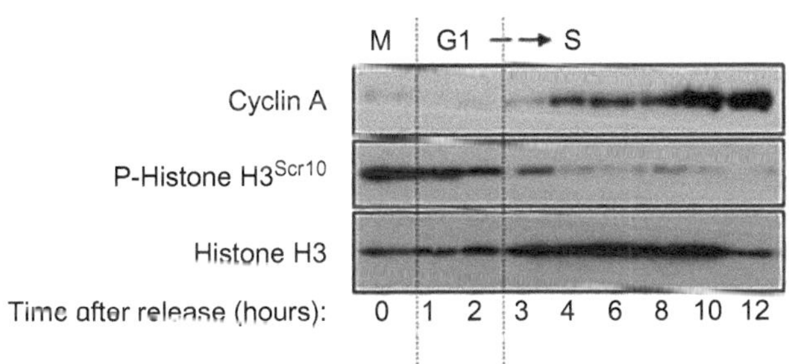

Fig. 8 Western blot of protein extracts obtained from mESCs (CGR8 cell line) synchronized with nocodazole. Cells were lysed in 1 × Laemmli buffer and proteins were fractionated by SDS-PAGE, followed by western blot with the indicated antibodies for specific cell cycle phases (mitosis: p-Histone $H3^{Ser\ 10}$; 09-797, Millipore; S-phase: Cyclin A; C4710, Sigma). A peak of histone H3 synthesis is observed in S-phase, as in somatic cells.

4. Conclusions

The protocols detailed in this chapter describe methods to study the DNA damage response in embryonic cells. Embryonic cells display several features common to cancer cells, including a high proliferation rate and a relaxed DNA damage response. Exploring the molecular mechanisms governing the DNA damage response in embryonic cells can be highly instrumental to understand the biology of cancer cells and can be useful to identify novel targets to sensitize cancer cells to therapeutic treatments.

Acknowledgments

Work in D.M. laboratory is funded by grants from "Fondation ARC pour la Recherche sur le Cancer", MSD Avenir, World Wide Cancer Research and Région Occitanie.

References

Aparicio Casado, T., & Gautier, J. (2021). DNA damage response in Xenopus laevis cell-free extracts. *Methods in Molecular Biology (Clifton, N.J.)*, *2267*, 103–144. https://doi.org/https://doi.org/10.1007/978-1-0716-1217-0_8.

Bakkenist, C. J., & Kastan, M. B. (2003). DNA damage activates ATM through inter-molecular autophosphorylation and dimer dissociation. *Nature*, *421*(6922), 499–506.

Blackford, A. N., & Jackson, S. P. (2017). ATM, ATR, and DNA-PK: The trinity at the heart of the DNA damage response. *Molecular Cell*, *66*(6), 801–817. https://doi.org/https://doi.org/10.1016/j.molcel.2017.05.015.

Blow, J. J., & Laskey, R. A. (1986). Initiation of DNA replication in nuclei and purified DNA by a cell-free extract of Xenopus eggs. *Cell*, *47*, 577–587.

Briano, R. A. (1983). A reproducible technique for breaking glass micropipettes over a wide range of tip diameters. *Journal of Neuroscience Methods*, *9*(1), 31–34. https://doi.org/https://doi.org/10.1016/0165-0270(83)90106-1.

Byun, T. S., Pacek, M., Yee, M. C., Walter, J. C., & Cimprich, K. A. (2005). Functional uncoupling of MCM helicase and DNA polymerase activities activates the ATR-dependent checkpoint. *Genes & Development*, *19*(9), 1040–1052.

Carbone, A., Fioretti, F. M., Fucci, L., Ausió, J., & Piscopo, M. (2012). High efficiency method to obtain supercoiled DNA with a commercial plasmid purification kit. *Acta Biochimica Polonica*, *59*(2), 275–278.

Ciccia, A., & Elledge, S. J. (2010). The DNA damage response: Making it safe to play with knives. *Molecular Cell*, *40*(2), 179–204. https://doi.org/https://doi.org/10.1016/j.molcel.2010.09.019.

Conn, C. W., Lewellyn, A. L., & Maller, J. L. (2004). The DNA damage checkpoint in embryonic cell cycles is dependent on the DNA-to-cytoplasmic ratio. *Developmental Cell*, *7*(2), 275–281.

Costanzo, V., & Gautier, J. (2004). Xenopus cell-free extracts to study DNA damage checkpoints. *Methods in Molecular Biology*, *241*, 255–267.

Dasso, M., & Newport, J. W. (1990). Completion of DNA replication is monitored by a feedback system that controls the initiation of mitosis in vitro: Studies in Xenopus. *Cell*, *61*(5), 811–823.

Falck, J., Mailand, N., Syljuasen, R. G., Bartek, J., & Lukas, J. (2001). The ATM-Chk2-Cdc25A checkpoint pathway guards against radioresistant DNA synthesis. *Nature*, *410*(6830), 842–847. https://doi.org/https://doi.org/10.1038/35071124.

Gonnot, F., Langer, D., Bourillot, P.-Y., Doerflinger, N., & Savatier, P. (2019). Regulation of cyclin E by transcription factors of the naïve pluripotency network in mouse embryonic stem cells. *Cell Cycle (Georgetown, Tex.)*, *18*(20), 2697–2712. https://doi.org/https://doi.org/10.1080/15384101.2019.1656475.

Hackett, J. A., & Surani, M. A. (2014). Regulatory principles of pluripotency: From the ground state up. *Cell Stem Cell*, *15*(4), 416–430. https://doi.org/https://doi.org/10.1016/j.stem.2014.09.015.

Kermi, C., Lo Furno, E., & Maiorano, D. (2017). Regulation of DNA replication in early embryonic cleavages. *Genes*, *8*(1). https://doi.org/https://doi.org/10.3390/genes8010042.

Kermi, C., Prieto, S., van der Laan, S., Tsanov, N., Recolin, B., Uro-Coste, E., et al. (2015). RAD18 is a maternal limiting factor silencing the UV-dependent DNA damage checkpoint in Xenopus embryos. *Developmental Cell*, *34*(3), 364–372. https://doi.org/https://doi.org/10.1016/j.devcel.2015.06.002.

Kumagai, A., & Dunphy, W. G. (2000). Claspin, a novel protein required for the activation of Chk1 during a DNA replication checkpoint response in Xenopus egg extracts. *Molecular Cell*, *6*(4), 839–849. https://doi.org/https://doi.org/10.1016/s1097-2765(05)00092-4.

Lin, Y., Ha, A., & Yan, S. (2019). Methods for studying DNA single-Strand break repair and signaling in Xenopus laevis egg extracts. *Methods in Molecular Biology (Clifton, N.J.)*, *1999*, 161–172. https://doi.org/https://doi.org/10.1007/978-1-4939-9500-4_9.

Lohka, M. J., & Masui, Y. (1983). Formation in vitro of sperm pronuclei and mitotic chromosomes induced by amphibian ooplasmic contents. *Science*, *220*, 719–721.

Mechali, M., & Harland, R. M. (1982). DNA synthesis in a cell-free system from Xenopus eggs: Priming and elongation on single-stranded DNA in vitro. *Cell*, *30*(1), 93–101.

Menut, S., Lemaitre, J. M., Hair, A., & Méchali, M. (1999). DNA replication and chromatin assembly using Xenopus egg extracts. In J. D. Richter (Ed.), *Advances in molecular biology: A comparative methods approach to the study of Oocytes and embryos* Oxford University Press.

Murray, A. W. (1991). Cell cycle extracts. *Methods in Cell Biology*, *36*, 581–605.

Newport, J., & Dasso, M. (1989). On the coupling between DNA replication and mitosis. *Journal of Cell Science. Supplement*, *12*, 149–160.

Newport, J., & Kirschner, M. (1982a). A major developmental transition in early Xenopus embryos: I. characterization and timing of cellular changes at the midblastula stage. *Cell*, *30*(3), 675–686.

Newport, J., & Kirschner, M. (1982b). A major developmental transition in early Xenopus embryos: II. Control of the onset of transcription. *Cell*, *30*(3), 687–696.

Nieuwkoop, P. D., & Faber, J. (1994). *Normal table of Xenopus laevis*. New York: Garland Publishing Inc.

Saldivar, J. C., Cortez, D., & Cimprich, K. A. (2017). The essential kinase ATR: Ensuring faithful duplication of a challenging genome. *Nature Reviews. Molecular Cell Biology*, *18*(10), 622–636.

Sannino, V., Pezzimenti, F., Bertora, S., & Costanzo, V. (2017). Xenopus laevis as model system to study DNA damage response and replication fork stability. *Methods in Enzymology*, *591*, 211–232. https://doi.org/https://doi.org/10.1016/bs.mie.2017.03.018.

Sive, H. L., Grainer, R. M., & Harland, R. M. (2000). *Early development of Xenopus laevis. A laboratory manual.* Cold Spring harbour Laboratory Press.

van der Laan, S., Golfetto, E., Vanacker, J. M., & Maiorano, D. (2014). Cell cycle-dependent expression of Dub3, Nanog and the p160 family of nuclear receptor coactivators (NCoAs) in mouse embryonic stem cells. *PLoS One*, *9*(4), e93663. https://doi.org/https://doi.org/10.1371/journal.pone.0093663.

van der Laan, S., Tsanov, N., Crozet, C., & Maiorano, D. (2013). High Dub3 expression in mouse ESCs couples the G1/S checkpoint to pluripotency. *Molecular Cell*, *52*(3), 366–379. https://doi.org/https://doi.org/10.1016/j.molcel.2013.10.003.

Walter, J., Sun, L., & Newport, J. (1998). Regulated chromosomal DNA replication in the absence of a nucleus. *Molecular Cell*, *1*(4), 519–529. https://doi.org/https://doi.org/10.1016/s1097-2765(00)80052-0.

Werbowetski-Ogilvie, T. E., Bossé, M., Stewart, M., Schnerch, A., Ramos-Mejia, V., Rouleau, A., et al. (2009). Characterization of human embryonic stem cells with features of neoplastic progression. *Nature Biotechnology*, *27*(1), 91–97. https://doi.org/https://doi.org/10.1038/nbt.1516.

Willis, J., DeStephanis, D., Patel, Y., Gowda, V., & Yan, S. (2012). Study of the DNA damage checkpoint using Xenopus egg extracts. *Journal of Visualized Experiments: JoVE*, *69*, e4449. https://doi.org/https://doi.org/10.3791/4449.

Ying, Q.-L., Wray, J., Nichols, J., Batlle-Morera, L., Doble, B., Woodgett, J., et al. (2008). The ground state of embryonic stem cell self-renewal. *Nature*, *453*(7194), 519–523. https://doi.org/https://doi.org/10.1038/nature06968.

CHAPTER SIX

Analysis of repair of replication-born double-strand breaks by sister chromatid recombination in yeast

Belén Gómez-González, Pedro Ortega, and Andrés Aguilera*

Centro Andaluz de Biología Molecular y Medicina Regenerativa (CABIMER), Universidad de Sevilla-CSIC-Universidad Pablo de Olavide, Seville, Spain

*Corresponding author: e-mail address: aguilo@us.es

Contents

Abstract

The repair of DNA double-strand breaks is crucial for cell viability and the maintenance of genome integrity. When present, the intact sister chromatid is used as the preferred repair template to restore the genetic information by homologous recombination. Although the study of the factors involved in sister chromatid recombination is hampered by the fact that both sister chromatids are indistinguishable, genetic and

Methods in Enzymology, Volume 661
ISSN 0076-6879
https://doi.org/10.1016/bs.mie.2021.08.010

molecular systems based on DNA repeats have been developed to overcome this problem. In particular, the use of site-specific nucleases capable of inducing DNA nicks that replication converts into double-strand breaks has enabled the specific study of the repair of such replication-born double strand breaks by sister chromatid recombination. In this chapter, we describe detailed protocols for determining the efficiency and kinetics of this recombination reaction as well as for the genetic quantification of recombination products.

1. Introduction

Double-strand breaks (DSBs) are among the most cytotoxic DNA lesions as a single unrepaired DSB is potentially lethal. Consequently, multiple options have evolved in eukaryotic cells to ensure that DSBs are repaired. However, not all possible DSB repair pathways result in the same outcome. Whereas some repair pathways accurately restore the DNA without consequences, the processing and repair of DSBs can also moderately or even extensively alter the genetic information, leading to mutations, loss of heterozygosity, or chromosome rearrangements that could ultimately result in cell death or tumorigenesis. Hence, to ensure accurate DSB repair is crucial for the maintenance of genome integrity. Despite that exogenous and endogenous agents can access and harm the DNA leading to various lesions including DSBs, one major source of genome instability is related to the DNA metabolism itself. Among DNA metabolic processes, evidence indicates that replication has a strong potential to cause alterations in the DNA, being S phase the most vulnerable phase of the cell cycle. Spontaneous DSBs occurring during DNA replication must therefore be carefully repaired to warrant fidelity. In accordance, the major repair pathway used during S phase is Sister Chromatid Recombination (SCR), a recombination reaction in which the genetic information from the intact sister chromatid is used as a template to restore the broken DNA (Johnson & Jasin, 2000; Kadyk & Hartwell, 1992).

Here, we describe the tools and methodology developed and used for the study of the repair of DSBs generated during DNA replication thanks to the use of site-specific nucleases that are able to produce nicks in the DNA before replication converts them into DSBs in the budding yeast *Saccharomyces cerevisiae*, vastly used as an eukaryotic model organism to study DNA DSB repair due to its easy growth and genetic manipulation.

1.1 The study of DSB repair mechanisms

There are two major DSB repair pathways: non-homologous DNA end joining (NHEJ), which re-ligates the two DNA ends together, and homologous recombination (HR), which uses the genetic information found in an homologous DNA sequence as a template to repair the break, resulting in either the unidirectional transfer of information (gene conversion) from the template (donor) to the broken molecule (recipient) or a reciprocal exchange of information between both molecules or crossover (reviewed in Pardo, Gómez-González, & Aguilera, 2009; Heyer, Ehmsen, & Liu, 2010). NHEJ can potentially ligate all kinds of double-stranded DNA ends and it is in large part mutagenic, in contrast to error-free HR. Indeed, HR restores the information of the broken molecule without altering its DNA sequence when using an identical copy as template, as is the case of a sister chromatid. This explains that, when available (in the S and G2 phases of the cell-cycle), the sister chromatid is the preferred donor molecule (Gonzalez-Barrera, Cortes-Ledesma, Wellinger, & Aguilera, 2003; Johnson & Jasin, 2000; Kadyk & Hartwell, 1992). Notwithstanding, the donor used for the HR repair reaction can be in the same locus of the homologous chromosome, leading to allelic recombination, or in any other location, leading to ectopic recombination. Thus, ectopic homologous DNA sequences or repeats, either in a different or the same DNA molecule whether a chromosome or a plasmid, can be used as alternative templates for the repair of DSBs. Apart from point mutations or loss of heterozygosity (LOH) as possible readouts of allelic recombination, ectopic recombination between DNA repeats can lead to important chromosomal reorganizations like deletions, duplications, inversions and translocations.

Diverse genetic systems based on DNA repeats have been developed and extensively used for the study of DSB repair by the different HR mechanisms: Single-Strand Annealing (SSA), Synthesis-Dependent Strand Annealing (SDSA), Double-Strand Break Repair (DSBR), and Break-Induced Replication (BIR) (Gómez-González, Ruiz, & Aguilera, 2011; Klein et al., 2019). Thus, constructs based on a duplication of non-identical tandem-repeats have been highly useful for the study of SCR. Despite the fact that the two sister chromatids resulting from replication are genetically indistinguishable, unequal SCR events occurring between the tandem-repeats and leading to triplications of the repeat can be genetically distinguishable. This rationale was used in the first yeast genetic assay generated to study spontaneous SCR (Fasullo & Davis, 1987), and has been used later applied to study DSB-induced SCR in yeast (Fasullo, Bennett, & Dave, 1999) and human cells (Johnson & Jasin, 2000; Puget, Knowlton, & Scully, 2005). However, an additional layer of complexity is required to physically study SCR and its kinetics of occurrence during time-course experiments. This was achieved with plasmid systems carrying either one (*pL2-HOr* system) or two (*pTINV* system) copies of the *LEU2* gene. In these cases, recombination intermediates from equal or unequal SCR events leading to reciprocal exchanges between the sister chromatids (Sister Chromatid Exchange, SCE) could be detected by Southern-blot after DNA-break induction by a site-specific endonuclease (Gonzalez-Barrera et al., 2003). Importantly, the studies with those systems support that unequal SCE accurately represents a reliable readout of natural SCR (Gonzalez-Barrera et al., 2003).

1.2 Repair of broken replication forks

During replication, advancing forks can encounter different sorts of obstacles that might cause their stalling or blockage (reviewed in Gómez-González & Aguilera, 2019). Specific protein-induced replication fork (RF) barriers have been used to study the fate of RFs after blockage (Lambert & Carr, 2013; Willis & Scully, 2021). Upon stalling, if replication does not resume, forks can collapse and break. This would be the case of a fork encountering a nick in the template. As a result, a one-ended DSB would be generated that could promote replication restart via BIR (see Anand, Lovett, & Haber, 2013; Yeeles, Poli, Marians, & Pasero, 2013 for reviews). However, once an incoming RF arrives from the opposite side, a two-ended DSB would be generated resembling a standard two-ended DSB. The proximity of the sister chromatids, which are held by cohesin rings, promotes such replication-born DSBs to be preferentially repaired with the sister chromatid (Cortés-Ledesma & Aguilera, 2006), thus maximizing the chances of restoring the genetic information without harmful consequences.

2. The use of a mini-HO site to study SCR induced by replication-born DSBs generated by the HO endonuclease

The yeast homothallic (HO) switching endonuclease is a site-specific endonuclease responsible for mating type switching that works by introducing a nick in both DNA strands and thus leading to a 4-bp staggered DSB into a 117 bp-target sequence (HO site) (Rudin, Sugarman, & Haber, 1989).

The HO endonuclease is involved in the cleavage of the yeast *MAT* locus in chromosome III. However, the expression of the HO endonuclease in a strain carrying the *MATa-inc* mutation, with a non-cleavable mutated HO site (Mascioli & Haber, 1980) allows the use of the HO site to induce DSBs elsewhere to study its repair without interference by mating type switching. Interestingly, the use of a mini-HO target site of 24 bp (mini-HO) lowers the efficiency of the HO endonuclease at that site (Gonzalez-Barrera et al., 2003). In this case, a large fraction of DSBs induced at this 24-bp mini-HO site would be replication-born as they would occur when the RF hit the nicks randomly generated in either DNA strand (Cortés-Ledesma & Aguilera, 2006). It is worth mentioning that the mini-HO site can also be occasionally cleaved in both strands at the same time causing replication-independent DSBs, although this happens in a low frequency (less than 2%) (Cortés-Ledesma & Aguilera, 2006).

3. The use of the FRT site and a mutant FLP to study replication-born DSB-induced SCR

The Flippase (FLP) endonuclease of the yeast 2 μ circle plasmid is involved in site-specific recombination induced by single-stranded cleavage and ligation reactions at the two 48-bp FLP recognition target (FRT) sites of the 2 μ circle (Sadowski, 1995). FLP recombinase action at the FRT leads to an intermediate FLP cleavage complex (FLP-cc), which is rapidly resolved by the ligation step (Lee, Whang, & Jayaram, 1996). However, a first step-arrest mutant version of FLP (FLPm) in which the histidine at position 305 is substituted by leucine (H305L), causes an irreversibly FLPm protein-bound nick that cannot be ligated (Tsalik & Gartenberg, 1998) similar to the TOP1cc formed after treatment with the DNA topoisomerase I poison camptothecin. Thus, the FLPm nicking enzyme can be used to introduce site-specific nicks at the FRT site that after replication would result in DSBs (Nielsen et al., 2009). However, in contrast to those produced by the HO endonuclease at the mini-HO site, FLPm-induced nicks are strand-specific. This allows, for instance, to target the nicks to the replication leading or lagging strands (Ortega, Gómez-González, & Aguilera, 2019). Furthermore, the frequency at which FLPm cleaves both strands is below 0.2%, 10-fold lower than that of the HO at the mini-HO site, making this a much better nickase with lower background levels of DSB induction (Ortega et al., 2019).

4. Rationale of the TINV recombination systems

The TINV recombination systems are based on a centromeric plasmid carrying the *URA3* gene as the selective genetic marker and two inverted repeats of the *LEU2* gene carrying different mutations (Gonzalez-Barrera et al., 2003; Ortega et al., 2019). One of the *LEU2* copies is 5′-truncated whereas the other one is a complete gene interrupted by the mini-HO sequences (TINV-HO system) or FRT (TINV-FRT system), which are inserted at the *LEU2 Eco*RI site (*leu2-HOr* or *leu2-FRT* alleles, respectively) (Fig. 1A) (Gonzalez-Barrera et al., 2003). These systems allow both the genetic detection of Leu + recombination products (Fig. 1A) that, although a good indirect indicator of unequal SCR products, can also arise via other HR events, as well as the physical detection of unequal SCR intermediates (Fig. 1B–D) preferentially arising from the SCE between the cut *leu2-HOr* repeat in one chromatid and the truncated *leu2*Δ*5′* repeat in the sister chromatid.

To study SCR, the endogenous *LEU2* gene must be deleted in the strain of interest to avoid ectopic recombination and undesired probe hybridization during physical analysis by Southern-blot (Gonzalez-Barrera et al., 2003). In the case of the TINV-HO, the yeast strain must also be *MATa-inc* to avoid HO cleavage at the *MAT* loci. The expression of the HO or FLPm endonucleases must be inducible, which is achieved by placing the corresponding genes under the *GAL* promoter. *GAL-HO* and *GAL-FLPm* constructs are available to be directly expressed from the genome in *ade3::GAL-HO* strains (Gonzalez-Barrera et al., 2003) or in *HPHMX6::GAL-FLPm* yeast strains (Ortega et al., 2019), respectively. Alternatively, the *GAL-HO* and *GAL-FLPm* constructs are also available in centromeric plasmids with different selective genetic markers to be co-transformed with the pTINV plasmid of interest (pTINV-HO or pTINV-FRT): pRS413GALHO (*HIS3*) (Muñoz-Galván, Jimeno, Rothstein, & Aguilera, 2013), pBISGALkFLP (*TRP1*) and pBIS-GALkFLP (*URA3*) (Tsalik & Gartenberg, 1998). Additionally, the *GAL*-HO construct has been inserted together with the TINV recombination system in the same centromeric plasmid allowing to perform experiments with single transformants: pTHGH (*URA3*) and pTHGH-2 (*TRP1*) (Ortega et al., 2019).

Time-course experiments are performed after HO or FLPm induction to determine the kinetics of DSB formation and HR repair. Southern-blot hybridization with a *LEU2* specific probe after XhoI-SpeI digestion allows

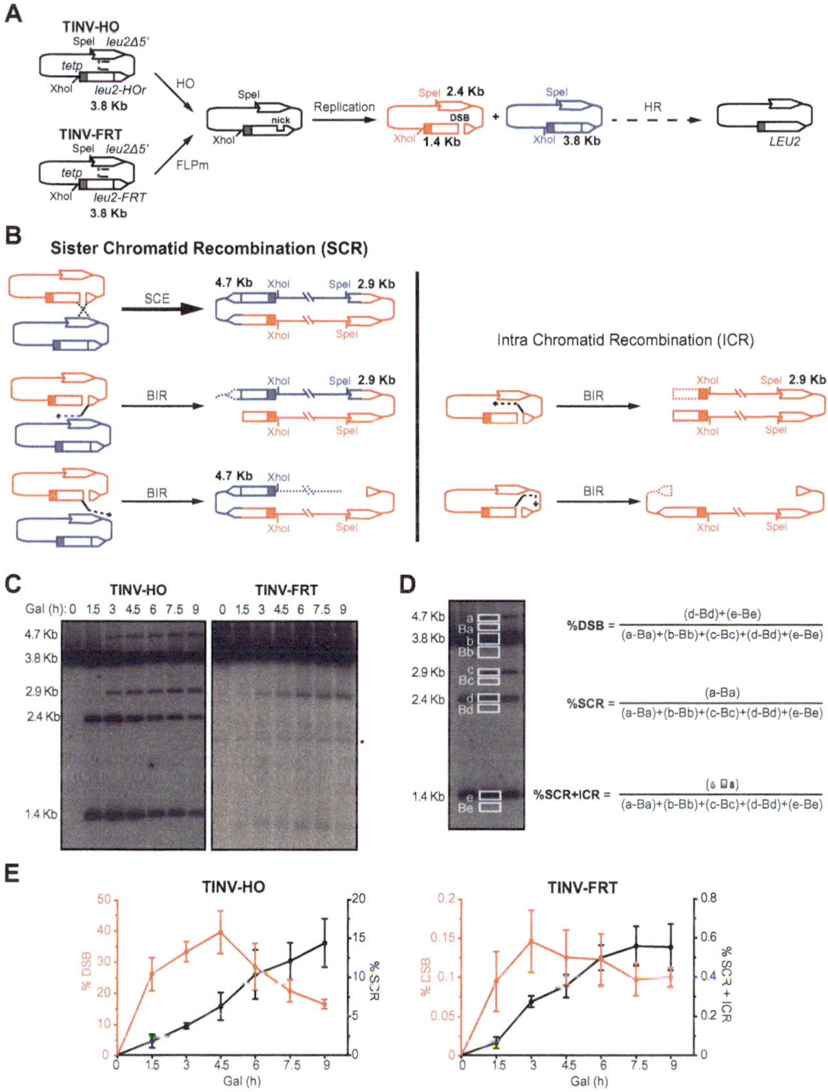

Fig. 1 Analysis of repair of replication-born DSBs with TINV plasmid systems. (A) Scheme of the TINV-based recombination plasmid systems (TINV-HO and TINV-FRT) with two inverted repeats of the *LEU2* gene, one 5′ truncated (*leu2Δ5′*) and the other one placed under the doxycycline-repressible promoter (*tet*p) and interrupted by the mini-HO or FRT sequences (*leu2-HOr* or *leu2-FRT* alleles, respectively). Induction of the expression of either HO or FLPm endonucleases creates a nick that is converted into a DSB in one of the chromatids (red) by replication, while the sister chromatid (blue) remains intact. All XhoI and SpeI restriction sites are indicated.

(Continued)

the visualization of the molecules of interest (Fig. 1C). A 3.8-Kb band corresponding to the linearized plasmid is observed in all time points. Break induction leads to 2.4- and 1.4-kb bands (Fig. 1A and C). At longer times, new 4.7- and 2.9-kb bands appear that indicate the occurrence of HR. This can happen by SCR using as a template either of the *leu2* copies located in the sister chromatid (equal SCR with the *leu2-HOr/leu2-FRT* or unequal SCR with the *leu2*△*5′*) or by Intra-Chromatid Recombination (ICR) using as a template the *leu2*△*5′* copy located within the same chromatid (Fig. 1B). Unequal SCR leads to the formation of recombination intermediates that correspond to the 4.7- and 2.9-kb bands generated after the XhoI-SpeI digestion. Importantly, the 4.7-kb band is specifically generated by SCR, and it is known to occur mostly by unequal SCE, with BIR being only a minor contributor (Cortes-Ledesma, Tous, & Aguilera, 2007). In contrast, the 2.9-kb band results also from either sister-chromatid BIR or

Fig. 1—Cont'd Fragments generated after DSB induction and XhoI-SpeI digestion, as detected by the *LEU2* probe shown as a line with asterisk are indicated with their corresponding sizes. HR repair can give rise to restoration of the *LEU2* gene Leu+recombinants. (B) Scheme of different recombination intermediates generated by either unequal Sister Chromatid Recombination (SCR), which can occur by Sister Chromatid Exchange (SCE) or by Break-Induced Replication (BIR) with the *leu2*△5′ copy in the intact sister chromatid, or by Intra-Chromatid Recombination (ICR) by BIR with the *leu2*△5′ copy within the same chromatid. The size of fragments generated after DSB induction and XhoI-SpeI digestion, as detected by the *LEU2* probe (shown in Fig. 1A) are indicated. Dashed lines depict putative fragments of undefined length depending on the extent of DNA synthesis of the BIR reaction. (C) Kinetics of appearance of DSBs and HR repair intermediates with the TINV-HO and TINV-FRT systems as determined by Southern blot hybridization with the *LEU2* probe depicted in Fig. 1A. The different-sized molecules are generated after XhoI-SpeI digestion and either HO (TINV-HO system) or FLPm (TINV-FRT system) induction in galactose for the indicated time-points. Bands resulting from unspecific hybridization are indicated with an asterisk. These illustrative gels correspond to experiments published in Ortega et al. (2019). (D) Diagram for the quantification of the percentage of each kind of molecule (DSBs, SCR intermediates or SCR +ICR intermediates). The signal from same-sized quantification squares over each band of interest (a, b, c or d), from which the corresponding Background (B) signal must be subtracted, is divided by the sum of the signal of each all other bands. (E) Results of the mean and SEM values from the quantification of the bands obtained in a wild-type strain transformed with from either pTINV-HO or pTINV-FRT plasmids. The percentage of molecules with DSBs is shown in red in both systems whereas the percentage of molecules that constitute an intermediate for HR reactions is shown in black. SCR intermediates were quantified in the TINV-HO system and SCR+ICR intermediates in the TINV-FRT system. The data correspond to independent experiments published in Ortega et al. (2019).

Intra-Chromatid Recombination (ICR) occurring between the two inverted repeats located within the broken chromatid (Fig. 1B), both reactions known to occur at low frequency (Cortes-Ledesma et al., 2007; Cortés-Ledesma & Aguilera, 2006; Gonzalez-Barrera et al., 2003). Quantifying the amount of the different-sized DNA molecules allows to infer the efficiency and kinetics of each HR reaction (Fig. 1D and E).

5. Reagents, materials and equipment

5.1 Reagents

- Glucose-containing minimal media (SC): 1.7 g/L amino acid-free yeast nitrogen base, 5 g/L ammonium sulfate, 20 g/L glucose. Add 20 g/L agar before autoclaving for solid media. Autoclave and supplement with the required amino acid. All amino acids should be added in standard concentrations except for leucine and adenine, which should be added three times more concentrated
- Raffinose-containing minimal media (SRaf): 1.7 g/L amino acid-free yeast nitrogen base, 5 g/L ammonium sulfate. Autoclave and supplement with the required amino acid. All amino acids should be added in standard concentrations except for leucine and adenine, which should be added three times more concentrated and with 2% raffinose (Thermo Fisher Scientific, cat # 15815188) from a 20% autoclaved stock solution
- 20% galactose (Sigma-Aldrich, cat # G0750) filtered stock solution
- 5 mg/mL doxycycline stock (Sigma-Aldrich, cat # D9891). Store in the dark at 4 °C for up to 2 weeks
- 1 M spermidine stock solution (Sigma-Aldrich, cat # S2626). Store at −20 °C. Make a fresh stock solution of this reagent every month
- 0.5 M spermine stock solution (Sigma-Aldrich, cat # S3256). Store at −20 °C
- Nuclei-isolating buffer (NIB) pH 7.2: 17% (*w*/*v*) glycerol, 50 mM (3-[N-morpholino] propanesulfonic acid) sodium salt (MOPS, Sigma-Aldrich, cat # M1254) pH 7.5, 150 mM CH_3CO_2K (Sigma-Aldrich, cat # P1190), 2 mM $MgCl_2$, 500 μM spermidine, 150 μM spermine. Autoclave and store at 4 °C in the dark for up to 6 months
- Zymolyase 20T (15 mg/mL) (USBiological)
- RNase A (10 mg/mL) (Sigma-Aldrich, cat # R6513)
- 3 M sodium acetate
- Phenol solution (Sigma-Aldrich, cat # P4557)

- Phenol:Chloroform:Isoamyl alcohol solution (25:24:1) (Sigma-Aldrich, cat # P3803)
- Isopropanol
- Milli-Q water
- 1 × TE: Tris–HCl 10 mM, EDTA 1 mM pH 8
- 100% ethanol
- Ethidium bromide (10 mg/mL) (Sigma-Aldrich, cat # E1510)
- Agarose
- 1 × TBE: 90 mM Tris base, 90 mM boric acid, 2 mM EDTA
- 1-kb DNA ladder (Invitrogen, cat # 10787018 or similar)
- 10 × Loading buffer
- XhoI and SpeI-HF restriction enzymes (New England Biolabs, cat # R0146L and R3133L, respectively)
- 0.25 M HCl
- Denaturation solution: 0.5 M NaOH, 1.5 M NaCl
- 20 × SSC: 0.3 M NaCl, 0.03 M trisodium citrate pH 7
- dATG solution: 0.5 mM dATP, 0.5 mM dTTP, and 0.5 mM dGTP
- Specific *LEU2* oligonucleotides (5′-GTTCCACTTCCAGATGAGGC-3′ and 5′-TTAGCAAATTGTGGCTTGA-3′)
- $\alpha 32^{P}$-dCTP (1 mCi [10 mCi/mL], 3000 Ci/mmol) (PerkinElmer, cat # NEG513H250UC)
- Klenow (Roche, cat # 1008412001)
- Hexanucleotide mix (Roche, cat # 11277081001)
- Taq DNA polymerase (MyTaq™ DNA polymerase) (Bioline, cat # BIO-21105)
- NucleoSpin Gel and PCR extraction kits (Macherey-Nagel, cat # 740609.50)
- Sephadex G50-TE (GE Healthcare, cat # 11524875): Dissolve 5 g Sephadex G50 in 75 mL 1 × TE, autoclave, and store at 4 °C
- Hybridization solution: 0.5 M phosphate buffer pH 7, 7% SDS
- Wash solution: 0.1 × SSPE, 5 mM EDTA, 0.5% SDS
- 20 × SSPE: 3 M NaCl, 200 mM sodium phosphate, 20 mM EDTA pH 7.4

5.2 Materials

- 50 mL tubes
- 1.5 mL microtubes
- 2 mL microtubes

- Fluorometer cuvettes
- Temperature controlled shaker
- Temperature controlled incubator
- Phase lock gel (PLG) heavy 2 mL tubes (Quantabio, cat # 2302830)
- Hybond-XL hybridization nylon membrane (GE-Healthcare, cat # RPN303S)
- Whatman filter paper
- Filter paper
- Petri dishes
- Fluorescent gel ruler

5.3 Equipment

- Centrifuge for 50 mL tubes
- Centrifuge for 1.5 mL tubes
- Fluorometer apparatus
- Block heater or water bath
- Magnetic stirrer
- Electrophoresis apparatus
- Electrophoresis power supply
- UV transilluminator
- 254 nm UV crosslinker
- Hybridization oven
- Hybridization bottles
- Vortex
- Nanodrop
- PhosphorImager Fujifilm FLA-5100 (Fujifilm)
- ImageGauge software (Fujifilm)

6. Protocols

6.1 Replication-born DSB induction

1. Zig-zag streak the yeast transformant of interest on an SC plasmid-selective media plate for 3–4 days at the selected temperature (30 °C for non-thermosensitive strains).
2. Inoculate a single colony in 5 mL liquid SC plasmid-selective media and incubate overnight on a shaker at the appropriate temperature (30 °C for non-thermosensitive strains).

3. Dilute the culture in 50 mL of the same SC plasmid-selective media and let it grow for two rounds of duplication. This usually takes around 6 h. Collect the cultures at OD600 of 0.6–0.8, before lag phase.
4. Dilute the culture in 350 mL for a of SRaf plasmid-selective media and incubate overnight at the appropriate temperature.

   Tip: To avoid any putative interference of transcription from the *TET* promoter placed in the *leu2* repeat to be cut (*leu2-HOr* or *leu2-FRT*), add doxycycline to a final concentration of 5 μg/mL.

   Tip: For the genetic determination of recombination frequencies, steps 2–4 can be avoided and the different colonies of the yeast transformant of interest can be directly inoculated in 5 mL SRaf plasmid-selective media with doxycycline (5 μg/mL) if possible. Make sure to use colonies of the same size to be able to assume that the frequencies are estimated from the same number of cell divisions.
5. Wait until OD600 of 0.4–0.5, add galactose to a final concentration of 2% and keep shaking to induce the expression of either the HO or FLPm endonuclease.
6. The time of incubation will depend on the goal of the experiment (genetic determination of recombination frequencies or physical analysis of DSBs and SCR intermediates), as explained below.

6.2 Genetic determination of recombination frequencies

To determine recombination frequencies, galactose-induction should be kept during 5 h for HO and 6 h for FLPm. After that, add glucose to a final concentration of 2% to stop the endonuclease expression and perform serial dilutions in H_2O to be plated on either SC plasmid-selective media, from which we can estimate the total number of cells; or SC plasmid-selective media lacking leucine, from which we can estimate the number of recombinant cells.

6.3 Physical analysis of DSBs and SCR intermediates

To determine the kinetics of DSB formation and repair by HR, the plasmid DNA is extracted at different time points after galactose addition (usually 0, 1.5, 3, 4.5, 6, 7.5 and 9 h). At every time-point, harvest 50 mL of the cell culture in 50 mL tubes, centrifuge at 4 °C (3000 *g*, 2 min), remove the supernatant and wash the pellet with cold milli-Q H_2O. Centrifuge again (3000 *g*, 2 min), remove the supernatant and freeze each pellet at −80 °C. In addition, take 1 mL of the cell culture for measuring OD in the fluorometer apparatus with the appropriate cuvettes. Write down the ODs for every time-point as these values will be the reference to estimate the volume containing the same number of cells at every time-point. Once all time-point samples are collected, proceed with the adjusted volumes as follows:

6.3.1 DNA extraction

1. Add 1 mL of milli-Q H_2O to the frozen pellet and transfer the previously estimated volume of cells into a 2 mL microtube, centrifuge (3000 *g*, 1 min), remove supernatant and resuspend the pellet in 400 μL NIB by vortexing.
2. For cell wall digestion, add 80 μL of zymolyase 20T (15 mg/mL) and incubate during 35 min at RT inverting the tubes every 10 min.
3. To stop the zymolyase action, add 1.5 mL of milli-Q H_2O to each tube, centrifuge (3000 *g*, 2 min), remove supernatant and resuspend the pellet in 720 μL of 1 × TE by pipetting and a subsequent vigorous vortexing. From now on, do not use the vortex anymore.
4. For cell lysis, add 80 μL 10% SDS and incubate on ice during 30 min inverting the tubes every 10 min during the incubation.
5. For protein removal, add 800 μL phenol, mix vigorously, centrifuge (16,000 *g*, 15 min) and transfer the clear upper phase into a 2 mL microtube. Add 800 μL Phenol:Chloroform:Isoamyl alcohol solution, mix vigorously, centrifuge (16,000 *g*, 5 min) and transfer the clear upper phase into a new 2 mL PLG heavy tube. Repeat this last step at least once more in order to obtain a clean sample.
6. For nucleic acid precipitation, add 160 μL sodium acetate 3 M and 800 μL isopropanol, invert the tubes several times, centrifuge (16,000 *g*, 15 min) and remove supernatant. Add 700 μL of cold 70% ethanol without mixing, centrifuge (16,000 g, 10 min), remove supernatant and resuspend the pellet in 500 μL 1 × TE.
7. To remove RNA, add 0.5 μL RNase A (10 mg/mL), resuspend the pellet and incubate at 37 °C during 30–60 min.
8. To remove RNAse A, add 500 μL Phenol:Chloroform:Isoamyl alcohol solution and transfer both phases into a 2 mL PLG heavy tube, mix vigorously, centrifuge (16,000 *g*, 5 min) and transfer the clear upper phase into a new 2 mL microtube.
9. Precipitate the DNA by adding 100 μL sodium acetate 3 M and 500 μL isopropanol. Centrifuge (16,000 *g*, 15 min) and remove supernatant. Add 700 μL of cold 70% ethanol without mixing, centrifuge (16,000 *g*, 10 min), remove supernatant and resuspend the pellet in 200 μL 1 × TE.

10. Quantify the DNA in each preparation with a DNA fluorimeter and use either 5 μg or 10–20 μg of DNA, for HO and FLPm inductions, respectively, for digestion with 50 U of XhoI and SpeI-HF restriction enzymes (2 h at 37 °C).
11. Precipitate the digested DNA by adding 100 μL sodium acetate 3 M and 500 μL isopropanol. Invert the tubes several times, centrifuge (16,000 *g*, 15 min) and remove supernatant. Add 700 μL of cold 70% ethanol without mixing, centrifuge (16,000 *g*, 10 min), remove supernatant and resuspend the pellet in 30 μL 1 × TE.

Tip: Make sure to let the pellet dry before dissolving in 1 × TE avoiding water and/or ethanol rests.

Tip: If the extracted DNA pellet is not properly dissolved, use pre-warmed the 1 × TE and incubate the pellet in it for 30–60 min at 30 °C.

6.3.2 Gel electrophoresis

12. Prepare a 0.8% agarose gel by dissolving 2.8 g agarose in 350 mL boiling 1 × TBE with a stir bar in a magnetic stirrer. Let it cool down to 65 °C in a water bath or a temperature-controlled incubator. Add ethidium bromide (to a 0.3 μg/mL final concentration) and then pour the solution in a 24 × 20 cm gel-casting tray to allow agarose polymerization (usually during 2 h). Then, place the gel at a gel stage containing a suitable volume of 1 × TBE and clean the wells with a syringe or a pipette.

Tip: The agarose must be perfectly diluted in order to avoid non-dissolved agar clumps.

13. Load 5 μL of a 1-kb DNA ladder in a well and the digested DNA samples (30 μL with 3.3 μL of 10 × Loading buffer each) in the rest of the wells. Run at constant low voltage (50 V, c.a. 1 V/cm) overnight. Stop the electrophoresis when the 1-kb band from the 1-kb DNA ladder reaches the end of the gel (usually after 20 h).
14. Take a picture of the ethidium bromide staining in a UV transilluminator with a fluorescent gel ruler.

6.3.3 Transfer of DNA to a nylon membrane

15. Immerse the gel in 0.25 M HCl during 10 min and rinse twice with distilled H_2O. Then, immerse the gel in denaturation solution during 30 min and rinse twice with distilled H_2O. Last, immerse the gel in 20 × SCC during 5–15 min prior its transfer.
16. Transfer the gel in standard Southern-blot conditions using Hybond-XL hybridization nylon membrane (GE-Healthcare) in 20 × SCC solution overnight.

17. UV crosslink the DNA to the membrane by subjecting both sides to 70,000 μJ/cm^2 UV light using a 254 nm UV crosslinker.
18. Rinse the membranes with 2× SSC to eliminate salt excess and let them dry on a Whatman filter paper.

6.3.4 Random primer radiolabeling

19. Amplify a 218-pb fragment of the *LEU2* gene by standard PCR with a Taq DNA polymerase and specific oligonucleotides (5′-GTTCCACT TCCAGATGAGGC-3′ and 5′-TTAGCAAATTGTGGCTTGA-3′). Load the PCR product in an agarose gel, run an electrophoresis and purify the DNA from the agarose gel with a NucleoSpin Gel and PCR extraction kit following manufacturer instructions. To obtain an ultrapure probe, repeat the purification with the NucleoSpin Gel and PCR extraction kit following manufacturer instructions (PCR clean up protocol) and measure the final DNA concentration in the nanodrop.
20. Boil 200 ng of the purified DNA probe in 36.5 μL of milli-Q H_2O (100 °C during 10 min). Then, centrifuge the sample (16,000 *g*, 30 s) and quickly transfer the tube to ice.
21. Add 5 μL of hexanucleotide mix, 5 μL of 0.5 mM dATG solution, 1 μL of Klenow and 50 μCi of 32^{P}-dCTP. Mix by pipetting up and down and incubate for 1 h at 37 °C.
22. Purify from unincorporated labeled nucleotides through a Sephadex G50 column.

6.3.5 Southern-blot hybridization

23. Prehybridize the hybond-XL membrane in a hybridization bottle with 10 mL of Hybridization Solution under rotation in a hybridization oven at 65 °C for at least 30 min. Place the DNA side of the membrane facing inwards.
24. Add the radioactive probe (previously boiled during 10 min at 100 °C for its denaturation) to 10 mL pre-warmed hybridization solution and incubate overnight under rotation in a hybridization oven at 65 °C.
25. Remove the radioactive solution and wash the filters twice during 5 min with 50 mL pre-warmed wash solution under rotation in a hybridization oven at 65 °C and then perform an additional wash during 45 min. Let the membranes dry on a Whatmann filter paper and detect the radioactive signals using a PhosphorImager Fujifilm FLA-5100.

7. Data analysis

To determine the recombination frequency of a particular yeast strain, the average of at least three experiments, each performed with an independent transformant, should be calculated. From each transformant, we use the median value from six different colonies. Student's *t*-tests can be used to determine the statistical differences between the average recombination frequencies of different strains.

After the physical detection of the radioactive signals corresponding to the DSB and HR intermediates in the PhosphorImager Fujifilm FLA-5100, the ImageGauge software can be used for quantification. The exposition time in the PhosphorImager Fujifilm FLA-5100 should be optimized for each membrane to obtain non-saturated images. The analysis of the signals is performed by drawing quantification squares over each band (Fig. 1D). In addition, a second square of the same size is selected below each band to quantify the background from the same lane. Such signal of the background should be subtracted from the signal of each band in the gel. Then, the resulting signal corresponding to the each of the bands of the gel is used to estimate the percentage of molecules of interest at each time point of study. The percentage of molecules with DSBs is calculated by dividing the signal corresponding to the 2.4- and 1.4-kb bands by the sum of the signal of each all other bands (Fig. 1D and E). Similarly, the percentage of either molecules that constitute an intermediate for SCR or SCR + ICR reactions is calculated by dividing the signal corresponding to the 4.7- or 2.9-kb bands, respectively, by the sum of the signal of each all other bands (Fig. 1D and E).

Acknowledgments

Research in A.A.'s lab is funded by the Spanish Ministry of Science and Innovation, the European Research Council and the Junta de Andalucía, plus the European Union (FEDER).

References

Anand, R. P., Lovett, S. T., & Haber, J. E. (2013). Break-induced DNA replication. *Cold Spring Harbor Perspectives in Biology*, *5*(12), a010397. https://doi.org/10.1101/cshperspect.a010397.

Cortés-Ledesma, F., & Aguilera, A. (2006). Double-strand breaks arising by replication through a nick are repaired by cohesin-dependent sister-chromatid exchange. *EMBO Reports*, 7(9), 919–926. https://doi.org/10.1038/sj.embor.7400774.

Cortes-Ledesma, F., Tous, C., & Aguilera, A. (2007). Different genetic requirements for repair of replication-born double-strand breaks by sister-chromatid recombination and break-induced replication. *Nucleic Acids Research*, *35*(19), 6560–6570. https://doi.org/10.1093/nar/gkm488.

Fasullo, M., Bennett, T., & Dave, P. (1999). Expression of Saccharomyces cerevisiae MATa and MAT alpha enhances the HO endonuclease-stimulation of chromosomal rearrangements directed by his3 recombinational substrates. *Mutation Research*, *433*(1), 33–44. https://doi.org/10.1016/S0921-8777(98)00059-7.

Fasullo, M. T., & Davis, R. W. (1987). Recombinational substrates designed to study recombination between unique and repetitive sequences in vivo. *Proceedings of the National Academy of Sciences of the United States of America*, *84*(17), 6215–6219. https://doi.org/10.1073/pnas.84.17.6215.

Gómez-González, B., & Aguilera, A. (2019). Transcription-mediated replication hindrance: A major driver of genome instability. *Genes and Development*, *33*(15–16). https://doi.org/10.1101/gad.324517.119.

Gómez-González, B., Ruiz, J. F., & Aguilera, A. (2011). Genetic and molecular analysis of mitotic recombination in Saccharomyces cerevisiae. In H. Tsubouchi (Ed.), *DNA recombination. Methods in molecular biology (methods and protocols), vol 745*. Humana Press. https://doi.org/10.1007/978-1-61779-129-1_10.

Gonzalez-Barrera, S., Cortes-Ledesma, F., Wellinger, R. E., & Aguilera, A. (2003). Equal sister chromatid exchange is a major mechanism of double-strand break repair in yeast. *Molecular Cell*, *11*(6), 1661–1671. https://doi.org/10.1016/S1097-2765(03)00183-7.

Heyer, W. D., Ehmsen, K. T., & Liu, J. (2010). Regulation of homologous recombination in eukaryotes. *Annual Review of Genetics*, *44*, 113–139. https://doi.org/10.1146/annurev-genet-051710-150955.

Johnson, R. D., & Jasin, M. (2000). Sister chromatid gene conversion is a prominent double-strand break repair pathway in mammalian cells. *The EMBO Journal*, *19*(13), 3398–3407. https://doi.org/10.1093/emboj/19.13.3398.

Kadyk, L. C., & Hartwell, L. H. (1992). Sister chromatids are preferred over homologs as substrates for recombinational repair in Saccharomyces cerevisiae. *Genetics*, *132*(2), 387–402.

Klein, H. L., Bačinskaja, G., Che, J., Cheblal, A., Elango, R., Epshtein, A., et al. (2019). Guidelines for DNA recombination and repair studies: Cellular assays of DNA repair pathways. *Microbial Cell*, *6*(1). https://doi.org/10.15698/mic2019.01.664.

Lambert, S., & Carr, A. M. (2013). Replication stress and genome rearrangements: Lessons from yeast models. *Current Opinion in Genetics & Development*, *23*(2), 132–139. https://doi.org/10.1016/j.gde.2012.11.009.

Lee, J., Whang, I., & Jayaram, M. (1996). Assembly and orientation of Flp recombinase active sites on two-, three- and four-armed DNA substrates: Implications for a recombination mechanism. *Journal of Molecular Biology*, *257*(3), 532–549. https://doi.org/10.1006/jmbi.1996.0183.

Mascioli, D. W., & Haber, J. E. (1980). A cis-acting mutation within the MAT(a) locus of Saccharomyces cerevisiae that prevents efficient homothallic mating-type switching. *Genetics*, *94*(2), 341–360.

Muñoz-Galván, S., Jimeno, S., Rothstein, R., & Aguilera, A. (2013). Histone H3K56 acetylation, Rad52, and non-DNA repair factors control double-strand break repair choice with the sister chromatid. *PLoS Genetics*, *9*(1), e1003237. https://doi.org/10.1371/journal.pgen.1003237.

Nielsen, I., Bentsen, I. B., Lisby, M., Hansen, S., Mundbjerg, K., Andersen, A. H., et al. (2009). A Flp-nick system to study repair of a single protein-bound nick in vivo. *Nature Methods*, *6*(10), 753–757. https://doi.org/10.1038/nmeth.1372.

Ortega, P., Gómez-González, B., & Aguilera, A. (2019). Rpd3L and Hda1 histone deacetylases facilitate repair of broken forks by promoting sister chromatid cohesion. *Nature Communications*, *10*(1), 1–15. https://doi.org/10.1038/s41467-019-13210-5.

Pardo, B., Gómez-González, B., & Aguilera, A. (2009). DNA repair in mammalian cells: DNA double-strand break repair: How to fix a broken relationship. *Cellular and Molecular Life Sciences*, *66*(6), 1039–1056. https://doi.org/10.1007/s00018-009-8740-3.

Puget, N., Knowlton, M., & Scully, R. (2005). Molecular analysis of sister chromatid recombination in mammalian cells. *DNA Repair*, *4*(2), 149–161. https://doi.org/10.1016/j.dnarep.2004.08.010.

Rudin, N., Sugarman, E., & Haber, J. E. (1989). Genetic and physical analysis of double-strand break repair and recombination in Saccharomyces cerevisiae. *Genetics*, *122*(3), 519–534.

Sadowski, P. D. (1995). The Flp recombinase of the 2-μm plasmid of *Saccharomyces cerevisiae*. *Progress in Nucleic Acid Research and Molecular Biology*, *51*, 53–91. https://doi.org/10.1016/S0079-6603(08)60876-4.

Tsalik, E. L., & Gartenberg, M. R. (1998). Curing Saccharomyces cerevisiae of the 2 micron plasmid by targeted DNA damage. *Yeast*, *14*(9), 847–852. https://doi.org/10.1002/(SICI)1097-0061(19980630)14:9 < 847::AID-YEA285 > 3.0.CO;2-9.

Willis, N. A., & Scully, R. (2021). Measurement of homologous recombination at stalled mammalian replication forks. In A. Aguilera, & A. Carreira (Eds.), *Methods in molecular biology: Vol. 2153. Homologous recombination*. New York, NY: Humana. doi:10.1007/978-1-0716-0644-5_23.

Yeeles, J. T., Poli, J., Marians, K. J., & Pasero, P. (2013). Rescuing stalled or damaged replication forks. *Cold Spring Harbor Perspectives in Biology*, *5*(5), a012815. https://doi.org/10.1101/cshperspect.a012815.

CHAPTER SEVEN

Determining the kinetics of break-induced replication (BIR) by the assay for monitoring BIR elongation rate (AMBER)

Liping Liu[a], Neal Sugawara[b], Anna Malkova[a,*], and James E. Haber[b,*]
[a]Department of Biology, University of Iowa, Iowa City, IA, United States
[b]Department of Biology, Rosenstiel Basic Medical Sciences Research Center, Brandeis University, Waltham, MA, United States
*Corresponding authors: e-mail address: anna malkova@uiowa.edu; haber@brandeis.edu

Contents

Abstract

A detailed understanding of how homologous recombination proceeds at the molecular level in vivo requires the ability to detect in real time the appearance of specific intermediates of DNA repair. The most detailed analysis of double-strand break (DSB) repair in eukaryotes has come from the study of budding yeast, using an inducible

ISSN 0076-6879
https://doi.org/10.1016/bs.mie.2021.09.004

site-specific HO endonuclease to initiate recombination synchronously in nearly all cells of the population. Polymerase chain reaction (PCR) and chromatin immunoprecipitation (ChIP) methods have been used to visualize the timing of the DSB, its resection by 5′ to 3′ exonucleases, the binding of the Rad51 recombinase and the pairing of the Rad51 filament with a homologous donor sequence. PCR has also been used to identify the next key step: the initiation of new DNA synthesis to extend the invading stand and copy the donor template. In break-induced replication (BIR), there appears to be a very long delay between strand invasion and this primer extension step. Here we describe an alternative method, an assay for monitoring BIR elongation rate (AMBER) based on digital droplet PCR that yields a much earlier time of initial DNA synthesis. We suggest that previous methods have failed to recover the initial long, single-stranded primer extension product that is readily detected by AMBER.

1. Introduction

A DNA double strand break (DSB) is one of the most harmful DNA lesions that threatens genomic integrity. DSBs can be repaired either by non-homologous end joining (NHEJ), which ligates the two ends of the DSB together, or by homologous recombination (HR), which uses the homologous DNA—on a sister chromatid, a homologous chromosome or at an ectopic location—as a template to repair the break site. Several HR pathways have been identified, including synthesis-dependent annealing (SDSA) and double Holliday Junction (dHJ)-mediated gene conversions, in which both ends of DSB can find homology and patch up the break with sequences copied from the donor (reviewed in Haber, 2016; Heyer, 2015; Jasin & Rothstein, 2013). An alternative HR pathway, break-induced replication (BIR), arises when there is homology on only one end of the DSB, as happens at eroded telomeres or at stalled and broken replication forks (reviewed in Anand, Lovett, & Haber, 2013; Kockler, Osia, Lee, Musmaker, & Malkova, 2021; Llorente, Smith, & Symington, 2008). Previous studies have identified key molecular steps in Rad51-dependent BIR. Strand invasion creates a migrating D-loop in which the synthesis of leading and lagging strands during BIR are asynchronous, with very long (tens of kilobase pairs) single-stranded DNA (ssDNA) accumulating behind the migrating D-loop (Fig. 1) (Donnianni & Symington, 2013; Saini et al., 2013; Wilson et al., 2013). In addition to nearly all the proteins required for normal DNA replication (where leading- and lagging-strand synthesis are coupled), BIR requires several additional proteins including the 3′ to 5′ helicase, Pif1, and the nonessential DNA polymerase δ subunit, Pol32 (Lydeard et al., 2010; Lydeard, Jain, Yamaguchi, & Haber, 2007; Wilson et al., 2013). One surprising aspect of BIR is the apparently long interval between the

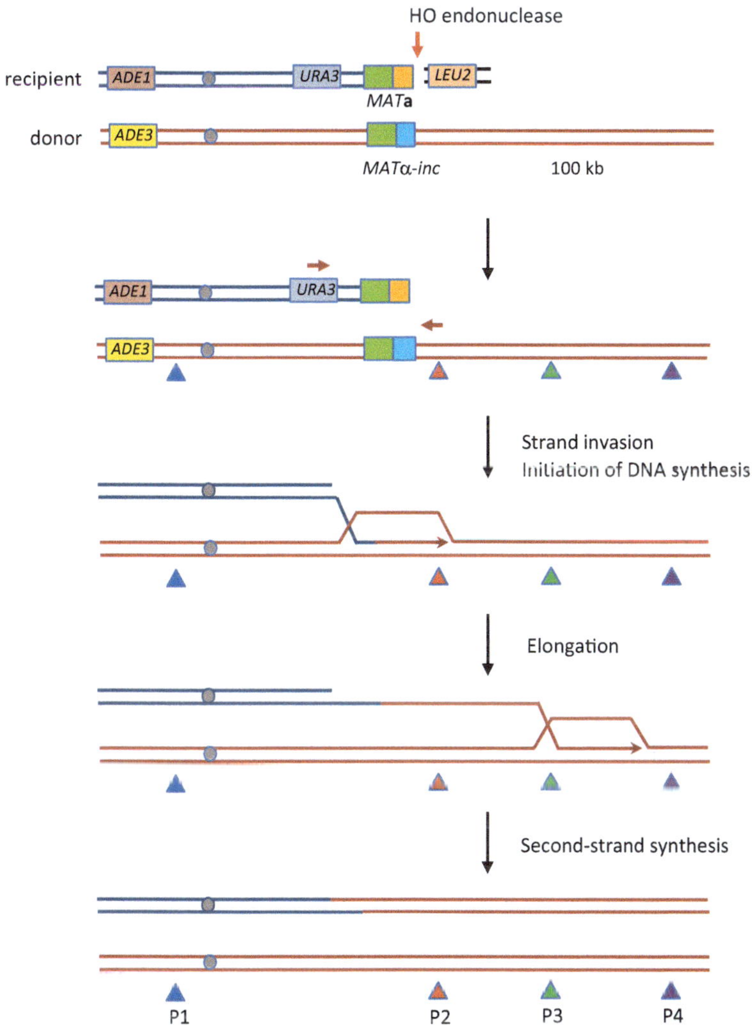

Fig. 1 BIR system and schematic of BIR synthesis detected by AMBER. Different steps of BIR including strand invasion and initiation, D-loop extension and finish are indicated in this schematic. BIR requires the resection of **a**-specific sequences (gold box) before strand invasion can be initiated between the homologous (green) *MAT* sequences. The triangles indicate the position of primer sets with probe used for ddPCR. Primer set P1, which is not involved in BIR, serves as a control. BIR is expected to complete ≥1860 bp of synthesis to copy through the position of P2 (initiation), ≥22 kb to copy through P3 (elongation) and close to 100 kb to copy through P4 (completion). The horizontal red arrows create a 5-kb PCR product that requires 1106 bp of BIR synthesis.

detection of Rad51 bound to the donor template (presumably during strand invasion and the creation of a D-loop) and the time at which a novel primer extension product could be detected; this delay appears to be under the control of a recombination execution checkpoint (Jain et al., 2009). Moreover, earlier studies had failed to see a further delay in achieving the completion of BIR, which involved the synthesis of as much as 100 kb of DNA, as visualized by Southern blot (Malkova, Naylor, Yamaguchi, Ira, & Haber, 2005). Recently we have developed a new technique to monitor the progression of BIR, which allows us to determine the timing of the initiation of new DNA synthesis, the rate of BIR, and the distance that BIR synthesis progresses in the absence of important BIR players (Liu et al., 2021). Here we summarize this method and compare it with previous methods.

2. Experimental system

A yeast disomic system (AM1153, a derivative of AM1003 (Deem et al., 2008)) is used in this experiment (Fig. 1). This system contains a truncated chromosome III, deleted for sequences beyond *MAT***a**, and a full-length chromosome III carrying *MATα-inc*, a mutation that prevents HO endonuclease cleavage. Homologous sequences at *HML* and *HMR* that could repair the DSB by ectopic gene conversion are deleted. The full-length chromosome serves as the template for BIR repair, copying 100 kb to the end of the chromosome. BIR is initiated by the galactose-inducible expression of HO (*GAL::HO*) and cleavage of *MAT***a**. Several primer sets specifically recognize the template sequences distal to the site of strand invasion are used to monitor BIR synthesis (Fig. 1, see legend for positions). The signal from those primer sets are normalized to the *ACT1* locus (on ChrVI) to determine relative copy number. Before BIR, the copy number of the P2, P3, and P4 sites is 1, relative to *ACT1*; after BIR, they increase, up to ~2×, depending on the efficiency of repair.

3. Collecting cells from the time course during BIR

3.1 Equipment

i. 30 °C incubator with agitation.
ii. Benchtop centrifuge with 50-mL conical tubes adaptors.
iii. 2-L flasks.
iv. 50-mL conical tubes.
v. Testing tubes.

3.2 Media for cell culture

i. YEPD: 1% yeast extract, 2% peptone, 2% dextrose, with 0.04 g adenine per liter media, adjusted to pH 5.5.
ii. YEP-raffinose: 1% yeast extract, 2% peptone, 3% raffinose, with 0.04 g adenine per liter media, pH 5.5.
iii. 20% Galactose, sterilized by filtration.

3.3 Procedure

1. Cells are first grown in 5 mL YEPD liquid media overnight to reach $\sim 2 \times 10^7$ cells per mL.
2. Add the cells from step 1 to ~300 mL YEP-Raffinose media by 1:100 dilution, to a starting culture at $\sim 2 \times 10^5$ cells per mL.
3. Incubate at 30 °C with agitation for about 14 h to reach a density of 5×10^6 cells per mL.
4. Collect 10–30 mL cells from the culture by centrifuging at 3000 rpm for 2 min and remove all liquid. This pellet represents time = 0. Store at −80 °C for later use.
5. Add galactose to the final concentration of 2% to the rest of the cultures to induce expression of HO endonuclease.
6. Collect ~30 mL samples from the culture every 1–2 h, and after centrifugation at 3000 rpm for 2 min, save the samples at −80 °C.

4. DNA preparation

4.1 Equipment

i. 37 and 60 °C water baths.
ii. Microcentrifuge.
iii. 1.5 mL Eppendorf tubes.
iv. Pipettes.
v. Rainin tips.
vi. Vortex.

4.2 Reagents

i. *Spheroplasting buffer:* 0.4 M sorbitol, 0.4 M KCl, 40 mM sodium phosphate buffer pH 7.2, 0.5 mM $MgCl_2$. Sterilize by filtration and store at 4 °C. Good for use within a month.
ii. *Zymolyase buffer:* 0.1 g/mL 20T zymolyase (MP Biomedicals, #08320921) dissolved in 2% glucose, 50 mM Tris–HCl, pH 7.5, sterilize by filtration and store at 4 °C. (Good for use within a month).

iii. *10× CutSmart buffer:* 0.5 M potassium acetate, 0.2 M Tris-acetate, 0.1 M magnesium acetate, 1 mg/mL bovine serum albumin (BSA) pH7.9 at 25 °C. Store at −20 °C.
iv. *10% Sodium dodecyl sulfate (SDS) buffer:* Keep at room temperature.
v. *10% Triton X-100 buffer:* Sterilize by filtration and keep at room temperature.
vi. *10 mg/mL RNase:* boil for 10 min and cool down on bench for 1 h. Store at −20 °C.
vii. *10 mg/mL proteinase K:* dissolve 10 mg proteinase K powder in 1 mL water (make fresh before use).
viii. 3 M sodium acetate buffer, pH 5.2.
ix. *Phenol:chloroform:isoamyl alcohol (25:24:1)* (Fisher, #15593-049).
x. *Isopropanol.*
xi. *80% Ethanol.*
xii. *RNase and DNase-free water.*

4.3 Procedure

1. Take out cells from −80 °C and rapidly thaw cells by incubating at 37 °C water bath for ~2 min.
2. During incubation, make digestion buffer using 1 mL spheroplasting buffer per sample and adding 5 μL zymolyase buffer.
3. Pellet cells by centrifugation at 3000 rpm for 1 min and remove all liquid.
4. Resuspend cells with 1 mL of digestion buffer per sample.
5. Digest cells by incubating at 37 °C for 10–15 min, with gentle mixing every 5 min.
6. After digestion, briefly shake by hand to suspend cells and then keep them on ice.
7. Count the cells by hemocytometer and take about 5×10^5 cells from each sample (~10–50 μL) and centrifuge at 3000 rpm for 2 min and remove liquid.
8. Resuspend cells in 360 μL 1 × CutSmart buffer.
9. Add 4 μL 10% SDS; mix by vortexing.
10. Incubate at 60 °C for 15 min. Vortex every 5 min.
11. Cool solution for ~5 min on ice. The solution is turbid.
12. Add 46 μL 10% Triton X-100 to neutralize SDS and mix by gentle vortexing. The solution should become clear.

13. Add 20 μL of 10 mg/mL RNAse, mix by vortexing, then incubate at 37 °C for 30 min.
14. Add 20 μL of 10 mg/mL Proteinase K. Mix by vortexing and incubate at 60 °C for 30 min.
15. Cool on ice for 5 min.
16. Add 500 μL phenol:chloroform:isoamyl alcohol (25:24:1) then vortex for 30 s.
17. Centrifuge at 13,000 rpm for 15 min.
18. Carefully transfer upper layer liquid to a new tube.
19. Add 500 μL phenol:chloroform:isoamyl alcohol (25:24:1) and mix by inverting then repeat steps 17 and 18.
20. Add 1/10 volume (~35 μL) of 3 M sodium acetate (pH 5.2) and 1 volume of isopropanol (~350 μL). Mix thoroughly by inverting. Incubate at room temperature for 30 min.
21. Precipitate DNA by centrifuging at 13,000 rpm for 20 min and drain the liquid as much as possible without touching the pellet.
22. Add 200 μL 0.2 M NaCl to dissolve the DNA pellet completely. Then add 800 μL of pre-cooled 100% EtOH and mix thoroughly by inverting. Incubate at −20 °C for 20 min and centrifuge at 13,000 rpm for 20 min.
23. Add 500 μL 80% EtOH and centrifuge for another 5 min at 13,000 rpm. Then remove all liquid.
24. Dry the pellet then add 100 μL water to dissolve DNA.
25. DNA is stored at −20 °C for later use.

5. Note

1. Buffers including Spheroplasting buffer, Zymolyase buffer, and 10% triton X-100 can become contaminated by bacteria, fungi or molds, and therefore need to be well stored and checked before use.
2. After digestion in step 5, check the cells under microscope to make sure that cells are efficiently digested. The digested cells usually look smaller than un-digested ones.
3. In step 7, you can make several aliquots and save them at -80 °C after quick centrifugation and removal of the liquid. Those samples after digestion can be used for other experiments without notable difference for months.

4. It is very important to make sure samples are mixed thoroughly in this experiment.
5. In step 13, to exclude potential DNase contamination (due to product quality), the RNase buffer should be boiled at 100 °C for 10 min, then cooled down on bench for 1 h before use. This doesn't impair RNase activity.
6. DNA pellets in step 21, 22, and 23, may be not visible by eye, therefore labeling the outer side of the Eppendorf tube before centrifuging will be useful.
7. Step 22 is optional.

6. BIR synthesis detected by ddPCR

6.1 Equipment and supplies

i. QX200 Droplet Digital PCR (ddPCR) System and related supplies including DG8 Cartridges (Biorad, #1864008), ddPCR 96 well PCR plates (Biorad, #12001925), PCR plate heat seal foil (Biorad, #1814040), ddPCR Reader oil (Biorad, #1863004), DG8 Gaskets (Biorad, #1863009).
ii. 50 and 200 μL multiple channel pipettes.
iii. Rainin micropipette tips (1000, 200, and 10 μL) (NOTE: using Rainin tips for ddPCR is important to get good quantity and quality of droplets).
iv. 8-strip PCR tubes.

6.2 Reagents

i. Droplet generation oil (Biorad, #1863005).
ii. 2 × Supermix for probe (Biorad, #186-3026).
iii. RNase and DNase-free water.
iv. Primer sets with probe (20 ×) ordered from IDT.

6.3 Procedures

1. DNA samples, isolated as described above, are quantified by Qubit or pico-green then diluted to 0.1 ng/μL.
2. PCR samples are mixed using the following recipe: 10 μL supermix for probe, 6 μL water, 1 μL primer set for target locus, 1 μL primer set for the *ACT1* locus, 2 μL diluted DNA sample. Mix thoroughly by gentle vortexing for 20 s, then briefly centrifuge to collect all liquid.

3. The PCR mix is then used to generate ddPCR droplets using the QX200 Droplet Digital PCR (ddPCR) System.
4. The droplets are then transferred to 96 well plates and sealed with foil.
5. PCR is then conducted using the following program: step 1: 95 °C for 10 min; step 2: 94 °C for 30 s (2 °C change per seconds), 60 °C for 1 min 20 s (2 °C change per second); step 3, repeat step 2 for 39 ×; step 4: 98 °C for 10 min; step 5: 12 °C for 30 min.
6. After carrying out PCR, the samples are analyzed using the QX200 Droplet Digital PCR (ddPCR) System. An example is shown in Fig. 2.

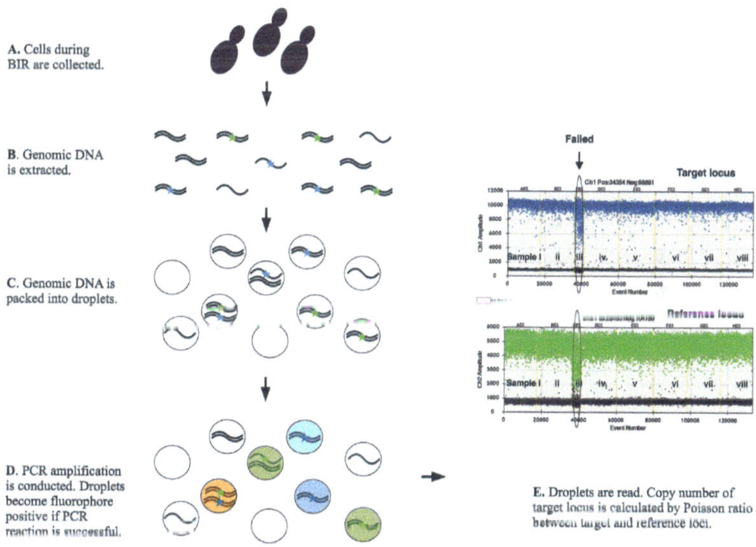

Fig. 2 Schematic of AMBER experiment. (A) Genomic DNA is extracted from cells undergoing BIR. (B) Stars represent target locus (blue) or reference locus (green). (C) Genomic DNA is diluted and mixed with dNTPs, primer sets, Taq enzyme and buffer, and then packed into droplets [~10,000–20,000 droplets per reaction, with each droplet containing 0 or 1 copy of template (target, reference, or both) DNA]. (D) After the PCR reaction, droplets that underwent successful PCR are colored (blue = target locus, green = reference locus, and yellow = both, clear = neither target locus nor reference locus). (E) All droplets are then analyzed by a QX200 droplet reader. Positive droplets from successful experiments are separated from negative droplets (Samples i, ii and iv-viii), while droplets from a failed experiment (sample iii) as circled in the figure are not clearly separated. Sample failures often result from dust in pipette tips or other problems in droplet formation and not from DNA preparation issues.

7. Note

1. In step 2, don't vortex harshly.
2. Extension time in Step 5 can be adjusted based on specific experiments.
3. After carrying out the PCR reaction, if not scanning right away, keep the plate in 4 °C fridge for 1–2 days but covered to avoid light.

8. Results

Using the AM1153 strain, a derivative of AM1003, which contains the *URA3* gene inserted 3 kb centromere proximal to *MAT***a**, we collected DNA samples by the procedure described above and analyzed the kinetics of BIR repair by the conventional PCR method, by densitometry of PCR products on agarose gels (Fig. 3A). This previously described method (Donnianni & Symington, 2013; Jain et al., 2009; Lydeard et al., 2007; Lydeard et al., 2010; Malkova et al., 2005), involves using one primer that lies in *URA3* on the donor chromosome and the second primer that lies distal to *MAT-Z* on the broken chromosome (see Table 1). Together they allow the formation of a 5-kb PCR product after BIR has synthesized 1106 bp of new DNA. Products were quantitated by densitometry on agarose gels relative to PCR products using dilution standards of DNA from an isolate of AM1153 that was repaired by BIR. A parallel assay was carried out using primers located at a reference locus (*ADE3*, 1778 bp product, primers listed in Table 1). The intensities of the BIR product were normalized to the intensities from the reference PCR products and plotted in Fig. 3A. A distinct increase in signal can be detected between 4 and 6 h, reaching a maximum value at 10 h. All PCRs were carried out using a processive, hot start Taq, KOD Xtreme polymerase (activation: 94 °C, 2.0 min, cycling: 98 °C for 10 s, 55 °C for 20 s, 68 °C for 2.0 or 5.0 min, final extension: 68 °C 6.0 min). For the BIR product PCR 5.0 min extensions for 29 cycles were used (for the reference, 2.0 min extensions for 24 cycles).

DNA synthesis during BIR by AMBER was determined by the copy number increase at each position detected by different primer sets. The primer set P1, which is located at the *EMC1* gene locus, blue triangle in Fig. 1, represents a quality control, which should be 2× relative to *ACT1* (which is on chromosome VI) before and after BIR. Primer sets (with fluorophore probes) P2, P3, and P4 (see legend of Figs. 1 and 3 for specific locations) are used to detect BIR synthesis at the beginning, middle, and

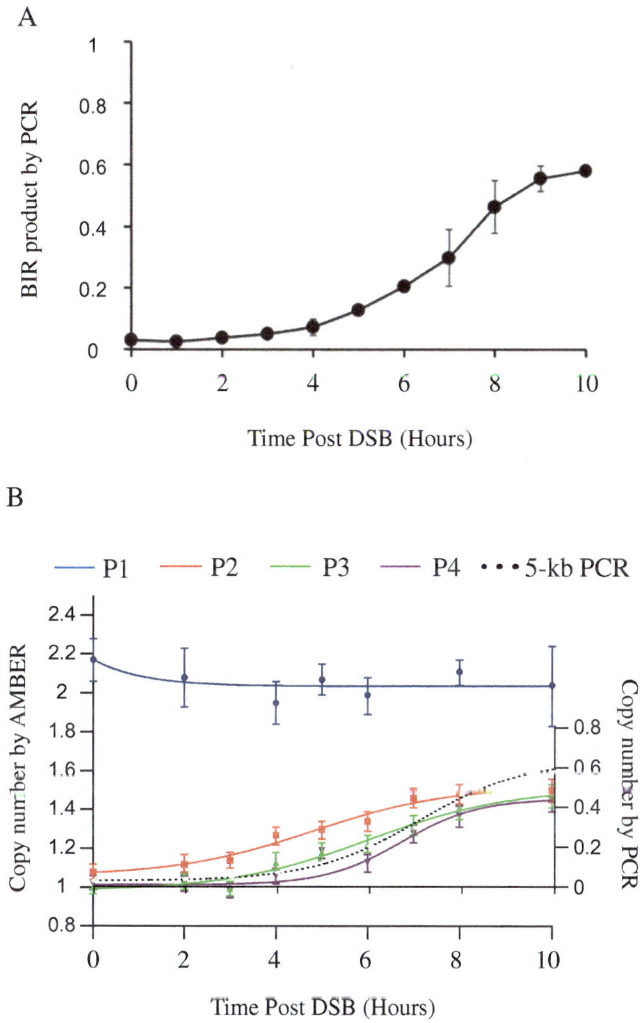

Fig. 3 DNA synthesis during BIR is detected by (A) PCR-based primer extension using primers indicated in Fig. 1, or (B) by AMBER using primer sets indicated in Fig. 1. (A) Conventional PCR assay of new BIR synthesis of at least 1106 bp. Vertical bars represent ranges of two experiments. (B) AMBER assay of new BIR synthesis. P1 is located at the *EMC1* gene locus, left arm of chromosome III. P2, P3, and P4 detect BIR synthesis of ≥1860 bp, ≥22 kb, and close to 100 kb respectively. Dotted line indicates the result of the PCR assay in (A) for comparison.

end, respectively. Copy number at these positions are normalized to 1 before BIR and increase to ≤2× following DSB and repair (Fig. 1). Evidence of DNA synthesis was defined as an increase in donor DNA copy number of at least 1.1× normalized to the value of the *ACT1* control locus. BIR synthesis

Table 1 Primers used in these experiments.

Primer set name	Dye	Type	Sequences
P-*ACT1*	HEX	Forward	GCCTTCTACGTTTCCATCCA
		Reverse	GGTAGAGAGAAACCAGCGTAAA
		Probe	TTCCGGTGATGGTGTTACTCACGT
P1	FAM	Forward	CCAGGACGTGTCTTCACTAAAC
		Reverse	CATATCAAGGCCGCCGATAAA
		Probe	ACCGTGGTATCCTTAGCGTCTAGAGC
P2	FAM	Forward	GTTGATGTGCTTTCTGAATGTGTA
		Reverse	CAGATTGGACGGAGGACTTAAT
		Probe	TGGATGGTAGTTTGGATACCATAAGTGACG
P3	FAM	Forward	GGCACAGAGACTACGTTATGG
		Reverse	CGGATGGCCTCTTGGTAAAT
		Probe	TCCCAGACAAAGACAGCACCGTC
P4	FAM	Forward	TGCGTCTGGAATGTTGCT
		Reverse	ACGAGTCAAGGCGGTAATTT
		Probe	TCAGGCGTGTGAGCATCTTCTTGT
Primers for PCR-based assay (BIR product)			GAAGAGGACTATTTGCAAAGGGAAG GTTAAGATAAGAACAAAGAAGGATGCT
Primers for PCR-based assay (*ADE3* reference)			GACATCTCCAGAGCACAACAG GCGGCAAGAGCGTATAAGTA

through P2 was detected by ddPCR around 3 h post-DSB induction (Fig. 3B). Synthesis through P3 and P4 was detected starting after 1–3 h later compared to P2 position, respectively, and increased by 10 h to about 1.5 × (Fig. 3B), which is lower than the final level of BIR efficiency (~80%) as detected by genetic methods, which indicates that some cells complete BIR later, due to some asynchrony of this process in the yeast cell population. A comparison of the PCR analysis to ddPCR (dotted line) shows that there is at least an hour's apparent delay in the formation of product as assessed by the 5-kb PCR product, comparing the time to detect 1106 bp of new DNA synthesis by PCR and 1860 bp for P2.

9. Discussion

We describe a new method that allows one to detect ongoing BIR synthesis, using ddPCR. We note that ssDNA intermediates and dsDNA products are not directly distinguished by AMBER since a droplet that contains either a single-or double-stranded sequence will be scored as a positive ddPCR signal (Fig. 2). In principle, the ssDNA vs dsDNA status can be distinguished by treating the donor DNA by S1-nuclease before ddPCR (Liu et al., 2021). Importantly, the usage of AMBER allowed us to detect the appearance of the initial BIR synthesis product at different times along the donor chromosome, which in turn allowed us to estimate the rate of BIR progression. Previous analysis of BIR by PCR for the initial primer extension and by Southern blot to see the final full-length product failed to see a difference in their timing, whereas with ddPCR we find a 3-h delay between the initiation and completion of the BIR-dependent synthesis of 100 kb (Liu et al., 2021).

We believe that there are several reasons that could contribute to the higher efficiency of AMBER as compared to the previous techniques in detecting the intermediates of BIR synthesis. First, we believe that AMBER protocol of genomic DNA isolation allows better preservation of BIR intermediates, especially of newly synthesized ssDNA. This DNA isolation protocol is based on the D-loop extension (DLE) assay developed by Heyer's lab (Piazza et al., 2019; Piazza, Koszul, & Heyer, 2018), which allowed the authors to capture the initial BIR synthesis in both wild type cells and in some BIR mutants (e.g., *pol32Δ*). In particular, we used two buffers described in the DLE protocol (lysis buffer and DNA dissolving buffer (Piazza et al., 2018)), which seemed to be amenable to isolating and preserving ssDNA, possibly because they were devoid of SDS.

Our protocol also includes procedures to remove detergents from the solutions to make the prepared DNA compatible with ddPCR. Previously, we compared DNA extraction from the same BIR time-course experiment by AMBER DNA purification method versus classic glass bead phenol extraction protocol (Liu et al., 2021). We observed that while ddPCR of the former samples allowed resolution of copy number increase for different primer positions along the donor chromosome, ddPCR of the latter demonstrated copy number increases for all primer positions simultaneously and only at later time points, suggesting that ssDNA intermediates might be lost and only double-stranded DNA outcomes were recovered.

However, there must be other differences in the state of DNA besides preserving single-stranded intermediates, because—using the AMBER method for DNA preparation—our PCR assay (Fig. 3) still failed to reveal early synthesis of the DNA product. It is possible that the long ssDNA intermediates described for BIR are present, but broken, in this preparation, allowing ddPCR amplification but still preventing recovery of a 5-kb PCR product until the DNA has become fully double-stranded. It is interesting that the long delay in detecting BIR by PCR methods was eliminated in strains lacking the Sgs1 and Mph1 helicases (homologs of BLM and FANCM helicases, respectively) (Jain, Sugawara, Mehta, Ryu, & Haber, 2016). Possibly the conversion of ssDNA intermediates to a double-stranded product is accelerated in this double mutant.

In practice, monitoring BIR by PCR or qPCR has relied on the creation of new, recombination-specific junctions. In the case of classic PCR detection of these junctions is enabled by the presence of unique sequences on recipient (*URA3* in the case of AM1153 (Fig. 1)) and donor chromosome. The downside of using such heterology is that PCR fragments are usually large (5 kb in AM1153), which makes them less amendable for quantification. In the DLE assay (Piazza et al., 2018, 2019), qPCR requires digestion of the DNA by restriction enzymes followed by proximal ligation. This creates new junctions that could only be formed after BIR synthesis and smaller sizes of PCR fragments, but in theory can lead to inter-sample variability. In theory qPCR should be able to detect quantitative changes in a BIR event that increases the copy number of a pre-existing sequence (indeed, we have used qPCR to monitor the loss by resection of sequences around a DSB (Eapen, Sugawara, Tsabar, Wu, & Haber, 2012); but in practice we have not found it reproducible for monitoring $<2\times$ increases associated with BIR.

The ability of ddPCR to detect newly synthesized DNA without the need to create a novel recombination product offers the following advantages. It allows detecting BIR synthesis by copy number changes at any position, whereas previously used methods could only detect synthesis at the very beginning of BIR, based on creation of a new recombination product. Also, use of fluorescent probes in addition to primers allows one to determine the 95% confidence interval (CI) for the ratio of the target to the reference in each individual experiment, thus eliminating the need for technical repeats. In addition, the primer sets for target and control loci are mixed for ddPCR in the same reaction, which eliminates variations that can be caused by pipetting. Finally, the ddPCR software can automatically calculate the target to the reference loci ratios, producing a Poisson ratio between the loci with 95% upper and lower Poisson CIs, which avoids unnecessary technical repeats.

Together, we believe that AMBER is an effective approach to detect ongoing BIR synthesis. It will be interesting to compare the initiation and progression of BIR synthesis in different experimental systems since they may vary. For example, based on our previous results (Liu et al., 2021) we expect that the initiation and progression of BIR will be affected in situations where strand invasion occurs in the vicinity of highly transcribed genes. In fact, this could at least in part add to the delay in BIR synthesis initiation reported previously (Jain et al., 2009). In addition, we envision that the AMBER approach can be used in the future to follow other steps of DNA repair, including DSB end resection, template switching or formation of chromosomal rearrangements.

Funding

Work in the Haber lab has been funded by NIH grant R35GM127029. Work in the Malkova lab has been funded by NIH grants R35GM127006 and R01CA232425.

References

Anand, R. P., Lovett, S. T., & Haber, J. E. (2013). Break-induced DNA replication. *Cold Spring Harbor Perspectives in Biology*, *5*(12), a010397. https://doi.org/10.1101/cshperspect.a010397.

Deem, A., Barker, K., Vanhulle, K., Downing, B., Vayl, A., & Malkova, A. (2008). Defective break-induced replication leads to half-crossovers in *Saccharomyces cerevisiae*. *Genetics*, *179*(4), 1845–1860. https://doi.org/10.1534/genetics.108.087940.

Donnianni, R. A., & Symington, L. S. (2013). Break-induced replication occurs by conservative DNA synthesis. *Proceedings of the National Academy of Sciences of the United States of America*, *110*(33), 13475–13480. https://doi.org/10.1073/pnas.1309800110.

Eapen, V. V., Sugawara, N., Tsabar, M., Wu, W. H., & Haber, J. E. (2012). The *Saccharomyces cerevisiae* chromatin remodeler Fun30 regulates DNA end resection and checkpoint deactivation. *Molecular and Cellular Biology, 32*(22), 4727–4740. https://doi.org/10.1128/MCB.00566-12.

Haber, J. E. (2016). A life investigating pathways that repair broken chromosomes. *Annual Review of Genetics, 50*, 1–28. https://doi.org/10.1146/annurev-genet-120215-035043.

Heyer, W. D. (2015). Regulation of recombination and genomic maintenance. *Cold Spring Harbor Perspectives in Biology*, 7(8), a016501. https://doi.org/10.1101/cshperspect.a016501.

Jain, S., Sugawara, N., Lydeard, J., Vaze, M., Tanguy Le Gac, N., & Haber, J. E. (2009). A recombination execution checkpoint regulates the choice of homologous recombination pathway during DNA double-strand break repair. *Genes & Development, 23*(3), 291–303. https://doi.org/10.1101/gad.1751209.

Jain, S., Sugawara, N., Mehta, A., Ryu, T., & Haber, J. E. (2016). Sgs1 and Mph1 helicases enforce the recombination execution checkpoint during DNA double-strand break repair in *Saccharomyces cerevisiae*. *Genetics, 203*(2), 667–675. https://doi.org/10.1534/genetics.115.184317.

Jasin, M., & Rothstein, R. (2013). Repair of strand breaks by homologous recombination. *Cold Spring Harbor Perspectives in Biology, 5*(11), a012740. https://doi.org/10.1101/cshperspect.a012740.

Kockler, Z. W., Osia, B., Lee, R., Musmaker, K., & Malkova, A. (2021). Repair of DNA breaks by break-induced replication. *Annual Review of Biochemistry, 90*, 165–191. https://doi.org/10.1146/annurev-biochem-081420-095551.

Liu, L., Yan, Z., Osia, B. A., Twarowski, J., Sun, L., Kramara, J., et al. (2021). Tracking break-induced replication shows that it stalls at roadblocks. *Nature, 590*(7847), 655–659. https://doi.org/10.1038/s41586-020-03172-w.

Llorente, B., Smith, C. E., & Symington, L. S. (2008). Break-induced replication: What is it and what is it for? *Cell Cycle*, 7(7), 859–864. https://doi.org/10.4161/cc.7.7.5613.

Lydeard, J. R., Jain, S., Yamaguchi, M., & Haber, J. E. (2007). Break-induced replication and telomerase-independent telomere maintenance require Pol32. *Nature, 448*(7155), 820–823. doi:nature06047 [pii] https://doi.org/10.1038/nature06047.

Lydeard, J. R., Lipkin-Moore, Z., Sheu, Y. J., Stillman, B., Burgers, P. M., & Haber, J. E. (2010). Break-induced replication requires all essential DNA replication factors except those specific for pre-RC assembly. *Genes & Development, 24*(11), 1133–1144. doi:24/11/1133 [pii] https://doi.org/10.1101/gad.1922610.

Malkova, A., Naylor, M. L., Yamaguchi, M., Ira, G., & Haber, J. E. (2005). RAD51-dependent break-induced replication differs in kinetics and checkpoint responses from RAD51-mediated gene conversion. *Molecular and Cellular Biology, 25*(3), 933–944. https://doi.org/10.1128/MCB.25.3.933-944.2005.

Piazza, A., Koszul, R., & Heyer, W. D. (2018). A proximity ligation-based method for quantitative measurement of D-loop extension in *S. cerevisiae*. *Methods in Enzymology, 601*, 27–44. https://doi.org/10.1016/bs.mie.2017.11.024.

Piazza, A., Shah, S. S., Wright, W. D., Gore, S. K., Koszul, R., & Heyer, W. D. (2019). Dynamic processing of displacement loops during recombinational DNA repair. *Molecular Cell, 73*(6), 1255–1266. e1254. https://doi.org/10.1016/j.molcel.2019.01.005.

Saini, N., Ramakrishnan, S., Elango, R., Ayyar, S., Zhang, Y., Deem, A., et al. (2013). Migrating bubble during break-induced replication drives conservative DNA synthesis. *Nature, 502*(7471), 389–392. https://doi.org/10.1038/nature12584.

Wilson, M. A., Kwon, Y., Xu, Y., Chung, W. H., Chi, P., Niu, H., et al. (2013). Pif1 helicase and Poldelta promote recombination-coupled DNA synthesis via bubble migration. *Nature*. https://doi.org/10.1038/nature12585.

CHAPTER EIGHT

Genomic mapping of DNA-repair reaction intermediates in living cells with engineered DNA structure-trap proteins

Jingjing Liu[a,†], Qian Mei[a,b,c,d,e,†], Sadeieh Nimer[a,b,c,d], Devon M. Fitzgerald[a,b,c,d], and Susan M. Rosenberg[a,b,c,d,e,*]

[a]Department of Molecular and Human Genetics, Baylor College of Medicine, Houston, TX, United States
[b]Department of Biochemistry and Molecular Biology, Baylor College of Medicine, Houston, TX, United States
[c]Department of Molecular Virology and Microbiology, Baylor College of Medicine, Houston, TX, United States
[d]Dan L Duncan Comprehensive Cancer Center, Baylor College of Medicine, Houston, TX, United States
[e]Systems, Synthetic, and Physical Biology Program, Rice University, Houston, TX, United States
*Corresponding author: e-mail address: smr@bcm.edu

Contents

† These authors contributed equally.

Methods in Enzymology, Volume 661
ISSN 0076-6879
https://doi.org/10.1016/bs.mie.2021.09.015

Abstract

Diverse DNA structures occur as reaction intermediates in various DNA-damage and -repair mechanisms, most of which results from replication stress. We harness the power of proteins evolutionarily optimized to bind and "trap" specific DNA reaction-intermediate structures, to quantify the structures, and discern the mechanisms of their occurrence in cells. The engineered proteins also allow genomic mapping of sites at which specific DNA structures occur preferentially, using a structure-trapping protein and ChIP-seq- or Cut-and-Tag-like methods. Genome-wide identification of sites with recurrent DNA-damage intermediates has illuminated mechanisms implicated in genome instability, replication stress, and chromosome fragility. Here, we describe X-seq, for identifying sites of recurrent four-way DNA junctions or Holliday-junctions (HJs). X-seq uses an engineered, catalysis-defective mutant of *Escherichia coli* RuvC HJ-specific endonuclease, RuvCDefGFP. X-seq signal indicates sites of recombinational DNA repair or replication-fork stalling and reversal. We also describe methods for genomic mapping of 3′-single-stranded DNA ends with SsEND-seq, in *E. coli*. Both methods allow genomic profiling of DNA-damage and -repair intermediates, which can precede genome instability, and are expected to have many additional applications including in other cells and organisms.

1. Introduction

Non-B-form DNA structures can arise as reaction intermediates in DNA replication mishaps, and DNA damage and its repair. The reaction intermediates are repaired or removed by DNA damage-processing proteins to promote faithful genetic transmission. Engineered proteins that bind specific DNA structures that are intermediates in DNA-damage and -repair reactions have been used to visualize, quantify and map the intermediates in spontaneous and induced DNA damage processing in living cells (Henrikus et al., 2019; Renzette et al., 2005; Rogakou, Boon, Redon, & Bonner, 1999; Scully et al., 1997). The advantage of protein-based detectors is high substrate specificity, a result of evolutionary optimization, which allows the discrimination of different DNA intermediates. Engineering of catalytically dead "trap"-protein detectors takes a further step by blocking processing once the DNA intermediate is bound (Chen et al., 2017; Kotlajich et al., 2018; Mei et al., 2021; Schmutz, Timashev, Xie, Patel, & de Lange, 2017; Shee et al., 2013; Xia et al., 2016). Transient DNA intermediates thus become persistent and so are more likely to be captured. Chromatin immunoprecipitation followed by sequencing (ChIP-seq) is a well-established method for detecting DNA-protein interactions (Park, 2009), as are Cut-and-Run (Skene & Henikoff, 2017) and Cut-and-Tag (Kaya-Okur et al., 2019) technologies. Here, we describe two modified ChIP-seq approaches for mapping DNA intermediate structures bound by engineered proteins.

1.1 Overview of the RuvCDefGFP Holliday-junction trap and X-seq

Holliday junctions (HJs) are four-way DNA junctions (Fig. 1A). HJs are intermediates in homologous recombination, which is required for repair of double-strand breaks (DSBs) (Fig. 1Bi), and single-strand gaps (ssDNA gaps) (Fig. 1Bii), and also occur during replication-fork stalling and reversal (Dunderdale et al., 1991; Seigneur et al., 1998) by the pairing of newly synthesized DNA strands with each other, which creates a HJ (Fig. 1Biii).

We developed a HJ-trap protein, RuvCDefGFP, based on a catalytically inactive mutant of *E. coli* RuvC HJ-specific endonuclease (Xia et al., 2016). RuvCDefGFP possesses mutations that encode D7N and E66D amino acid

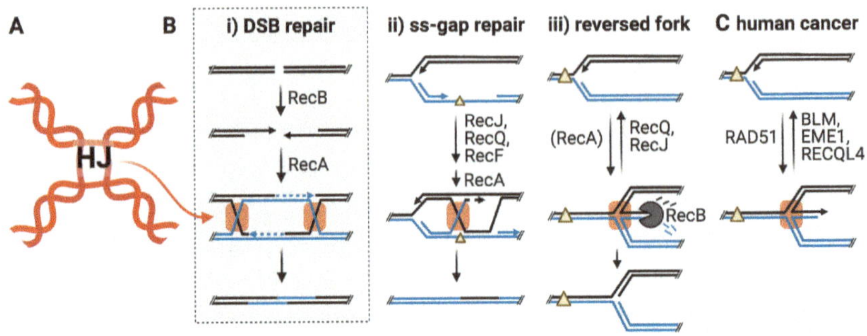

Fig. 1 Origins of recurrent HJs in *E. coli* can be distinguished by the proteins that promote or oppose their accumulation. The various origins of Holliday junctions (HJs) in *E. coli* cells can be distinguished genetically, by their requirements for or reduction by specific proteins shown. Orange boxes, HJs; lines, single DNA strands; dotted lines, newly synthesized DNA. (A) HJs can be formed by (Bi) homology-directed repair of DNA double-strand breaks (DSBs), which requires RecA recombinase, and has a unique requirement for RecBCD DNA exonuclease, which resects DSB ends and loads RecA only at DSB ends (Churchill, Anderson, & Kowalczykowski, 1999). (Bii) Formation of HJs during single-strand (ss)DNA gap repair requires RecA recombinase, RecQ helicase, RecJ 5′-ssDNA-dependent exonuclease, and RecF protein, which loads RecA at ssDNA gaps (Morimatsu & Kowalczykowski, 2003; Xia et al., 2016). (Biii) HJs that occur when stalled replication forks are reversed are prevented by RecQ and RecJ, and destroyed by RecBCD, and so are increased in mutants that lack any of these proteins (Xia et al., 2016). RecA is not required for fork reversal (Seigneur, Bidnenko, Ehrlich, & Michel, 1998), but its overproduction induces fork reversal (Xia et al., 2016). (C) Human counterparts of the reversed fork-promoting RAD51 (RecA ortholog), and reducing BLM, EME1 and RECQL4 proteins, inferred from cancer transcriptome data (Xia et al., 2016). *Adapted from Fitzgerald, D. M., & Rosenberg, S. M. (2021). Biology before the SOS response—DNA damage mechanisms at chromosome fragile sites.* Cells, 10.

substitutions, which inactivate the catalytic activity but not RuvC binding affinity for HJs. Based on a previous ChIP-seq method (Bonocora & Wade, 2015), we developed HJ mapping by X-seq (Mei et al., 2021; Xia et al., 2016), in which HJs are bound by RuvCDefGFP in cells that produce the RuvCDefGFP protein (Fig. 2A). The DNA is prepared, sheared by sonication, immunoprecipitated using anti-RuvC antibody, and then converted into sequencing libraries using ligation-based library preparation as shown in Fig. 2A. The DNA ends at HJs are blunted, A-tailed and eventually ligated to sequencing adapters while the HJ-RuvCDefGFP complex remains bound to complexes of antibody-IgG-Dynabeads with Protein A (Fig. 2A). This simplifies DNA purification between each step.

We have used X-seq to map spontaneous recurrent HJs in the *E. coli* genome and identified three major chromosome fragile sites (Mei et al., 2021).

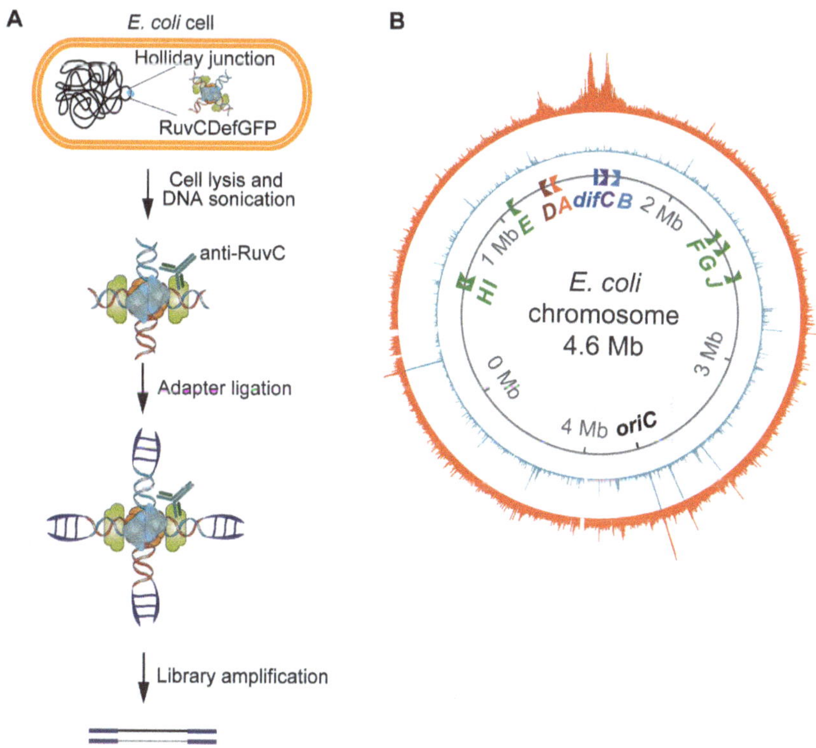

Fig. 2 HJ mapping by X-seq in *E. coli*. (A) Diagram of X-seq (Mei et al., 2021; Xia et al., 2016). RuvCDefGFP-bound HJs are immunoprecipitated with anti RuvC antibody. While RuvCDefGFP remains bound to immobilized antibody-protein A complex, DNA ends are blunted, A-tailed, and ligated with NEBNext adapters prior to elution. Adapter-ligated DNA fragments are converted into a sequencing library by PCR amplification, then sequenced. (B) X-seq detects three significant spontaneous HJ sites within the terminus region of the *E. coli* genome, one at the *TerA* unidirectional replication-barrier site, and two that flank the *dif* site of chromosome decatenation. All three form dependently on DSB repair-proteins RecA and RecBCD indicating their origins in DSB repair (see Fig. 1B). Replication terminus region at top, and origin, *oriC*, at the bottom. Orange, X-seq with RuvC antibody; sky blue, chromatin immunoprecipitation and sequencing (ChIP-seq) with nonspecific immunoglobulin G2a (IgG2a). Italics capital letters A–H, locations and orientations of *TerA-TerH* sites. *Panel (B): Data from Mei, Q., Fitzgerald, D. M., Liu, J., Xia, J., Pribis, J. P., Zhai, Y., et al. (2021). Two mechanisms of chromosome fragility at replication-termination sites in bacteria.* Science Advances, *7(25), eabe2846.*

These peaks (Fig. 2B), which represent recurrent HJs, occur primarily in the replication terminus region, and are formed by homology-directed repair of DNA DSBs, which, we found, occur via two separate mechanisms. One-ended DNA-break ends form near the *TerA* site, dependently on termination of runaway replication forks that cross the terminus region and would otherwise continue "backward" in the genome towards the origin (Fig. 2B). These DNA-break ends occur via replication-blocking Tus protein binding at the *TerA* unidirectional replication-terminus site, and so are Tus-dependent. Two other major HJ sites flank the site at which chromosomes are decatenated following replication, the *dif* site, apparently caused by DNA breakage and repair during chromosome decatenation (Fig. 2B). Both of these mechanisms may be useful models of human chromosome fragility (Fitzgerald & Rosenberg, 2021; Mei et al., 2021).

1.2 Overview of the ExID single-stranded DNA 3′-end trap and SsEND-seq

We created a trap protein for DNA with a 3′-single-stranded (ss)DNA end or overhang, ExID, for Exonuclease I Defective (Liu et al., n.d.). ExID is based on the protein encoded by the *sbcB15*- mutant form of *E. coli* Exonuclease I (Exo I), a 3′-specific ssDNA-dependent exonuclease that degrades ssDNA from a 3′ end (Lehman & Nussbaum, 1964). We have shown that ExID binds to recurrent induced and spontaneous 3′-ssDNA-ends, which, we found, are created in a replication-dependent, transcription-influenced manner (Liu et al., n.d.). The approach for genomic mapping of 3′-ssDNA ends with ExID is called Single-strand END-seq or SsEND-seq.

Fig. 3A and B shows schematic diagrams of SsEND-seq. In cells that carry a doxycycline-inducible gene encoding ExID-FLAG, 3′-ssDNA ends are bound by ExID-FLAG. The DNA is prepared, sheared by sonication, then immunoprecipitated using anti-FLAG antibody, and purified using the TaKaRa ChIP Elute kit, after which it is converted into a strand-specific sequencing library using the TaKaRa DNA SMART ChIP-Seq Kit. Sequencing then identifies sites of ExID enrichment: recurrent 3′-ssDNA ends (Fig. 3C) (Liu et al., n.d.).

We have used SsEND-seq to map 3′-ssDNA end sites that arise spontaneously in replicating *E. coli* (Liu et al., n.d.). Recurrent 3′-ssDNA end sites detected by SsEND-seq give rise to asymmetrical peaks whose steep sides correspond to 3′-ssDNA ends trapped by ExID, and the gentle slopes correspond to the random distribution of 5′ ends created by DNA sonication (Fig. 3B).

Detailed protocols for both X-seq and SsEND-seq in *E. coli* are presented below. In principle, these approaches can be applied to any organism or cell, as long as the trap protein can be produced in the organism and remain functional.

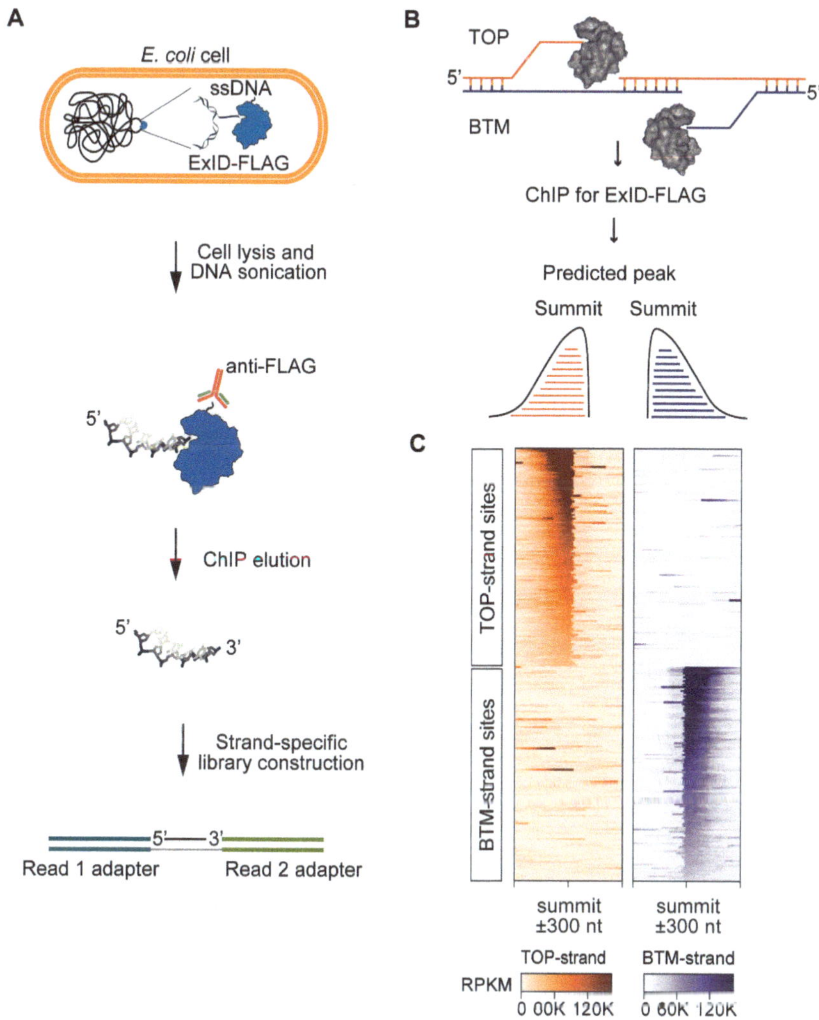

Fig. 3 SsEND-seq mapping of 3′-ssDNA-ends in *E. coli*. (A) Diagram of SsEND-seq. ExID-FLAG-bound 3′-ssDNA ends are immunoprecipitated with anti-FLAG antibody, followed by strand-specific library preparation using the Takara DNA SMART ChIP-Seq Kit, in which 5′ and 3′ ends of ssDNA fragments are tagged with known adapter sequences, then sequenced and a genomic map of recurrent 3′-ssDNA ends is made. (B) SsEND-seq detects asymmetrical peaks. SsEND-seq reads are mapped to the top (TOP, orange) or bottom (BTM, purple) strand, and form different asymmetrical peaks in which summits indicate recurrent DNA 3′ ssDNA termini that ExID traps. (C) SsEND-seq detects peaks of predicted asymmetry and strandedness (Liu et al. (n.d.)). Heatmaps of SsEND-seq signals around peak summits of ~220 recurrent 3′-ssDNA end sites. The signals were normalized by Reads Per Kilobase per Million mapped reads (RPKM), and sorted according to signal intensity. Signals are displayed along ±300-nt regions around and centered along the peak summits.

2. Considerations for experimental design with X-seq in *E. coli*

2.1 Interpretation of X-seq data

Before the experimental set-up, it is useful to consider in which cells to perform X-seq. Mapping HJs in genomes with X-seq indicates the presence and positions of four-way-junction structures but does not reveal how recurrent HJs formed, unless genetic tests are carried out (Fitzgerald & Rosenberg, 2021; Mei et al., 2021; Xia et al., 2016). In any organism or cell, HJs can be formed during homology-directed repair of DNA DSBs (Fig. 1Bi) and ssDNA gaps (Fig. 1Bii), or by replication-fork stalling and reversal (Fig. 1Biii and C). These sources can be distinguished definitively in *E. coli* by use of cells with null mutations in genes that encode the proteins that promote or reduce HJ formation via any of these mechanisms (Fitzgerald & Rosenberg, 2021; Mei et al., 2021; Xia et al., 2016) (Fig. 1B). Particularly, cells that lack the RecBCD double-strand end-dependent DNA exonuclease activity are defective for DSB repair and cannot form the DSB-repair HJs (Fig. 1Bi); so, *recB* or *recC* null mutants lack HJs (and X-seq signal) caused by DSB repair (Mei et al., 2021; Xia et al., 2016). By contrast, because RecBCD nuclease destroys reversed-fork HJs (Fig. 1Biii) (Michel, Ehrlich, & Uzest, 1997; Seigneur et al., 1998), fork-reversal HJs are more abundant in *recB* or *recC* deficient mutant cells and, with previous assays, were undetectable in wild-type cells, visible only in *recB*-null mutants (Michel et al., 1997; Seigneur et al., 1998). These opposite roles of RecBCD and these mutants can be used to distinguish HJs formed in DSB repair from those formed by fork reversal (Fig. 1Bi and iii) (Fitzgerald & Rosenberg, 2021; Mei et al., 2021; Xia et al., 2016). In addition, loss of RecQ or RecJ increased HJs, detected as RuvCDefGFP foci, formed by fork reversal (Xia et al., 2016). Similarly, HJ signal that is unaffected by RecBCD presence or absence, and instead requires RecF, which loads RecA recombinase onto ssDNA at ssDNA gaps, is likely to represent HJs formed during ssDNA gap repair (Fitzgerald & Rosenberg, 2021; Magner et al., 2007; Mei et al., 2021; Xia et al., 2016) . Thus, analysis of the source of recurrent HJs detected by X-seq can be accomplished with an isogenic set of RuvCDefGFP-expressing cells that carry no *rec* mutations ("wild-type") and each single null mutation of *recA*, *recB*, and *recF*, with *recQ* and *recJ* null mutant cells to shore up the conclusions. One such set

Table 1 Phenotypes of RuvCDefGFP-trapped HJ signals formed by various mechanisms in *E. coli*.

Processes that form HJs	*E. coli* strain genotype						
	WT	Δ*recA*	Δ*recB*	Δ*recF*	Δ*recJ*	Δ*recQ*	References
DSB repair	+	−	−	+	+	+	Mei et al. (2021) and Xia et al. (2016)
ssDNA gap repair	+	−	+	−	−	−	Xia et al. (2016)
Reversed forks (RecA overproduced)[a]	+	not tested	+	+	+++	+++	Xia et al. (2016)

[a]Observed in cells with overproduced RecA (Xia et al., 2016), except the Δ*recA* data for reversed forks, which are from pulsed-field gel electrophoresis, and were observed only in *recB* null-mutant cells without use of a HJ-trap protein (Michel et al., 1997; Seigneur et al., 1998).

is given in Section 4.1. A summary of these relationships between cell genotype and the appearance of detectable HJ structures that occur during repair of DSBs or ssDNA gaps, or those indicating reversed replication forks is given in Table 1.

2.2 Bioinformatic analysis of X-seq data

Illumina paired-end sequencing reads are aligned by BWA-MEM to the *E. coli* genome sequence of strain W3110 [National Center for Biotechnology Information Reference Sequence database accession: NC_007779.1] for strains given in Section 4.1, or to other reference sequences for other strain backgrounds used. Reads that have multiple primary hits or low-mapping quality are discarded using samtools. PCR duplicates are removed by Picard Tools MarkDuplicates. X-seq peaks are broad, and are difficult to detect with conventional peak-calling algorithms that are optimized for narrow peaks. To estimate the boundaries of X-seq peaks, we first divide the genome into 1-kb bins and count the reads in each bin. Next, numbers of reads are normalized to the median number of total reads. We then use change-point analyses to detect peak boundaries by detecting the change point in the mean for a minimum of two independent experiments and datasets. We use the cpt.mean function in the "changepoint" R package (Mei et al., 2021), with the PELT (Pruned Exact Linear Time) algorithm and Akaike's information criterion as penalty. The input is normalized reads in 1-kb bins.

3. Considerations for SsEND-seq in *E. coli*

3.1 Bioinformatic analysis of SsEND-seq data

We process SsEND-seq data as follows (Liu et al., n.d.). Paired-end sequencing reads are mapped to the *E. coli* K12 MG1655 genome (NCBI: NC_000913.3) using BWA-MEM (v0.7.12). PCR duplicates are removed by Picard Tool MarkDuplicates. Other reference sequences would be used for other strain backgrounds. Mapped reads are then divided into two classes, top-strand (reference sequence in the 5′-3′ direction) and bottom-strand (reference sequence in the 3′-5′ direction) reads. The TaKaRa DNA SMART ChIP-Seq Kit ligates the read-2 adapter (Fig. 3A) to 3′ ends of the original ssDNA fragments and ligates the read-1 adapter to the 5′ ends (Fig. 3A). It then converts ssDNA to duplex DNA. Therefore, the reads 2, aligned to the top strand, and their readmate, reads 1, aligned to the bottom strand, are classified as reads mapped to top strand, shown as orange signal in Fig. 3B and C. Similarly, the reads 2 aligned to the bottom strand, and reads 1 aligned to the top strand, are reads mapped to the bottom strand, shown as purple signal in Fig. 3B and C. Significant peaks (adjusted P-value $\leq$0.00001) are called by MACS2 (v1.4.2) on the top and bottom strands individually.

4. Materials for X-seq in *E. coli*

4.1 Bacterial growth, induction of RuvCDefGFP (RDG) and HJ-RDG crosslinking

- Orbital shaker
- LB medium: 0.5% NaCl, 0.5% Bacto-Yeast extract (Gibco), 1% Bacto-Tryptone (Gibco)
- Doxycycline
- Formaldehyde Solution, 37%
- Phosphate-buffered saline (PBS), pH 7.0–7.2
- 2.5 M glycine
- Bacterial strains (Table 2): Note that the 594 strain background is derived from the W3110 "wild-type" *E. coli* K12 strain and carries the following mutations: *galK2(Oc) galT22 rpsL179* IN*(rrnD-rrnE)1 lac3350* (Bachmann, 1996). It lacks the F conjugative plasmid and contains no lambda prophage.

Table 2 *E. coli* strains that can diagnose origins of HJ signal observed with X-seq.

Name	Relevant genotype	References
SMR19425 ("WT")	594 *hsdrk*$^-$ *mK*$^+$ Δ*att*λ::P_{N25}*tetR* FRT *zfe2512.7*:: $P_{N25tetO}$*ruv*CDef*gfp* FRT	Xia et al. (2016)
SMR19406	594 *hsdrk*$^-$ *mK*$^+$ Δ*att*λ::P_{N25}*tetR* FRT *zfe2512.7*:: $P_{N25tetO}$*ruv*CDef*gfp* FRT*cat*FRT Δ*recA*::FRTKanFRT	Xia et al. (2016)
SMR19407	594 *hsdrk*$^-$ *mK*$^+$ Δ*att*λ::P_{N25}*tetR* FRT *zfe2512.7*:: $P_{N25tetO}$*ruv*CDef*gfp* FRT*cat*FRT Δ*recB*::FRTKanFRT	Xia et al. (2016)
SMR19427	594 *hsdrk*$^-$ *mK*$^+$ Δ*att*λ::P_{N25}*tetR* FRT *zfe2512.7*:: $P_{N25tetO}$*ruv*CDef*gfp* FRT Δ*recF*::FRT	Xia et al. (2016)
SMR19392	594 *hsdrk*$^-$ *mK*$^+$ Δ*att*λ::P_{N25}*tetR* FRT *zfe2512.7*:: $P_{N25tetO}$*ruv*CDef*gfp* FRT*cat*FRT Δ*recQ1906*::FRT	Xia et al. (2016)
SMR19398	594 *hsdrk*$^-$ *mK*$^+$ Δ*att*λ::P_{N25}*tetR* FRT *zfe2512.7*:: $P_{N25tetO}$*ruv*CDef*gfp* FRT*cat*FRT Δ*recJ*::FRT	Xia et al. (2016)

4.2 DNA sonication

- Labquake rotator (Fisher Scientific)
- Bioruptor Pico Sonicator (Diagenode)
- 15 mL Bioruptor Pico Tube (Diagenode)
- Nanodrop (Thermo Scientific)
- FA lysis buffer:
  0.05 M HEPES
  150 mM NaCl
  1% Triton
  0.1% sodium deoxycholate
  0.1% SDS
  1 mM ethylenediaminetetraacetic acid (EDTA)
- Lysozyme (Fisher Scientific)

4.3 Immunoprecipitation of RDG

- Labquake rotator (Fisher Scientific)
- Magnetic rack for 1.5 mL microtube
- Anti-RuvC antibody 200 μg/mL (Santa Cruz sc-53,437)
- Dynabeads Protein A magnetic beads (Invitrogen 10002D)
- FA lysis buffer

4.4 Library construction

- Magnetic rack for 0.2 mL PCR strip
- Quick Blunting and Quick Ligation kits (NEB E0542L)
- Klenow Fragment (3′ → 5′ exo-) (NEB M0212L)
- NEBNext Multiplex Oligos for Illumina (NEB E7335L, E7500L, E7700L, E7730L)
- FA lysis buffer
- FA lysis buffer 500:
  0.05 M HEPES
  500 mM NaCl
  1% Triton
  0.1% sodium deoxycholate
  0.1% SDS
  1 mM EDTA
- Wash buffer:
  50 mM HEPES
  0.5 M NaCl
  1% Triton
  0.1% sodium deoxycholate
  0.1% SDS
  1 mM EDTA
- Deoxycholate buffer:
  10 mM TE, pH 8.0
  250 mM LiCl
  0.5% sodium deoxycholate
  0.5% Nonidet-P40
- ChIP elution buffer:
  50 mM Tris–HCl, pH 7.5
  1% SDS
  10 mM EDTA
- 10 mM Tris–HCl, pH 7.5
- 10 mM Tris–HCl, pH 8.0
- TE buffer, pH 7.5
- 10 μM ChIP_p1 Primer: 5′-AATGATACGGCGACCACCGAGATC TACAC
- 10 μM ChIP_p2 Primer: 5′-CAAGCAGAAGACGGCATACGAGAT

4.5 Sequencing-library amplification and bead-based size selection

- Thermocycler
- Agilent TapeStation (or equivalent)
- Nanodrop (Thermo Scientific)
- NEBNext Multiplex Oligos for Illumina (NEB E7335L, E7500L, E7700L, E7730L)

5. Methods for X-seq in *E. coli*

5.1 Overview

The protocol can be completed in 5 days as follows:

Day 0 and 1: Bacterial growth, induction of RDG and HJ-RDG crosslinking.

Day 2: DNA sonication and immunoprecipitation of RDG.

Day 3: Library construction.

Day 4: Sequencing-library amplification and bead-based size selection.

5.2 Bacterial growth, induction of RDG and HJ-RDG crosslinking

1. Start a 3 mL LB shaking overnight culture for cells that carry the doxycycline-inducible *ruv*CDef*gfp* expression cassette.
2. Dilute saturated overnight cultures 1:500 (400 μL) into 20 mL LB medium with 100 ng/mL doxycycline to induce the production of RDG, then grow for 3 h at 37 °C shaking until the OD_{600} is 0.4~0.8
3. Add 550 μL 37% formaldehyde (~1% final concentration) to cross-link HJs with RDG. Incubate at room temperature for 20 min.
4. Add 5 mL 2.5 M glycine to quench the formaldehyde.
5. Transfer cells to centrifuge bottles and centrifuge at 7000 rpm for 7 min at 4 °C using a refrigerated centrifuge.
6. Resuspend the cell pellet in 1 mL PBS and transfer cells to a 1.5 mL microtube.
7. Centrifuge at 7000 rpm for 2 min at 4 °C.
   Discard the supernatant.
   Store cell pellets overnight at −20 °C.

5.3 DNA sonication

1. Resuspend cell pellet in 0.5 mL FA lysis buffer containing 4 mg/mL lysozyme.
2. Incubate for 30 min at 37 °C with rotation on a Labquake rotator.
3. Chill cell lysate on ice for 5 min.
4. Add ~750 g sonication beads to a 15 mL Bioruptor Pico tube.
   Wash beads once with 0.9 mL PBS.
5. Transfer lysate to the prepared Bioruptor Pico Tube.
   Sonicate in a 30 s on/off continuous pulse for 30 min.
6. Transfer sonicated lysate to a 15 mL tube.
   Centrifuge at 11,000 rpm for 20 min at 4 °C.
   Transfer the supernatant to a new 15 mL tube and discard cell debris (pellet).
7. Measure the DNA concentration of the lysate with these steps:
   - take a 15 μL sample of lysate and boil at >95 °C for 10 min to decrosslink
   - spin and take the supernatant
   - purify the supernatant with a PCR purification kit
   - use Nanodrop to measure the concentration of purified DNA
8. Adjust the lysate to ~150 ng/μL.
9. Split diluted lysate into two 1.5 mL microtubes with 1 mL in each tube.
   Store lysates at −80 °C or proceed to the next step.

5.4 Immunoprecipitation of RDG

1. Transfer 50 μL Dynabeads to a 1.5 mL microtube.
   Place tubes on magnetic rack for 1 min and discard supernatant.
2. Add 1 mL FA lysis buffer to wash Dynabeads.
   Pipette up and down 10 times.
   Place tubes on magnetic rack for 1 min and discard supernatant.
3. Add 50 μL anti-RuvC antibody to Dynabeads.
   Incubate at room temperature with rotation on a Labquake rotator for 30 min.
4. Centrifuge briefly to bring all the liquid to the bottom of the tube.
   Place tubes on magnetic rack for 1 min and discard supernatant.
5. Add 1 mL FA lysis buffer to wash antibody-coated Dynabeads.
   Pipette up and down 10 times.
   Place tubes on magnetic rack for 1 min and discard supernatant.
6. Add 1 mL normalized cell lysates.
   Incubate overnight at 4 °C with rotation.

5.5 Library construction

1. Centrifuge briefly to bring all the liquid to the bottom of the tube.
   Place tubes on magnetic rack for 1 min and discard supernatant.
2. Resuspend Dynabeads in 200 µL FA lysis buffer and transfer to a 0.2 mL PCR strip.
   Place PCR strip on magnetic rack for 1 min and discard supernatant.
3. Add 200 µL FA lysis buffer.
   Pipette up and down 10 times.
   Place PCR strip on magnetic rack for 1 min and discard supernatant.
4. Add 200 µL 10 mM Tris–HCl, pH 7.5.
   Pipette up and down 10 times.
   Place PCR strip on magnetic rack for 1 min and discard supernatant.
5. Repeat step 4 once.
6. Prepare blunting solution as follows (per sample):
   10 µL 10 × quick blunting buffer
   10 µL 1 mM dNTP Mix
   80 µL dH2O
   2 µL blunt enzyme mix
   Make a master mix for the number of samples planned.
7. Add ~100 µL blunting solution to each tube and incubate at room temperature for 30 min. Resuspend with pipette every 10 min.
8. Place PCR strip on magnetic rack for 1 min and discard supernatant.
9. Add 200 µL FA lysis buffer.
   Pipette up and down 10 times.
   Place PCR strip on magnetic rack for 1 min and discard supernatant.
   Repeat this step once.
10. Add 200 µL 10 mM Tris–HCl, pH 8.0.
    Pipette up and down 10 times.
    Place PCR strip on magnetic rack for 1 min and discard supernatant.
    Repeat this step once.
11. Prepare A-tailing solution master mix as follows (per sample):
    10 µL 10 × NEB buffer 2
    2 µL 100 mM dATP
    88 µL dH2O
    2 µL Klenow fragment (3′ → 5′exo–)
    Make a master mix for the number of samples planned.
12. Add ~100 µL A-tailing solution and incubate at 37 °C for 30 min.
    Resuspend with pipette every 10 min.

13. Place tubes on magnetic rack for 1 min and discard supernatant.
14. Add 200 μL FA lysis buffer.
    Pipette up and down 10 times.
    Place PCR strip on magnetic rack for 1 min and discard supernatant.
    Repeat this step once.
15. Add 200 μL 10 mM Tris–HCl, pH 7.5.
    Pipette up and down 10 times.
    Place PCR strip on magnetic rack for 1 min and discard supernatant.
    Repeat this step once.
16. Dilute the NEBNext Adapter for Illumina (15 μM) 10-fold in 10 mM Tris–HCl, pH 7.5 to a final concentration of 1.5 μM.
17. Prepare adapter ligation solution master mix as follows (per sample):
    50 μL 10 × ligase buffer
    50 μL H2O
    2 μL quick ligase
    1 μL diluted NEBNext Adapter
    Make a master mix for the number of samples planned.
18. Add ~100 μL adapter ligation solution and incubate at room temperature for 15 min.
    Resuspend with pipette every 10 min.
19. Add 2 μL USER enzyme and incubate at 37 °C for 15 min.
    Resuspend with pipette every 10 min.
20. Place PCR strip on magnetic rack for 1 min and discard supernatant.
21. Add 200 μL FA lysis buffer.
    Pipette up and down 10 times.
    Place PCR strip on magnetic rack for 1 min and discard supernatant.
    Repeat this step once.
22. Add 200 μL FA lysis buffer 500.
    Pipette up and down 10 times.
    Place PCR strip on magnetic rack for 1 min and discard supernatant.
23. Add 200 μL wash buffer.
    Pipette up and down 10 times.
    Place PCR strip on magnetic rack for 1 min and discard supernatant.
24. Add 200 μL deoxycholate buffer.
    Pipette up and down 10 times.
    Place PCR strip on magnetic rack for 1 min and discard supernatant.
25. Add 200 μL TE buffer.
    Pipette up and down 10 times.
    Place PCR strip on magnetic rack for 1 min and discard supernatant.

26. Add 50 μL ChIP elution buffer and incubate at 70 °C for 15 min.
Place PCR strip on magnetic rack for 1 min.
27. Transfer the supernatant to a 1.5 mL microtube and boil at >95 °C for 10 min.
28. Use 1.1 × Ampure XP beads to clean the DNA according to the manufacturer's protocol.
29. Elute DNA with 24 μL water.
Store at 4 °C or proceed to the next step.

5.6 Sequencing-library amplification and bead-based size selection

1. Prepare 1st round of library amplification solution as follows (per sample):
20 μL adapter-ligated DNA
25 μL NEBNext Q5 Hot Start HiFi PCR Master Mix
2.5 μL 10 μM Universal PCR Primers/i5 Primer
2.5 μL 10 μM Index Primers/i7 Primer
2. Run the following PCR protocol:

98 °C	30 s	
98 °C	10 s	8 cycles
65 °C	75 s	
65 °C	5 min	
4 °C	∞	

3. Add 25 μL (0.5 ×) resuspended AMPure XP beads.
Mix well by pipetting up and down 10 times.
Incubate at room temperature for 5 min.
4. Place PCR strip on the magnetic rack until the supernatant is clear.
Carefully transfer the supernatant to a new well.
Discard beads that contain the large fragments.
5. Add 20 μL (0.4 ×) AMPure XP beads to the supernatant.
Mix well by pipetting up and down 10 times.
Incubate at room temperature for 5 min.
6. Put PCR strip on the magnetic rack until the supernatant is clear.
Carefully remove and discard the supernatant.

7. Add 200 μL of freshly prepared 80% ethanol while the plate is still on the magnetic rack.
   Incubate at room temperature for 30 s and carefully remove and discard the supernatant.
8. Repeat step 7 once.
9. Air dry beads for 5 min while the plate is on the magnetic rack.
10. Remove PCR strip from the magnetic rack.
    Add 23 μL water.
    Mix well by pipetting up and down.
11. Incubate at room temperature for 2 min.
    Put PCR strip on the magnetic rack until the solution is clear.
    Transfer 20 μL of the supernatant to a new PCR tube.
12. Prepare 2nd round of library amplification solution as follows (per sample):
    20 μL size-selected DNA
    25 μL NEBNext Q5 Hot Start HiFi PCR Master Mix
    2.5 μL 10 μM ChIP_p1 Primer
    2.5 μL 10 μM ChIP_p2 Primer
13. Run the following PCR protocol:

98 °C	30 s	
98 °C	10 s	15 cycles
65 °C	75 s	
65 °C	5 min	
4 °C	∞	

14. Clean the DNA with 45 μL (0.9 ×) Ampure XP beads according to the manufacturer's protocol.
15. Quantify the library with Nanodrop.
16. Proceed to next-generation sequencing.

6. Materials for SsEND-Seq in *E. coli*

6.1 Bacterial growth, ExlD-FLAG induction and crosslinking to 3′-single-stranded DNA ends

- Orbital shaker
- LB medium: 0.5% NaCl, 0.5% Bacto-Yeast extract (Gibco), 1% Bacto-Tryptone (Gibco)

Table 3 *E coli* strain that produces ExlD-3xFLAG protein.

Name	Relevant genotype	Reference
SMR24776	MG1655 Δ*araBAD567* Δ*att*λ::P_{BAD} Δ*ypeC*::P_{N25}*tetR* FRT (*mntH-nupC*)::$P_{N25tetO}$ExlD-3xFLAG FRT	Liu et al. (n.d.)

- Doxycycline
- Formaldehyde Solution, 37%
- Phosphate-buffered saline (PBS), pH 7.0–7.2
- 4.5 M Tris, pH 8.0
- Bacterial strain (Table 3): Note that the MG1655 strain background is the first sequenced *E. coli* K12 wild-type strain (Blattner et al., 1997). It lacks the F conjugative plasmid and contains no lambda prophage.

6.2 DNA sonication

- Labquake rotator (Fisher Scientific)
- Bioruptor Pico Sonicator (Diagenode)
- 1.5 mL Bioruptor Pico microtube (Diagenode)
- FA lysis buffer:
  0.05 M HEPES
  150 mM NaCl
  1% Triton
  0.1% sodium deoxycholate
  0.1% SDS
  1 mM EDTA
- Lysozyme (Fisher Scientific)

6.3 Immunoprecipitation of ExlD-FLAG

- 1.5 mL snaplock microfuge tube (Axygen MCT-150-C-S)
- Labquake rotator (Fisher Scientific)
- Magnetic rack for 1.5 mL microtube
- ANTI-FLAG M2 antibody (Sigma-Aldrich F1804)
- Dynabeads Protein G magnetic beads (Invitrogen 10003D)
- Wash buffer:
  50 mM HEPES
  0.5 M NaCl
  1% Triton
  0.1% sodium deoxycholate
  0.1% SDS
  1 mM EDTA

- Deoxycholate buffer:
  10 mM TE, pH 8.0
  250 mM LiCl
  0.5% sodium deoxycholate
  0.5% Nonidet-P40
- PBS with 0.02% Tween 20
- TaKaRa ChIP Elute Kit
- QuantStudio™ 3 Real-Time PCR (or equivalent)
- DEPC (Diethyl pyrocarbonate)-treated water (Fisher Bioreagents)
- BIO-RAD iTaq Universal SYBR Green Supermix
- 100 μM qPCR Primer 1: 5′-ACGGCATGATTGAAATATCGA GTTC
- 100 μM qPCR Primer 2: 5′-GGTACTGGATTTTCGTGAGACAG

6.4 Strand-specific library construction

- Vacufuge (Eppendorf)
- TaKaRa DNA SMART ChIP-Seq Kit
- Thermocycler
- Agilent TapeStation (or equivalent)

6.5 DNA polyacrylamide gel size selection and DNA extraction

- Vertical electrophoresis cells
- 30% Acrylamide/Bis Solution, 29:1 (Bio-Rad 161,015)
- 10 × TBE:
  1 M Tris-base, pH 7.5
  1 M boric acid
  20 mM EDTA
- 10% Ammonium persulfate (APS, store aliquotes at −20 °C)
- *N,N,N′,N′*-Tetramethylethylenediamine (TEMED)
- DNA extraction buffer:
  10 mM Tris–HCl, pH 8.0
  0.3 M NaCl
  1 mM EDTA
- Ethidium bromide
- 3 M sodium acetate
- Glycogen (20 mg/mL)
- Isopropyl alcohol
- KAPA Library Quantification kit (Roche)

7. Methods for SsEND-seq in *E. coli*

7.1 Overview

The protocol can be completed in 6 days as follows:

Day 0 and 1: Bacterial growth, ExID-FLAG induction and crosslinking to 3′-single-stranded DNA ends.

Day 2: DNA sonication, and immunoprecipitation of ExID-FLAG steps 1 through 8.

Day 3: The rest of the steps of immunoprecipitation of ExID-FLAG.

Day 4: Strand-specific library construction.

Day 5: DNA polyacrylamide gel size selection and DNA extraction.

7.2 Bacterial growth, ExID-FLAG induction and crosslinking to 3′-single-stranded DNA ends

1. Start 3 mL LB overnight cultures for cells that carry the doxycycline-inducible ExID-FLAG expression cassette.
2. Dilute saturated overnight cultures 1:100 putting 400 μL in 40 mL LB medium with 1 ng/mL doxycycline to induce the production of ExID-FLAG, then grow for 2 h shaking at 37 °C to OD600 0.5 ~ 1.0.
3. Take 1 mL of the sample into cuvette to measure the OD600.
4. Add 1.1 mL 37% formaldehyde (~1% final concentration) to cross-link chromatin and proteins.
   Incubate at room temperature for 20 min.
5. Take 12.5 mL of cells at OD600 = 1, or normalize from lower density from 12.5 mL to as much as 25 mL cells.
6. Add 2.5 mL ~ 5 mL 4.5 M Tris, pH 8.0 (750 mM final concentration) to quench the formaldehyde.
7. Transfer cells to centrifuge bottles and centrifuge at 7000 rpm for 7 min at 4 °C using a refrigerated centrifuge.
8. Resuspend the cell pellet in 1 mL PBS and transfer cells to a 1.5 mL microtube.
9. Centrifuge at 7000 rpm for 2 min at 4 °C.
   Discard the supernatant and store cell pellet overnight at −80 °C.
   Notes:

- Because 3′-ssDNA ends are mostly DNA-replication intermediates, we use rich (LB) medium, which allows multifork replication, and a low concentration of doxycycline that induces low levels of ExID-FLAG, to minimize the amount of non-DNA-bound ExID-FLAG.

7.3 DNA sonication

1. Resuspend cell pellet in 1 mL FA lysis buffer containing 4 mg/mL lysozyme.
2. Incubate for 1 h at 37 °C with rotation on a Labquake rotator.
   Chill cell lysate on ice for 5 min.
3. Transfer 125 μL lysate to a 1.5 mL Bioruptor Pico microtube.
   Sonicate in 30 s on/off continuous pulse for 15 min; wait for 5 min to cool the lysate, then sonicate again for 15 min.
4. Transfer sonicated lysate to a 1.5 mL microtube.
5. Repeat step 3 and 4.
6. Centrifuge the sonicated lysate at 11,000 rpm for 10 min at 4 °C.
   Transfer supernatant to a new tube and discard the cell debris pellet.
7. Take 15 μL for agarose gel electrophoresis analysis.
   Visually inspect aiming for a DNA smear between 100 and 400 bp.
   Store the rest at −80 °C or proceed to the next step.
   Notes:

- Sonication optimization might be necessary if a different cell density or a different lysate volume is desired.
- After sonication, agarose gel electrophoresis analysis can be performed to verify that DNA fragments range from 100 to 400 bp. Boil 15 μL lysate at 96 °C for 10 min to decrosslink lysate. Centrifuge at 11,000 rpm for 10 min at 4 °C. Transfer supernatant to a tube and add 1 μL RNase A solution. RNase A treatment reduces the background signal, improving visual assessment of shearing.

7.4 Immunoprecipitation of ExID-FLAG

1. Dilute 16 μL ANTI-FLAG M2 antibody into 400 μL PBS with 0.02% Tween 20.
   Prepare a master mix for the number of immunoprecipitation reactions planned.
2. Vortex the vial of Dynabeads for >30 s to resuspend.
   Transfer 100 μL Dynabeads to a 1.5 mL Snaplock microfuge tube.
3. Place tubes on magnetic rack for 1 min to separate beads from solution and discard supernatant.
4. Remove tubes from magnetic rack.
   Add ~416 μL of the prepared antibody master mix to the Dynabeads.
   Incubate at room temperature with rotation on a Labquake rotator for 1 h.

5. Centrifuge briefly to bring the liquid down to the bottom of the tube. Place tubes on magnetic rack for 1 min and discard supernatant.
6. Remove tubes from the magnetic rack.
   Wash the Dynabeads once with 400 μL PBS with 0.02% Tween 20. Rotate the tubes at room temperature for 2 min.
7. Centrifuge briefly to bring the liquid to the bottom of the tube. Place tubes on magnetic rack for 1 min and discard supernatant.
8. Remove tubes from the magnetic rack.
   Add 200 μL cell lysate and 200 μL PBS with 0.02% Tween 20. Incubate at 4 °C overnight with rotation.
9. Centrifuge briefly and place tubes on magnetic rack for 1 min. Discard supernatant.
10. Wash magnetic beads 3 times with 400 μL wash buffer.
    Between each wash, resuspend beads by rotating the tubes for 2 min at room temperature, then place the tubes on a magnetic rack for 1 min and discard supernatant.
11. Wash magnetic beads 3 times with 400 μL deoxycholate buffer.
    Between each wash, resuspend beads by rotating the tubes for 2 min at room temperature, then place the tubes on a magnetic rack for 1 min and discard supernatant.
12. Elute DNA using the TaKaRa ChIP Elute Kit according to its user manual.
13. Dilute 1 μL ChIPed ssDNA in 7.5 μL DEPC-treated water for qPCR quantification.
    Store the rest at −80 °C.
14. Prepare qPCR primer mix by diluting 10 μL of 100 μM qPCR primer 1 and 2 in 1.4 mL water.
15. Prepare qPCR reactions as follows (per sample):
    5 μL SYBR Green master mix
    2.5 μL diluted qPCR primer mix
    2.5 μL diluted ChIPed ssDNA
16. Run the following qPCR protocol:

50 °C	2 min	
95 °C	3 min	
95 °C	15 s	40 cycles
60 °C	30 s	

17. Proceed to the next step with samples of Ct value less than 28.
 Notes:
- Low-retention tubes are used throughout the DNA elution step to minimize DNA loss.
- Elute DNA first with 23 μL DEPC-treated water, then with 18 μL water to maximize DNA recovery.
- Quantitative PCR is performed to approximate the quantity of ChIPed ssDNA based on an empirical Ct value. Ct values <28 indicate adequate DNA for library construction.

7.5 Strand-specific library construction

1. Thaw ~40 μL ssDNA on ice.
 Vacufuge at 60 °C for 40 min to concentrate ssDNA.
2. Resuspend ssDNA in 10.5 μL DEPC-treated water and transfer to a PCR strip.
3. Prepare library with the Takara DNA SMART ChIP-Seq Kit according to its user manual.
4. Do 24-cycle PCR amplification for library construction.
5. Follow the single size selection protocol in the user manual.
 Elute library in 27 μL water.
6. Use TapeStation or equivalent for DNA analysis to ensure generation of a successful library with a visible DNA smear.
 Proceed to the next step with successful libraries or store the libraries at −20 °C.
 Notes:
- 24 PCR cycles are used because the amount of ChIPed ssDNA is usually less than 100 pg, which is the lower limit of the kit.
- There might be non-specific PCR products between 170 and 200 bp if using >18 cycles of PCR. If seen, proceed to the next step for DNA separation and purification from polyacrylamide gel.

7.6 DNA polyacrylamide gel size selection and DNA extraction

1. Make an 8% nondenaturing DNA polyacrylamide gel (0.75 mm-thick, 5 mL):

30% Acrylamide/Bis solution 29:1	1.35 mL
10 × TBE	0.5 mL
Water	3.15 mL
10% APS	52 μL
TEMED	3.125 μL

2. Thaw 25 μL library.
   Vacufuge at 60 °C for 15 min.
   Resuspend in 10 μL water.
3. Add 2 μL 6 × DNA loading dye.
   Load entire sample onto the 8% PAGE gel.
4. Separate by electrophoresis for ~1.5 h at 100 V until the xylene cyanol FF band reaches the bottom of gel.
5. Dilute 2 μL ethidium bromide in ~30 mL water.
   Submerge the gel in the staining solution for 5 min.
6. Cut out the smear 250–600 bp.
   Pierce the bottom of a 0.7 mL tube and place into a 1.5 mL tube and place gel slice into the pierced 0.7 mL tube.
7. Shred the gel slice by spinning at 13,000 rpm for 5 min.
   Discard the 0.7 mL tube.
8. Add 0.4 mL DNA extraction buffer.
   Incubate with rotation for >16 h at 4 °C.
9. Transfer gel and liquid into a Spin-X column and spin at 13,000 rpm for 5 min.
10. Add 40 μL 3 M sodium acetate, 2 μL glycogen (20 mg/mL), and 400 μL isopropyl alcohol to flow-through.
    Incubate at −80 °C for 30 min.
11. Centrifuge at 13,000 rpm for 10 min at 4 °C.
    Discard the supernatant without disturbing the pellet.
12. Rinse the pellet with 1 mL cold 70% ethanol.
    Centrifuge at 10,000 rpm for 5 min.
    Discard the supernatant carefully to avoid disturbing the pellet.
13. Air dry the pellet.
    Dissolve the nucleic acid pellet in 12 μL DEPC-treated water.
14. Use KAPA Library Quantification kit to measure library concentration.
15. Proceed to next-generation sequencing.
    Notes:

- 6 × DNA loading dye contains two different dyes, bromophenol blue and xylene cyanol FF, for visual tracking of DNA migration during electrophoresis. In an 8% nondenaturing PAGE gel, bromophenol blue migrates to a position near 45 bp, and xylene cyanol FF 160 bp.

8. Conclusions

X-seq and SsEND-seq are both powerful approaches for identifying genomic sites with recurrent DNA-damage and -repair intermediate structures, which are otherwise transient. An informative landscape of replication stress-induced DNA damage intermediates can be obtained using these two approaches along with other DNA intermediate-mapping tools, such as END-seq (Canela et al., 2016), which we used with X-seq in a previous study (Mei et al., 2021). However, it should be noted that rare, or non-recurrent sites cannot be seen with bulk sequencing-based methods because their signals are indistinguishable from background noise.

Acknowledgments

We thank K. M. Miller, who has helped push forward all of our collaborative work on the engineered-protein reagents, and P. J. Hastings for comments on the manuscript. This work was supported by the WM Keck Foundation, and US National Institutes of Health (NIH) grants R01-CA250905 to SMR and K. M. Miller, and R35-GM122598 and NIH Director's Pioneer Award DP1-AG072751 to SMR.

References

Bachmann, B. J. (1996). Derivations and genotypes of some mutant derivatives of *Escherichia coli* K-12. In *Escherichia coli and Salmonella: Cellular and molecular biology* (2nd ed., pp. 2460–2488). Washington, DC: ASM Press.

Blattner, F. R., Plunkett, G., III, Bloch, C. A., Perna, N. T., Burland, V., Riley, M., et al. (1997). The complete genome sequence of *Escherichia coli* K-12. *Science*, *277*, 1453–1474.

Bonocora, R. P., & Wade, J. T. (2015). ChIP-seq for genome-scale analysis of bacterial DNA-binding proteins. In *Bacterial transcriptional control* (pp. 327–340). Springer.

Canela, A., Sridharan, S., Sciascia, N., Tubbs, A., Meltzer, P., Sleckman, B. P., et al. (2016). DNA breaks and end resection measured genome-wide by end sequencing. *Molecular Cell*, *63*(5), 898–911.

Chen, L., Chen, J.-Y., Zhang, X., Gu, Y., Xiao, R., Shao, C., et al. (2017). R-ChIP using inactive RNase H reveals dynamic coupling of R-loops with transcriptional pausing at gene promoters. *Molecular Cell*, *68*(4), 745–757. e745.

Churchill, J. J., Anderson, D. G., & Kowalczykowski, S. C. (1999). The RecBC enzyme loads RecA protein onto ssDNA asymmetrically and independently of χ, resulting in constitutive recombination activation. *Genes & Development*, *13*(7), 901–911.

Dunderdale, H. J., Benson, F. E., Parsons, C. A., Sharpies, G. J., Lloyd, R. G., & West, S. C. (1991). Formation and resolution of recombination intermediates by *E. coli* RecA and RuvC proteins. *Nature*, *354*(6354), 506–510.

Fitzgerald, D. M., & Rosenberg, S. M. (2021). Biology before the SOS response—DNA damage mechanisms at chromosome fragile sites. *Cells*, *10*(9), 2275. https://doi.org/10.3390/cells10092275.

Henrikus, S. S., Henry, C., Ghodke, H., Wood, E. A., Mbele, N., Saxena, R., et al. (2019). RecFOR epistasis group: RecF and RecO have distinct localizations and functions in *Escherichia coli*. *Nucleic Acids Research*, *47*(6), 2946–2965.

Kaya-Okur, H. S., Wu, S. J., Codomo, C. A., Pledger, E. S., Bryson, T. D., Henikoff, J. G., et al. (2019). CUT&Tag for efficient epigenomic profiling of small samples and single cells. *Nature Communications*, *10*(1), 1–10.

Kotlajich, M. V., Xia, J., Zhai, Y., Lin, H.-Y., Bradley, C. C., Shen, X., et al. (2018). Fluorescent fusions of the N protein of phage Mu label DNA damage in living cells. *DNA Repair*, *72*, 86–92.

Lehman, I., & Nussbaum, A. (1964). The deoxyribonucleases of *Escherichia coli* V. On the specificity of exonuclease I (phosphodiesterase). *Journal of Biological Chemistry, 239*(8), 2628–2636.

Liu, J., Fitzgerald, D. M., Mei, Q., Nimer, S., Halliday, J. A., Perren, J. O., Rogers, C., Crum, M., Kozmin, S. G., Xia, J., Zhai, Y., Bates, D., Li, L., Hastings, P. J., Sung, P. M., Miller, K. M., Herman, C., & Rosenberg, S. M. n.d. DNA damage at sites of terminated transcripts manuscript in preparation.

Magner, D. B., Blankschien, M. D., Lee, J. A., Pennington, J. M., Lupski, J. R., & Rosenberg, S. M. (2007). RecQ promotes toxic recombination in cells lacking recombination intermediate-removal proteins. *Molecular Cell, 26*(2), 273–286.

Mei, Q., Fitzgerald, D. M., Liu, J., Xia, J., Pribis, J. P., Zhai, Y., et al. (2021). Two mechanisms of chromosome fragility at replication-termination sites in bacteria. *Science Advances*, 7(25), eabe2846.

Michel, B., Ehrlich, S. D., & Uzest, M. (1997). DNA double-strand breaks caused by replication arrest. *The EMBO Journal, 16*(2), 430–438.

Morimatsu, K., & Kowalczykowski, S. C. (2003). RecFOR proteins load RecA protein onto gapped DNA to accelerate DNA strand exchange: A universal step of recombinational repair. *Molecular Cell, 11*(5), 1337–1347.

Park, P. J. (2009). ChIP–seq: Advantages and challenges of a maturing technology. *Nature Reviews Genetics, 10*(10), 669–680.

Renzette, N., Gumlaw, N., Nordman, J. T., Krieger, M., Yeh, S. P., Long, E., et al. (2005). Localization of RecA in *Escherichia coli* K-12 using RecA–GFP. *Molecular Microbiology, 57*(4), 1074–1085.

Rogakou, E. P., Boon, C., Redon, C., & Bonner, W. M. (1999). Megabase chromatin domains involved in DNA double-strand breaks in vivo. *Journal of Cell Biology, 146*(5), 905–916.

Schmutz, I., Timashev, L., Xie, W., Patel, D. J., & de Lange, T. (2017). TRF2 binds branched DNA to safeguard telomere integrity. *Nature Structural & Molecular Biology, 24*(9), 734–742.

Scully, R., Chen, J., Ochs, R. L., Keegan, K., Hoekstra, M., Feunteun, J., et al. (1997). Dynamic changes of BRCA1 subnuclear location and phosphorylation state are initiated by DNA damage. *Cell, 90*(3), 425–435.

Seigneur, M., Bidnenko, V., Ehrlich, S. D., & Michel, B. (1998). RuvAB acts at arrested replication forks. *Cell, 95*(3), 419–430.

Shee, C., Cox, B. D., Gu, F., Luengas, E. M., Joshi, M. C., Chiu, L. Y., et al. (2013). Engineered proteins detect spontaneous DNA breakage in human and bacterial cells. *eLife, 2*, e01222.

Skene, P. J., & Henikoff, S. (2017). An efficient targeted nuclease strategy for high resolution mapping of DNA binding sites. *eLife, 6*, e21856.

Xia, J., Chen, L.-T., Mei, Q., Ma, C.-H., Halliday, J. A., Lin, H.-Y., et al. (2016). Holliday junction trap shows how cells use recombination and a junction-guardian role of RecQ helicase. *Science Advances, 2*(11), e1601605.

CHAPTER NINE

A mapping platform for mitotic crossover by single-cell multi-omics

Peter Chovanec and Yi Yin*
Department of Human Genetics, University of California, Los Angeles, CA, Unites States
*Corresponding author: e-mail address: yeastyin@g.ucla.edu

Contents

Abstract

Mitotic crossovers have the potential to cause large-scale genome rearrangements. Here, we describe high-throughput, single-cell, whole-genome sequencing methods for mapping crossovers genome-wide at scale. The methods are generalizable to various eukaryotes and to other end points requiring high-throughput, high-coverage single cell sequencing.

Methods in Enzymology, Volume 661
ISSN 0076-6879
https://doi.org/10.1016/bs.mie.2021.08.017

1. Introduction

DNA experiences various types of lesions that can lead to double-strand breaks (DSBs). DSBs can be repaired by error-prone end joining using limited sequence homology, or in an error-free manner using extensive homology by homologous recombination (HR) (Ceccaldi, Rondinelli, & D'Andrea, 2016). HR has two classes: local gene conversion and crossover (CO) (St Charles et al., 2012; Symington, Rothstein, & Lisby, 2014). The latter involves a linkage switch upstream and downstream of the break (Fig. 1A). Crossover can be a major source of large-scale genome rearrangement.

Mitotic COs can result in seven types of chromosome rearrangements depending on the HR partner (Fig. 1B–H). Classes1–2 occur between allelic sequences. The vast majority occur between sister chromatids generated by replication. Sister chromatid exchange (SCE, class1) is error-free. The minority, 1% in mammalian cell lines and 3% in yeast (Claussin et al., 2017; Johnson, Liu, & Jasin, 1999; Kadyk & Hartwell, 1992; van Wietmarschen et al., 2018), occur between homologs bearing SNPs, causing copy-neutral loss-of-heterozygosity (cnLOH, class2). Classes3–5 use non-allelic repeats. Non-allelic HR (NAHR, (Sudmant et al., 2015)) can cause copy number variations (CNV, class3) and inversions (class4) with intra-chromatid repeats, or translocations (class5) with inter-chromosomal repeats. Lastly, unresolved HR intermediates and other errors could cause aneuploidy (class6) or uniparental disomy (UPD, class7) (Andersen & Petes, 2012). Unlike mutations, these CO outcomes are difficult to map without bias by standard bulk whole-genome sequencing (WGS). For example, the 99% error-free SCE is almost entirely missed in tumor signature studies that detect "scars" in the genome (Alexandrov et al., 2020).

Genome-wide mapping of DNA rearrangements has the advantage of naturally following to the end product and complements other biochemical, cell biology and molecular biology approaches that study snapshots of HR intermediates (Canela et al., 2016; Crosetto et al., 2013; Khil, Smagulova, Brick, Camerini-Otero, & Petukhova, 2012; Mimitou, Yamada, & Keeney, 2017; Sriramachandran et al., 2020; Wu et al., 2021). Such studies are particularly lacking for mitotic HR. In *S. cerevisiae*, genome instability screens are typically done in haploid with phenotypic markers (Putnam, Pennaneach, & Kolodner, 2005; Putnam et al., 2016); yet the assay is limited in the possible types of events that can be studied. Losing a marker in a

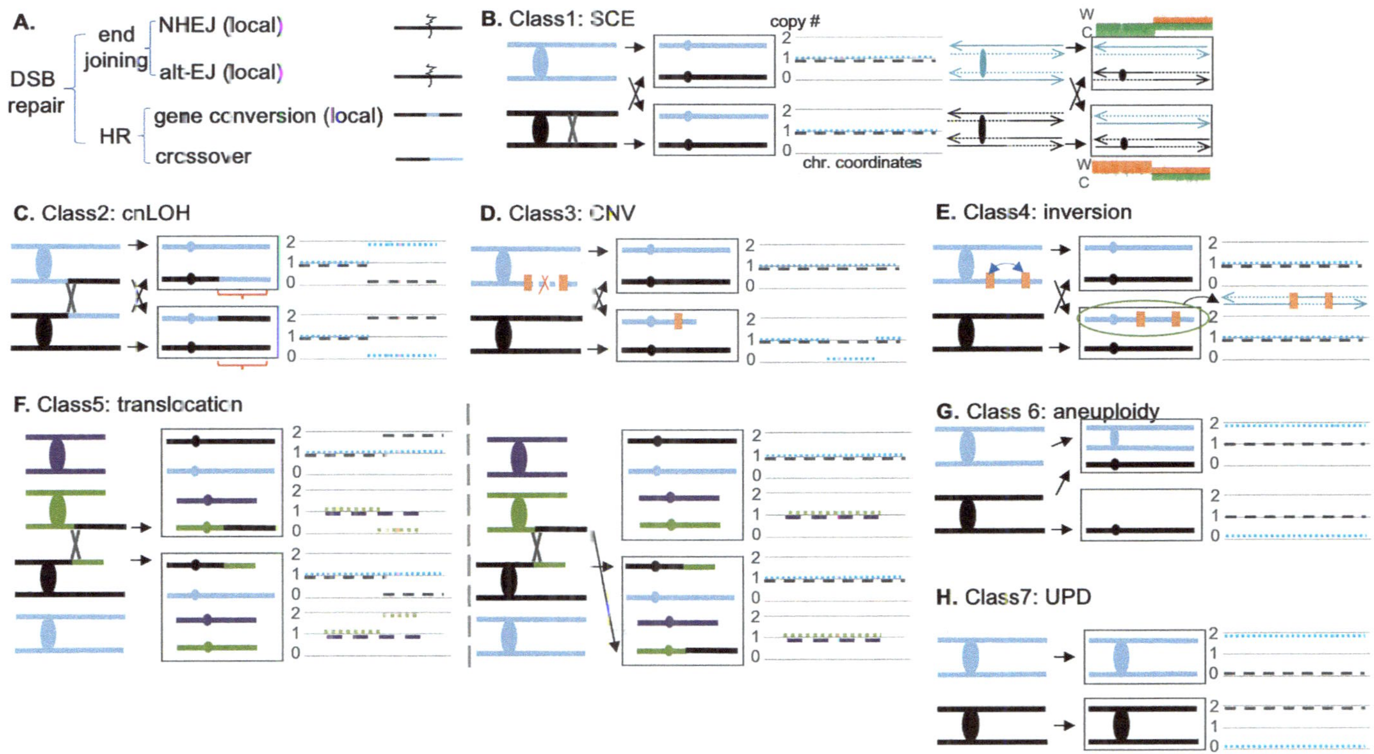

Fig. 1 See figure legend on next page.

diploid can be a result of mutation, cnLOH, deletion, monosomy, or UPD. Only mutations and deletions are possible in haploids subject to bias of survival. In addition, marker-based screens are confined to specific regions of the genome. A few studies focused on genome-wide mapping of mitotic HR in hybrid diploid yeast by analyzing cnLOH (Loeillet et al., 2020; Sampaio et al., 2020; St Charles et al., 2012; Sui et al., 2020; Yin, Dominska, Yim, & Petes, 2017; Yin & Petes, 2013). These global maps revealed the complexity of HR beyond simple explanations by current biochemical models, and uncovered variations not available in studies of induced DSB at defined loci, both regarding distinct types of rearrangements in various mutants, and non-random distributions of the events.

One can imagine extending such genome-wide mapping to various mutants and to human cells. However, such bulk sequencing assays are limiting because: (1) SNP-based analyses can only map inter-homolog HR; yet mitotic HR strongly prefers identical sister chromatids, particularly in mammalian cells; (2) the mapping requires cloning daughter cells experimentally, which is impossible for cells during development *in vivo* and is hard to scale up. Limited in throughput, roughly only 20 yeast mutants

Fig. 1 Mitotic crossover. A. Mitotic DSB repair. DSB can be repaired by end joining causing small indels, or by HR. While gene conversion involves local homology, crossover can cause extensive genome rearrangements. B—H. Types of mitotic crossover. Blue and black are homologs with SNPs, identical sister chromatids are of the same color, linked by centromere (oval). Homolog-specific copy number plots depict chromosomal alterations analyzed by WGS; for example, a chromosome with an inter-sister crossover still harbors one copy of the blue chromosome and one copy of the black chromosome. Abbreviations are defined in the text. B. SCE. Heterozygosity is maintained across the chromosome, but strand-switch can be detected by Strand-seq of the replication template strand (solid arrow). Without SCE, template strands continuously map to Watson ("W") or Crick ("C") strands (blue chromatids); with SCE, template strands have a switch (black chromatids). Reads mapped to the W and C strands are shown in orange and green, respectively. C. cnLOH. Inter-homolog crossover leads to reciprocal cnLOH in daughter cells (red bracket). D. CNV. Interstitial deletion between intra-chromatid repeats (orange boxes) as an example. E. inversion. Intra-chromatid repeats-mediated inversion detected by Strand-seq. F. translocation. Left: unbalanced translocation, recombined chromatids segregate apart, causing terminal CNV. Right: balanced translocation, recombined chromatids co-segregate, causing no change in copy number or heterozygosity, but change in Hi-C signal as now black chromosome is interacting with green chromosome. G. aneuploidy. Unresolved crossover intermediate is one cause of nondisjunction: monosomy and trisomy in reciprocal daughter cells. H. UPD. Reciprocal UPD shown as example. UPD can also be formed by duplication of monosomy or chromosome loss of trisomy. (For interpretation of the references to color in this figure legend, the reader is referred to the web version of this article.)

have been analyzed (Andersen, Sloan, Petes, & Jinks-Robertson, 2015; Andersen et al., 2016; Jinks-Robertson & Petes, 2021; Loeillet et al., 2020; Song, Dominska, Greenwell, & Petes, 2014; Yin & Petes, 2014, 2015; Zheng, Zhang, Wu, Mieczkowski, & Petes, 2016).

In Fig. 2A, we detail sequencing technologies that best map the seven types of mitotic CO detailed in Fig. 1. Global mapping of class1 (SCE) requires single-cell Strand-seq (Falconer et al., 2012), which monitors the strand switch of the replication template strand. Classes2–7 alter genetic information and are in theory detectable by bulk WGS; yet most can substantially benefit from single-cell assays, as they are not only copy-neutral but often reciprocal (like cnLOH and UPD-(Andersen & Petes, 2012; St Charles et al., 2012)). Single-cell WGS of daughter cells bypasses cloning and thus is highly scalable (Yin et al., 2019). Strand-seq uniquely detects error-free and mutational events, and provides a unifying tool for mapping all the seven types of CO outcomes without bias (Porubsky et al., 2021, 2020; Sanders et al., 2016).

Various methods of single-cell genome sequencing methods have been developed and applied to characterize genome instability events genome-wide. Technology development in single-cell WGS largely focuses on two aspects: minimizing amplification bias (Chen et al., 2017) and increasing throughput (Cusanovich et al., 2015; Vitak et al., 2017). Here, we describe single-cell DNA sequencing methods developed by our laboratory that combine improvements in both uniformity in genome amplification and throughput as scalable global assays that unbiasedly detect various types of CO outcomes (Yin et al., 2019).

The suite is called sci-L3: single-cell combinatorial indexing (sci) with linear amplification (L) and 3-level barcoding (3) (Fig. 2B). In a nutshell, sci-L3 minimizes amplification bias with *in vitro* transcription (IVT)-based linear amplification, enables exponential gains in throughput with 1 M single cells per experiment at low cost by a 3-level split-and-pool barcoding strategy, and the scheme generalizes to multiple assays beyond WGS such as target-seq and DNA/RNA co-assay. We applied sci-L3-WGS to map 86,786 meiotic COs and chromosome mis-segregations in >10k sperm and discovered whole-genome equational segregation in meiosis I in both fertile and infertile mice (Yin et al., 2019). Sci-L3-WGS bypasses cloning daughter cells and can map at scale rare spontaneous mitotic CO, the mapping is SNP-based and thus misses the abundant error-free SCE. However, the sci-L3 scheme can be easily adapted to Strand-seq with small modification of the protocol (Fig. 2C).

A.

	bulk WGS	sci-L3-WGS	sci-L3-Strand-seq	sci-L3-Hi-C
SCE	no	no	yes	no
cnLOH	hard**	yes	yes	yes
CNV	yes	yes	yes	yes
inversion	very hard*	very hard*	yes	hard***
translocation (balanced)	very hard*	very hard*	very hard	yes
translocation (unbalanced)	yes	yes	yes	yes
aneuploidy	yes	yes	yes	yes
UPD	hard**	yes	yes	yes

* requires junction-spanning reads
** reciprocal events require daughter cell cloning
***weaker signal in contact map than translocation

B.

- High-throughput barcoding with combinatorial indexing

- Uniform amplification with *in vitro* transcription

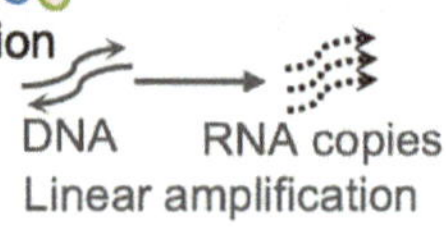

- Generalizable to multiple assays
  - sci-L3-WGS
  - sci-L3-target-seq
  - sci-L3-DNA/RNA co-assay

C. sci-L3-WGS → sci-L3-Strand-seq (+BrdU)
sci-L3-WGS → sci-L3-Hi-C (+proximity ligation)
sci-L3-DNA/RNA co-assay → sci-L3-ATAC/RNA (-nucleosome depletion)

Fig. 2 sci-L3 technologies enable high-throughput, global mapping of mitotic crossover. A. Sequencing assays required for detecting 7 types of mitotic crossover. Strand-seq and Hi-C uniquely detect error-free SCE and balanced translocations, respectively. B. Overview of sci-L3: single-cell combinatorial indexing ("sci"), linear amplification ("L") and 3-level ("3"). Cells are distributed into separate pools and barcoded for the 1st (colored edge), the 2nd (colored fill) and the 3rd round (different shapes). In-between each round, cells are pooled and sorted into new separate pools. A unique combination of three rounds of barcodes defines a single cell. We use IVT-based linear amplification to reduce exponential bias, achieve uniform coverage and prevent multiplicative errors. Sci-L3 enables single-cell WGS, target-seq or DNA and RNA co-assay. C. Expansion of sci-L3 toolset. By adding BrdU labeling or proximity ligation, sci-L3-WGS can be easily adapted to Strand-seq or Hi-C, respectively. By omitting nucleosome depletion, the sci-L3-DNA/RNA co-assay can be converted to an ATAC/RNA co-assay.

2. Overview of sci-L3-WGS

Fixed nuclei first undergo nucleosome depletion. The nuclei are then evenly distributed into X wells and barcoded for the 1st round by Tn5 insertion. The Tn5 transposomes fragment the genome and simultaneously insert a unique barcode for each well. The process of tagging the DNA with well-specific barcodes on average every 0.5–1.5 kb is termed "tagmentation." We then pool and split the nuclei into Y wells and ligate the 2nd round of barcodes and a T7 promoter, enabling subsequent IVT-based linear amplification. All of the nuclei are once again pooled together with X times Y number of barcode combinations after two rounds of split-and-pool. We then FACS sort or dilute to distribute the nuclei into a final round of wells. The number of nuclei per well is much smaller than X times Y such that the nuclei in the same well do not share the same barcode 1 and barcode 2 combinations. Nuclei are then lysed and subjected to downstream IVT, reverse transcription (RT) and second-strand synthesis (SSS) to amplify the genome in a linear fashion. A third round of barcodes is added during SSS. The combination of three rounds of barcodes define the single-cell origin of each amplified and sequenced molecule.

3. Preparing for a sci-L3 experiment

1. Prepare the following stock solutions:
   a) **QIAGEN Protease** (60 mg/mL; 2.8 mL)
      Add 2778 μL ddH_2O to 166.7 mg of lyophilized protease powder to make 60 mg/mL stock.
      Store at 4 °C. Note that the amount of water to add may vary by lot.
   b) **1× TE** (10 mM Tris-HCl pH 8.0, 1 mM EDTA; 1 mL)
      1 μL 1 M Tris-HCl pH 8.0
      2 μL 0.5 M EDTA pH 8.0
      988 μL ddH_2O
      Store at room temperature
   c) **1× Lysis buffer** (60 mM Tris-Ac pH 8.3, 2 mM EDTA pH 8.0, 15 mM DTT; 1 mL)
      150 μL 10× TAE pH 8.0

1 μL 0.5 M EDTA pH 8.0
15 μL 1 M DTT
10 μL 1 × TE
824 μL ddH_2O
Store at −20 °C

d) 10× thms buffer (50 mM $MgCl_2$, 100 mM Tris pH 8.0; 1 mL)
100 μL 1 M Tris-HCl pH 8.0
50 μL 1 M $MgCl_2$
850 μL ddH_2O
Store at room temperature

e) Tn5 Storage buffer - STB (50% glycerol, 0.5 × TE; 400 μL)
200 μL 1 × TE pH 8.0
200 μL glycerol
Store at room temperature

f) Glycine (2.5 M glycine; 1 mL)
Weigh out 187.7 mg of glycine powder and make up to 1 mL with ddH_2O
Store at room temperature

g) 1× LBT (1 × LB, 0.1% TritonX-100; 1 mL)
990 μL 1 × Lysis buffer
10 μL 10% TritonX-100
Prepare fresh on the day of experiment

2. Anneal oligonucleotides:

a) Prepare **annealing buffer** (10 mM Tris-HCl pH 8.0, 50 mM NaCl, 1 mM EDTA, pH 8.0; 1 mL)
10 μL 1 M Tris-HCl pH 8.0
10 μL 5 M NaCl
2 μL 0.5 M EDTA pH 8.0
978 μL ddH_2O
Store at room temperature

b) Resuspend:
lyophilized Tn5 mosaic end (ME) oligo and the Tn5 ME ligation oligo in annealing buffer to a 100 μM concentration. Mix the two oligos together 1:1 (50 μM).
lyophilized ligation hairpin oligo in annealing buffer to a 50 μM concentration

c) Anneal with gradual cooling:
95 °C 5 min, −0.1 °C/cycle, hold for 9 s/cycle, 700 cycles to 25 °C. This should take 2–3 h.

d) Dilute:

Tn5 mosaic end double stranded (MEDS) oligo to a final concentration of 1.5 μM with annealing buffer and store in −20 °C until ready to use

Hairpin ligation double-stranded oligo to a final concentration of 1.5 μM with annealing buffer and store in −20 °C until ready to use

4. Key resources table

Reagent or resource	Source	Identifier
Experimental Models: Cell Lines		
HEK293T	ATCC	CRL-3216
3T3	ATCC	CRL-1658
Oligonucleotides		
Barcoding oligos for sci-L3	(Yin et al., 2019)	Table S5
Software and Algorithms		
sciL3Pipe	This Paper	https://github.com/recombinationlab/sciL3Pipe

5. Materials and equipment

1. Reagents
   - QIAGEN Protease (QIAGEN, cat. no. 19155)
   - 1 M Tris-HCl pH 8.0 (Invitrogen, cat. no. 15568025)
   - 0.5 M EDTA pH 8.0 (Invitrogen, cat. no. AM9260G)
   - 10× TAE pH 8.0 (Invitrogen, cat. no. AM9869)
   - 1 *M* DTT (Sigma-Aldrich, cat. no. 646563-10X.5ML)
   - TritonX-100 (Sigma-Aldrich, cat. no. 93443-100ML)
   - 1 M $MgCl_2$ (Invitrogen, cat. no. AM9530G)
   - Glycerol (Sigma-Aldrich, cat. no. G5516-100ML)
   - DMEM complete media (Thermofisher Scientific, cat. no. 11054001)

- 37% Formaldehyde (Sigma-Aldrich, cat. no. F15587)
- Glycine (Sigma-Aldrich, cat. no. G7403-250G)
- IGEPAL® CA-630 (Sigma-Aldrich, cat. no. I8896)
- 10 × NEBuffer™ 2.1 (New England Biolabs, cat. no. B7202S)
- SDS, 20% Solution (Invitrogen, cat. no. AM9820)
- EZ-Tn5™ Transposase (Lucigen, cat. no. TNP92110)
- 5 M NaCl (Invitrogen, cat. no. AM9760G)
- dNTP set 10 mM each (Kapa Biosystems, cat. no. KK1017)
- PEG-4000 (Sigma-Aldrich, cat. no. 25322-68-3)
- T4 DNA ligase (Thermofisher Scientific, EL0011)
- DAPI (Thermofisher Scientific, cat. no. D3571)
- Ethanol, 200 proof, Molecular Biology Grade (FisherScientific, cat. no. 07-678-003)
- Bst 2.0 WarmStart® DNA Polymerase (New England Biolabs, cat. no. M0538)
- HiScribe T7 Quick High Yield RNA Synthesis Kit (New England Biolabs, cat. no. E2050S)
- SUPERase•In™ RNase Inhibitor (Thermofisher Scientific, cat. no. AM2696)
- SuperScript™ IV Reverse Transcriptase (Thermofisher Scientific, cat. no. 18090200)
- RNaseH (New England Biolabs, cat. no. M0297)
- RNaseA (Invitrogen, cat. no. AM2270)
- Q5® High-Fidelity DNA Polymerase (New England Biolabs, cat. no. M0491)
- NEBNext® Ultra™ II DNA Library Prep Kit for Illumina® (New England Biolabs, cat. no. E7645)
- Agencourt AMPure XP Magnetic Beads (Beckman Coulter, cat. no. A63881)
- Qubit ™ dsDNA HS Assay Kit (Thermofisher Scientific, cat. no. Q32851)

2. Equipment

- Eppendorf LoBind® twin. tec® PCR Plates, 96 well (Eppendorf, cat. no. 10049-108)
- Magnetic racks for 1.5 mL tubes (DynaMag-2; Thermofisher Scientific, cat no. 12321D)
- 1.5 mL DNA/RNA low retention microcentrifuge tubes (Eppendorf, cat. no. 022431021)
- 1.7 mL microcentrifuge tubes (Genesee Scientific, cat. no. 22-282)

- Axygen® 0.2 mL MAXYMum Recovery™ Thin Wall PCR Tubes with Flat Cap (Corning, cat. no. PCR-02-L-C)
- RNA Clean & Concentrator™-5 kit (Zymo Research, cat. no. R1016)
- DNA Clean & Concentrator™-5 kit (Zymo Research, cat. no. D4014)
- Benchtop microcentrifuge (Eppendorf, cat. no. 5420)
- Refrigerated centrifuge (Eppendorf, cat. no. 5424R)
- Eppendorf ThermoMixer® C (Eppendorf, cat. no. 5382000023)
- Eppendorf SmartBlock™ 1.5 mL (Eppendorf, cat. no. 5360000038)
- Eppendorf SmartBlock™ PCR 96 (Eppendorf, cat. no. 5306000006)
- Hemacytometer
- Qubit Fluorometer 4 (Thermofisher Scientific, cat. no. Q33240)
- Single channel Pipettes (Rainin, P2, P20, P200, P1000)
- 8-well and 12-well Multi-channel pipettes (Rainin, P20, P200)
- Low retention tips (Rainin)
- Reagent Reservoirs (FisherScientific, cat. no. 21381093)

6. Step-by-step method details

6.1 Crosslinking and nucleosome depletion

Timing: 2 h 30 min

1. Crosslink cells in 10 mL DMEM Complete medium with 406 μL 37% formaldehyde (final concentration 1.5%) at room temperature (21–25 °C) for 10 min while gently inverting the tube
2. Add 800 μL 2.5 M glycine and incubate on ice for 5 min. After, pellet cells at 500 g for 5 min at 4 °C and wash cells in 1 mL 1 × LBT

Critical: Ensure that the 2.5 M glycine solution used is no older than a month.

3. Resuspend pellet in 1 mL 1 × LBT with 10 μl 10% IGEPAL CA-630 (0.1% final) and incubate on ice for 20 min. After, pellet nuclei at 500 g for 5 min at 4 °C and wash once with 1 × NEBuffer2.1 (diluted to 1 × with ddH_2O).
4. Resuspend pellet in 776 μL 1 × NEBuffer 2.1 + 24 μL 10% SDS and place the tubes in a thermomixer set to 42 °C with vigorous shaking (500 rpm) for 30 min
5. Add 180 μL 10% Triton-X and place the tubes in a thermomixer set to 42 °C with vigorous shaking (500 rpm) for 30 min

Note: During the incubation, to save time, assemble the transposome as described in step 7.

6. Spin at 500*g* for 5 min at 4 °C and wash once in 1 mL LBT. After the last wash, leave about 20 μL behind, gently resuspend the pellet and count nuclei using a hemocytometer. Make up volume with 1 × LBT to about 20,000 cells/μL

6.2 First level indexing with tagmentation

Timing: 1 h

7. Assemble transposome (for 24 wells):

a) Dilute EZ-Tn5 with STB:

14 μL EZ-Tn5 (~1 μM)

7.5 μL STB

b) Combine 0.8 μL diluted EZ-Tn5 with 0.6 μL annealed Tn5 MEDS (1.5 μM) in a 96-well LoBind plate and incubate at room temperature for 30 min.

Critical: Here, we prepare volumes for at least 2 additional wells to account for pipetting errors.

Note: Here, we prepare 24 reactions for 24 first round barcodes, with each barcode located within an individual well. Depending on application, it may be desirable to use more first round barcodes.

Note: During assembly, the transposome dimerizes to a final concentration of ~0.25 μM. The assembled transposome complex can be stably stored at −20 °C for up to a year.

8. Preheat the thermomixer (or thermocycler) to 55 °C

9. Prepare the $MgCl_2$ solution:

6.5 μL H_2O ×26: 169 μL H_2O

0.7 μL 50 mM $MgCl_2$ ×26: 18.2 μL $MgCl_2$

10. Into each well of a new 96-well LoBind plate, add 7.2 μL of the $MgCl_2$ solution (final $MgCl_2$ concentration of 3.22 mM, accounting for EDTA in the lysis buffer and STB), along with 1.5 μL nuclei in LBT (approximately 30,000) and 1.19 μL of the assembled transposome. Incubate at 55 °C for 20 min in a thermomixer (or can also use a thermocycler).

11. After tagmentation, set the thermomixer at 20 °C. Leave the lid open to cool (takes about 30 min to cool from 55 °C to 20 °C).

6.3 Second level indexing with ligation

Timing: 1 h

12. Prepare RBS solution (for 64 wells):

24 μL 10 mM dNTP, 48 μL 10× thms buffer, 96 μL ddH2O, 144 μL LBT.

Note: Here, we prepare 64 reactions for 64 s round barcodes, with each barcode located within an individual well. Depending on application, it may be desirable to use more second round barcodes.

13. Pool cells from all well into a trough containing 500 μL of LBT and transfer to a 1.5 mL conical tube. Wash wells and trough with an additional 500 μL LBT and transfer into the same conical tube

14. Spin at 500 g at 4 °C for 8 min and resuspend the nuclei pellet in 312 μL RBS. Add 40 μL into each well of a new 8-well strip

15. Prepare ligation mix (for 64 wells):

48 μL 10 × T4 ligase buffer, 38.4 μL PEG-4000, 8 μL T4 DNA ligase

Note: For 64 wells, we prepare enough ligation mix for 80 wells to account for pipetting errors and loss between transfers. A single well contains 0.6 μL 10 × T4 ligase buffer, 0.48 μL PEG-4000, 0.1 μL T4 DNA ligase.

16. Add 11 μL of ligation mix into each well of a new 8-well strip

17. Prepare a new 96-well LoBind plate with 0.8 μL (1.5 μM) annealed double-stranded ligation hairpin oligos

18. Using a multichannel pipette, add 4.9 μL of nuclei in RBS and 1.2 μL of the ligation mix into the first 8 columns (64 wells) of the 96-well LoBind plate containing the double-stranded ligation hairpin oligos. Mix by pipetting up and down

19. Incubate in a thermomixer for 30 min at 20 °C

20. During ligation, prepare a 96-well LoBind plates for FACS by adding 3 μL of 1 × LBT into each well

21. After ligation, combine all well in a trough (approximately 630 μL). Add 70 μL 50 μg/mL DAPI, transfer to an appropriate FACS tube and keep on ice until ready to sort

6.4 FACS

Timing: 1–2 h

22. Sort 100–300 nuclei per well straight into 1 × LBT using DAPI, gating for 2C cells and excluding doublets

Note: Each sorting event is a droplet of approximately 3–5 nL of FACS buffer, depending on the size of the nozzle. To keep the salt concentration low, we recommend keeping the total volume sorted into each well to be less than 1 μL.

23. After sorting, spin down plate at 500 g for 5 min at 4 °C.

Pause Point: sorted nuclei can be frozen at −80 °C at this point for up to a year.

6.5 Nuclei lysis

Timing: 5–9 h

24. Lyse sorted nuclei

   a) By incubating the plate at 75 °C for 30 min, followed by cooling to 4 °C

   b) Into each well add 1 μL of freshly diluted QIAGEN protease (8 mg/mL diluted with ddH_2O, final concentration 2 mg/mL)

   c) Incubate for 4–8 h at 55 °C, followed by 30 min at 75 °C to heat-inactivate the protease. For overnight incubations, the plate can stay at 4 °C

Pause Point: lysed nuclei can be frozen at −80 °C at this point for up to a year.

6.6 Gap extension and linear amplification by *in vitro* transcription

Timing: 11–17 h

Note: We recommend processing no more than 32 wells of samples (approximately 9600 cells) at a time as the subsequent amplification step involves RNA and is subject to sample degradation.

25. Prepare gap extension mix (for 24 samples):

   50 μL H_2O, 17.5 μL 10 × thms buffer, 8.76 μL of dNTP, 8.74 μL of Bst WarmStart 2.0 polymerase

Note: For 24 samples, we prepare enough gap extension mix for 25 samples to account for pipetting errors. A single well contains 2 μL H_2O, 0.7 μL 10 × thms buffer, 0.35 μL 10 mM dNTPs, 0.35 μL Bst WarmStart 2.0 polymerase.

26. Add 3.4 μL into three 8-well strips and incubate for 5 min at 68 °C

27. Prepare the HiScribe T7 *in vitro* transcription mix (for 24 samples):

   48 μL H_2O, 48 μL T7 RNA Polymerase mix, 240 μL 10 mM rNTP mix

Note: A single well contains 2 μL H_2O, 2 μL T7 RNA Polymerase mix and 10 μL 10 mM rNTP mix.

28. Add 13.9 μL of the T7 mix into each tube for a final IVT reaction volume of 20 μL.

29. Incubate at 37 °C for 10–16 h

Critical: RT and SSS steps need to be performed right after IVT and take approximately 4 h.

Note: Before proceeding to the RT step, set up 24 × 1.5 mL LoBind tubes, 24 × Zymo spin IC columns, 24 × 1.7 mL microcentrifuge tubes and 24 × low retention PCR tubes.

6.7 Reverse transcription of linearly amplified samples

Timing: 2 h

30. Add 2.22 μL 0.5 M EDTA into each tube to terminate IVT

31. Purify the IVT RNA using the RNA Clean & Concentrator-5 kit. To each sample:

  a) Add 52 μL RNA Binding Buffer and mix well by pipetting up and down

  b) Add 95 μL of 100% ethanol, mix well by pipetting up and down and transfer to a Zymo-Spin IC column and spin at 13,000 g for 30 s. Discard the flow-through

  c) Add 400 μL RNA Prep Buffer and spin at 13,000 g for 30 s. Discard the flow-through

  d) Add 700 μL RNA Wash Buffer and spin at 13,000 g for 30 s. Discard the flow-through

  e) Add 500 μL RNA Wash Buffer and spin at 13,000 g for 2 min to ensure complete removal of ethanol from the column silica matrix

  f) Transfer each column to a new 1.5 mL LoBind tube and add 18 μL 0.1 × TE (pH 8.0) directly onto the column matrix. Wait 2 min to allow RNA to re-dissolve, after which spin at 13,000 *g* for 1 min

32. Prepare the RT mix (for 24 samples):

  a) RT mix A: 48 μL 10 mM dNTP, 14.4 μL RT RNA primers (50 μM), 12 μL Superase-In

  Note: A single tube will contain 2 μL 10 mM dNTP, 0.6 μL RT RNA primers (50 μM), 0.5 μL Superase-In.

  b) RT mix B: 144 μL 5 × SuperScript IV buffer, 36 μL 0.1 M DTT, 24 μL SUPERase-In, 24 μL SuperScript IV

  Note: A single tube will contain 6 μL 5 × SSIV buffer, 1.5 μL 0.1 M DTT, 1 μL SUPERase-In, 1 μL SuperScript IV.

33. Distribute 3.1 μL of RT mix A into low retention PCR tubes

34. Add approximately 18 μL of purified IVT RNA and mix by pipetting up and down
35. Incubate for 1 min at 70 °C, followed by 20 s at 90 °C in a thermocycler to denature and remove any RNA secondary structure. After denaturation, place samples into a 4 °C water bath for fast cooling. If processing more than 8 samples, start cooling after 10 s at 90 °C. After approximately 30 s in the 4 °C water bath, transfer samples to a 96-well ice block
36. Add 9.4 μL of RT mix B into each tube, mix well by pipetting up and down followed by a quick pulse spin
37. Incubate in a thermocycler with the following RT program:

25 °C 1 min
37 °C 1 min
55 °C 15 min
60 °C 10 min
65 °C 12 min
70 °C 8 min
75 °C 5 min
80 °C 10 min
25 °C hold

6.8 RNase treatment

Timing: 40 min

38. Prepare a mix of 12.5 μL RNaseH (0.5 μL per sample) and 7.5 μL RNaseA (0.3 μL per sample).
39. Add 0.8 μL of the RNase mix to each sample
40. Incubate at 37 °C for 30 min to eliminate the RNA
41. Thaw 24 SSS primers during incubation

6.9 Third level indexing with second-strand synthesis (SSS)

Timing: 1 h

42. Add 1 μL of a unique SSS barcoding primer into each tube
43. Prepare the Q5 master mix for SSS (24 samples):

661.5 μL H_2O, 490 μL 5 × Q5 buffer, 490 μL Q5 GC enhancer, 24.5 μL Q5 polymerase.

Note: The Q5 master mix for a single sample contains 27 μL H_2O, 20 μL 5 × Q5 buffer, 20 μL Q5 GC enhancer, 1 μL Q5 polymerase.

44. Add 67 μL of the Q5 master mix to each RT sample (100 μL reaction volume).
45. Incubate in a thermocycler with the following SSS program:
    98 °C 40 s
    58 °C 30 s
    60 °C 30 s
    65 °C 30 s
    70 °C 30 s
    72 °C 6 min
    4 °C hold
46. Purify the resulting double stranded DNA using the DNA Clean & Concentrator-5 kit:
    a) To each SSS reaction, add 500 μL DNA binding buffer and mix well by pipetting up and down. Load onto a Zymo-Spin IC Column and spin at 13,000 g for 30 s. Discard the flow-through
    b) Wash twice with 200 μL DNA wash buffer and spin at 13,000 g for 30 s. Discard the flow-through
    c) Spin at 13,000 g for 2 min
    d) Transfer each column to a new 1.5 mL LoBind tube and add 10 μL 0.1 × TE. Wait 2 min to allow DNA to re-dissolve, after which spin at 13,000 g for 1 min
47. Qubit 0.8 μL of the sample using the dsDNA high sensitivity kit

6.10 Library preparation

Timing: 2 h 30 min

48. Bring up dsDNA (20–100 ng) after SSS to 50 μL with 0.1 × TE
49. Add 7 μL End-repair buffer and 3 μL End-repair enzyme (from the NEBNext Ultra II kit, please see manufacturer's protocol for additional details) and mix well by pipetting up and down
50. Incubate for 30 min at 20 °C and then 30 min at 65 °C
51. Dilute sequencing adaptors with 10 mM Tris-HCl, 10 mM NaCl depending on sample concentration.
    a) >200 ng, no dilution (15 μM working adaptor concentration)
    b) 50–200 ng, 1:10 dilution (1.5 μM working adaptor concentration)
    c) <50 ng, 1:20 dilution (0.6 μM working adaptor concentration)
52. Add 2.5 μL diluted adaptor, with 30 μL ligation master mix and 1 μL ligation enhancer and mix well by pipetting up and down
53. Incubate for 15 min at 20 °C

54. Add 3 µL of USER enzyme and mix well by pipetting up and down
55. Incubate for 15 min at 37 °C
56. Perform an AMPure XP magnetic bead cleanup:
    a) Prepare 80% ethanol by diluting with ddH_2O

Note: Always prepare fresh 80% ethanol just before performing cleanup.

    b) Add 140 µL (1.4 ×) AMPure XP beads and mix by pipetting up and down at least 10 times
    c) Incubate at room temperature for 5 min, after which place on a magnetic stand for another 3 min

Note: During incubations, prepare the PCR reaction and pre-heat the thermocycler.

    d) Remove supernatant and wash with 200 µL 80% ethanol twice, briefly incubating 30–60 s each time. Do not remove tubes from the magnet during the washes
    e) After a second wash, remove any residual liquid and air dry the beads until they turn a matte brown color

Critical: Do not over dry beads.

    f) Elute with 17 µL 0.1 × TE, take off the magnet and mix well by pipetting up and down at least 10 times. Incubate for 2 min
    g) Place back onto the magnetic rack and take 16 µL of the supernatant into the PCR reaction
57. For the PCR reaction:
    a) Combine in an 8-well strip, 2 µL Universal primer, 2 µL Indexed primer, 20 µL 2 × NEBNext Ultra II Q5 Master Mix, 16 µL purified DNA (40 µL reaction).
    b) PCR cycling conditions:

       Initial denaturation 98 °C 30 s.

       4 cycles of:
       - Denaturation 98 °C 10 s.
       - Annealing/Extension 65 °C 75 s.

       Final extension 65 °C 5 min

       Hold 4 °C
58. After PCR add 34.6 µL (0.86 ×) AMPure XP magnetic beads and repeat step 56, eluting with 10–30 µL 0.1 × TE
59. Qubit 0.8 µL of the sample using the dsDNA high sensitivity kit

7. Expected outcomes

With single human or mouse cells, we expect the dsDNA concentration to be at about 50 ng/µL with Qubit quantification at step 47 with

100–300 sorted per well. The concentration and size distribution of the final library depend on how much was used as input for library preparation. We recommend quantifying the molar concentration of the sequencing library using only fragments between 200 and 400 bp with BioAnalyzer or TapeStation.

8. Quantification and statistical analysis

To process raw fastq files into single cell BAM files for bespoke downstream analysis, we have created a Snakemake workflow available at https://github.com/recombinationlab/sciL3Pipe. The major step of the workflow include:

1. Because of the bi-directional T7 transcription and the orientation independent ligation of sequencing adaptors, the final library contains reads with barcodes on either read 1 (R1) or reads 2 (R2). The first step involves identifying the read that contains the RT primer sequence, swapping orientation if sequence present in R2. Subsequently, the SSS barcode (third level index) is extracted and used to generate a separate fastq file for each barcode. The splitting by barcode enables parallel execution of all the downstream step
2. The next step involves identifying the Tn5 and ligation barcodes (first and second level indexes) and trimming off all the barcodes using the Tn5 MEDS sequence. Add barcodes are retained within the read names
3. Trimmed reads are aligned using BWA-MEM
4. The output BAM file is finally split into individual cell BAM files using the barcodes saved within the read names

The automated workflow also generates several QC metrics, including genome coverage and collisions if a mixture of cells from two species was used within the experiment.

To run the workflow:

1. Clone the sciL3Pipe repository from github:
   git clone https://github.com/recombinationlab/sciL3Pipe
2. Make sure the required dependencies are installed:
   a. Conda
   b. Python 3
   c. Snakemake
3. The workflow requires a JSON file containing locations and names of the input fastq files. This can be generated using fastq2json.py using the following command:
   python fastq2json.py –fastq_dir /path/to/fastq/directory

4. Users are expected to generate or download their own BWA-MEM genome index. The path to the local index needs to be updated in the configs/config.yaml file
5. Finally, the workflow can be executed using the provided shell run scripts:
   sh run_pipeline_sge.sh (to run on an SGE cluster) or sh run_pipeline_local.sh (to run locally).

Acknowledgments

Studies of mitotic crossover by single-cell sequencing in our laboratory are supported by grants from the National Institutes of Health (R35GM142511) and Damon Runyon Cancer Research Foundation (DFS-43-20).

References

Alexandrov, L. B., Kim, J., Haradhvala, N. J., Huang, M. N., Tian Ng, A. W., Wu, Y., et al. (2020). The repertoire of mutational signatures in human cancer. *Nature*, *578*(7793), 94–101.

Andersen, S. L., & Petes, T. D. (2012). Reciprocal uniparental disomy in yeast. *Proceedings of the National Academy of Sciences of the United States of America*, *109*(25), 9947–9952.

Andersen, S. L., Sloan, R. S., Petes, T. D., & Jinks-Robertson, S. (2015). Genome-destabilizing effects associated with top1 loss or accumulation of top1 cleavage complexes in yeast. *PLoS Genetics*, *11*(4), e1005098.

Andersen, S. L., Zhang, A., Dominska, M., Moriel-Carretero, M., Herrera-Moyano, E., Aguilera, A., et al. (2016). High-resolution mapping of homologous recombination events in rad3 hyper-recombination mutants in yeast. *PLoS Genetics*, *12*(3), e1005938.

Canela, A., Sridharan, S., Sciascia, N., Tubbs, A., Meltzer, P., Sleckman, B. P., et al. (2016). DNA breaks and end resection measured genome-wide by end sequencing. *Molecular Cell*, *63*(5), 898–911.

Ceccaldi, R., Rondinelli, B., & D'Andrea, A. D. (2016). Repair pathway choices and consequences at the double-Strand break. *Trends in Cell Biology*, *26*(1), 52–64. https://doi.org/10.1016/j.tcb.2015.07.009.

Chen, C., Xing, D., Tan, L., Li, H., Zhou, G., Huang, L., et al. (2017). Single-cell whole-genome analyses by linear amplification via transposon insertion (LIANTI). *Science*, *356*(6334), 189–194.

Claussin, C., Porubský, D., Spierings, D. C., Halsema, N., Rentas, S., Guryev, V., et al. (2017). Genome-wide mapping of sister chromatid exchange events in single yeast cells using Strand-seq. *eLife*, *6*, e30560. https://doi.org/10.7554/eLife.30560.

Crosetto, N., Mitra, A., Silva, M. J., Bienko, M., Dojer, N., Wang, Q., et al. (2013). Nucleotide-resolution DNA double-strand break mapping by next-generation sequencing. *Nature Methods*, *10*(4), 361–365.

Cusanovich, D. A., Daza, R., Adey, A., Pliner, H. A., Christiansen, L., Gunderson, K. L., et al. (2015). Multiplex single cell profiling of chromatin accessibility by combinatorial cellular indexing. *Science*, *348*(6237), 910–914.

Falconer, E., Hills, M., Naumann, U., Poon, S. S. S., Chavez, E. A., Sanders, A. D., et al. (2012). DNA template strand sequencing of single-cells maps genomic rearrangements at high resolution. *Nature Methods*, *9*(11), 1107–1112.

Jinks-Robertson, S., & Petes, T. D. (2021). Mitotic recombination in yeast: What we know and what we don't know. *Current Opinion in Genetics & Development*, *71*, 78–85.

Johnson, R. D., Liu, N., & Jasin, M. (1999). Mammalian XRCC2 promotes the repair of DNA double-strand breaks by homologous recombination. *Nature, 401*(6751), 397–399.

Kadyk, L. C., & Hartwell, L. H. (1992). Sister chromatids are preferred over homologs as substrates for recombinational repair in Saccharomyces cerevisiae. *Genetics, 132*(2), 387–402.

Khil, P. P., Smagulova, F., Brick, K. M., Camerini-Otero, R. D., & Petukhova, G. V. (2012). Sensitive mapping of recombination hotspots using sequencing-based detection of ssDNA. *Genome Research, 22*(5), 957–965.

Loeillet, S., Herzog, M., Puddu, F., Legoix, P., Baulande, S., Jackson, S. P., et al. (2020). Trajectory and uniqueness of mutational signatures in yeast mutators. *Proceedings of the National Academy of Sciences of the United States of America, 117*(40), 24947–24956.

Mimitou, E. P., Yamada, S., & Keeney, S. (2017). A global view of meiotic double-strand break end resection. *Science, 355*(6320), 40–45. https://doi.org/10.1101/067199.

Porubsky, D., Ebert, P., Audano, P. A., Vollger, M. R., Harvey, W. T., Marijon, P., et al. (2021). Fully phased human genome assembly without parental data using single-cell strand sequencing and long reads. *Nature Biotechnology, 39*(3), 302–308.

Porubsky, D., Sanders, A. D., Taudt, A., Colomé-Tatché, M., Lansdorp, P. M., & Guryev, V. (2020). breakpointR: An R/Bioconductor package to localize Strand state changes in Strand-seq data. *Bioinformatics, 36*(4), 1260–1261.

Putnam, C. D., Pennaneach, V., & Kolodner, R. D. (2005). Saccharomyces cerevisiae as a model system to define the chromosomal instability phenotype. *Molecular and Cellular Biology, 25*(16), 7226–7238.

Putnam, C. D., Srivatsan, A., Nene, R. V., Martinez, S. L., Clotfelter, S. P., Bell, S. N., et al. (2016). A genetic network that suppresses genome rearrangements in Saccharomyces cerevisiae and contains defects in cancers. *Nature Communications, 7*, 11256.

Sampaio, N. M. V., Ajith, V. P., Watson, R. A., Heasley, L. R., Chakraborty, P., Rodrigues-Prause, A., et al. (2020). Characterization of systemic genomic instability in budding yeast. *Proceedings of the National Academy of Sciences of the United States of America, 117*(45), 28221–28231.

Sanders, A. D., Hills, M., Porubský, D., Guryev, V., Falconer, E., & Lansdorp, P. M. (2016). Characterizing polymorphic inversions in human genomes by single-cell sequencing. *Genome Research, 26*(11), 1575–1587.

Song, W., Dominska, M., Greenwell, P. W., & Petes, T. D. (2014). Genome-wide high-resolution mapping of chromosome fragile sites in Saccharomyces cerevisiae. *Proceedings of the National Academy of Sciences of the United States of America, 111*(21), E2210–E2218.

Sriramachandran, A. M., Petrosino, G., Méndez-Lago, M., Schäfer, A. J., Batista-Nascimento, L. S., Zilio, N., et al. (2020). Genome-wide nucleotide resolution mapping of DNA replication patterns, single-strand breaks, and lesions by GLOE-Seq. *Molecular Cell, 78*(5), 975–985.e7.

St Charles, J., Hazkani-Covo, E., Yin, Y., Andersen, S. L., Dietrich, F. S., Greenwell, P. W., et al. (2012). High-resolution genome-wide analysis of irradiated (UV and γ-rays) diploid yeast cells reveals a high frequency of genomic loss of heterozygosity (LOH) events. *Genetics, 190*(4), 1267–1284.

Sudmant, P. H., Rausch, T., Gardner, E. J., Handsaker, R. E., Abyzov, A., Huddleston, J., et al. (2015). An integrated map of structural variation in 2,504 human genomes. *Nature, 526*(7571), 75–81.

Sui, Y., Qi, L., Wu, J.-K., Wen, X.-P., Tang, X.-X., Ma, Z.-J., et al. (2020). Genome-wide mapping of spontaneous genetic alterations in diploid yeast cells. *Proceedings of the National Academy of Sciences of the United States of America, 117*(45), 28191–28200.

Symington, L. S., Rothstein, R., & Lisby, M. (2014). Mechanisms and regulation of mitotic recombination in Saccharomyces cerevisiae. *Genetics, 198*(3), 795–835.

van Wietmarschen, N., Merzouk, S., Halsema, N., Spierings, D. C. J., Guryev, V., & Lansdorp, P. M. (2018). BLM helicase suppresses recombination at G-quadruplex motifs in transcribed genes. *Nature Communications*, *9*(1), 271.

Vitak, S. A., Torkenczy, K. A., Rosenkrantz, J. L., Fields, A. J., Christiansen, L., Wong, M. H., et al. (2017). Sequencing thousands of single-cell genomes with combinatorial indexing. *Nature Methods*, *14*(3), 302–308.

Wu, W., Hill, S. E., Nathan, W. J., Paiano, J., Callen, E., Wang, D., et al. (2021). Neuronal enhancers are hotspots for DNA single-strand break repair. *Nature*, *593*(7859), 440–444.

Yin, Y., Dominska, M., Yim, E., & Petes, T. D. (2017). High-resolution mapping of heteroduplex DNA formed during UV-induced and spontaneous mitotic recombination events in yeast. *eLife*, 6. https://doi.org/10.7554/eLife.28069.

Yin, Y., Jiang, Y., Lam, K.-W. G., Berletch, J. B., Disteche, C. M., Noble, W. S., et al. (2019). High-throughput single-cell sequencing with linear amplification. *Molecular Cell*, *76*(4), 676–690. e10.

Yin, Y., & Petes, T. D. (2013). Genome-wide high-resolution mapping of UV-induced mitotic recombination events in Saccharomyces cerevisiae. *PLoS Genetics*, *9*(10), e1003894.

Yin, Y., & Petes, T. D. (2014). The role of Exo1p exonuclease in DNA end resection to generate gene conversion tracts in Saccharomyces cerevisiae. *Genetics*, *197*(4), 1097–1109.

Yin, Y., & Petes, T. D. (2015). Recombination between homologous chromosomes induced by unrepaired UV-generated DNA damage requires Mus81p and is suppressed by Mms2p. *PLoS Genetics*, *11*(3), e1005026.

Zheng, D.-Q., Zhang, K., Wu, X.-C., Mieczkowski, P. A., & Petes, T. D. (2016). Global analysis of genomic instability caused by DNA replication stress in Saccharomyces cerevisiae. *Proceedings of the National Academy of Sciences of the United States of America*, *113*(50), E8114–E8121.

CHAPTER TEN

Characterization of DNA-PK-bound end fragments using GLASS-ChIP

Rajashree A. Deshpande and Tanya T. Paull*
The Department of Molecular Biosciences, The University of Texas at Austin, Austin, TX, United States
*Corresponding author: e-mail address: tpaull@utexas.edu

Contents

Abstract

Endonucleolytic cleavage of DNA ends by the human Mre11-Rad50-Nbs1 (MRN) complex occurs in a manner that is promoted by DNA-dependent Protein Kinase (DNA-PK). A method is described to isolate DNA-PK-bound fragments released from chromatin in human cells using a modified Gentle Lysis and Size Selection chromatin immunoprecipitation (GLASS-ChIP) protocol. This method, combined with real-time PCR or next-generation sequencing, can identify sites of MRN endonucleolytic cutting adjacent to DNA-PK binding sites in human cells.

1. Introduction

In eukaryotes, double-strand breaks in DNA are repaired through two pathways, non-homologous end-joining (NHEJ) and homologous recombination (HR), each of which requires a dedicated set of repair factors

Methods in Enzymology, Volume 661
ISSN 0076-6879
https://doi.org/10.1016/bs.mie.2021.08.014

(Pannunzio, Watanabe, & Lieber, 2018; Symington, 2016). The Mre11/Rad50/Nbs1 (MRN) complex and the CtIP protein are central to the HR pathway because of their importance in 5′ resection: the removal of a few hundred base pairs (or more) from the 5′ strand at a double-strand break, a process required for subsequent Rad51-mediated strand invasion into an intact template for replication-based repair (Jasin & Rothstein, 2013; Paull, 2018; Wright, Shah, & Heyer, 2018). In contrast, NHEJ does not rely on a separate template but instead resolves breaks by direct ligation of the broken ends, often with small deletions or occasional insertions at the break points (Pannunzio et al., 2018). The DNA-dependent protein kinase catalytic subunit (DNA-PKcs) is an important regulator of this process along with the Ku70/Ku80 (Ku) heterodimeric complex that recruits the kinase to DNA ends (Blackford & Jackson, 2017). The fate of double-strand breaks with respect to these pathways has often been presented as a competition between the MRN and DNA-PK complexes; however, recent work suggests that the presence of DNA-PK on DNA ends actually promotes the initial endonucleolytic processing of DNA by MRN (Deshpande et al., 2020). This result as well as other observations showing that NHEJ proteins arrive at double-strand break sites earlier than other factors (Jones & Forsburg, 2021; Kim et al., 2005; Kochan et al., 2017; Whelan & Rothenberg, 2021; Wu, Topper, & Wilson, 2008; Yang et al., 2018) and results showing a fast NHEJ phase of repair preceding HR (Lobrich et al., 2010; Mao, Bozzella, Seluanov, & Gorbunova, 2008) suggest that the decision process involves NHEJ factor recruitment followed by HR factors if the NHEJ pathway is unsuccessful.

We previously demonstrated CtIP-dependent MRN endonucleolytic cutting of DNA-PK-bound ends in vitro with purified recombinant components using ensemble and single-molecule approaches (Deshpande et al., 2020). In that work we observed MRN/CtIP-dependent removal of DNA-PK from DNA ends and showed that this requires all components of MRN, CtIP, and DNA-PK (DNA-PKcs and Ku). We also demonstrated that fragments of DNA bound by DNA-PK are generated by MRN in human cells using an inducible ER-*Asi*SI system (Iacovoni et al., 2010) to induce DSBs at specific sites in genomic DNA (Fig. 1). In this DSB inducible via *Asi*SI (DIvA) system, the restriction enzyme AsiSI is fused to the estrogen receptor hormone-binding domain and can be induced to enter the nucleus where it generates DSBs at sequence-specific sites (5′-GCGATCGC-3′) upon 4-hydroxytamoxifen (4-OHT) treatment. There are approximately 1000 AsiSI restriction sites in the human genome

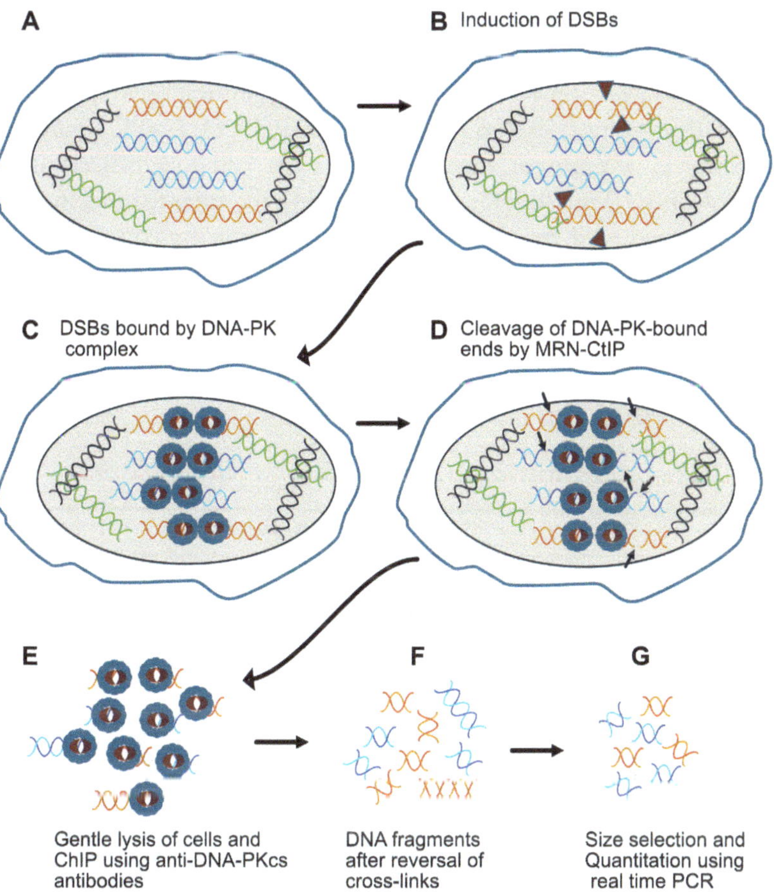

Fig. 1 Overview of GLASS-ChIP method. (A) U2OS cells with HA-AsiSI-ER expression, 10–15 million cells per treatment. (B) AsiSI cuts (DSBs) (brown triangles) are induced by addition of 4OHT which causes nuclear localization of AsiSI. (C) DSBs are bound by the Ku heterodimer (brown) and DNA-PKcs (blue). (D) MRN, in cooperation with CtIP, cleaves the DNA bound by the DNA-PK complex (black arrows). (E, F) GLASS-ChIP is carried out to collect these fragments using anti-DNA-PKcs pS2056 antibodies. (G) The smaller (<300 bp) DNA fragments are size selected and quantified using qPCR.

although approximately 100–200 DSBs are actually generated by the AsiSI enzyme in human cells and a subset of these undergo resection (Vítor, Huertas, Legube, & de Almeida, 2020; Zhou & Paull, 2021). Because DNA-PK is bound to the ends that are cleaved by MRN, it is possible to recover these ends by crosslinking and immunoprecipitation of DNA-PK (Deshpande et al., 2020). In contrast to standard chromatin immunoprecipitation (ChIP), these fragments are released from the chromatin and thus can be recovered from the supernatant rather than the pellet fraction that contains the vast majority of the chromatin. This protocol describes the preparation of cells for this procedure, crosslinking, immunoprecipitation, and size selection of fragments to efficiently purify the DNA-PK bound DNA fragments, in a protocol called Gentle Lysis and Size Selection ChIP (GLASS-ChIP) (Fig. 2). This assay, excluding the time for making appropriate cell lines, cell treatment and harvesting, can be completed within 3–4 days.

2. Key resource table

Item	Source	Identifier
NU7441	Fisher	371210R
4OHT	Sigma	H7904
Formaldehyde (37% stock) (See Note 1)	Fisher	F79-500
Igepal (NP-40)	Sigma	I3021
Sodium deoxycholate	Sigma	D6750
Protease inhibitor tablets	Fisher Scientific	A32955
Anti-DNA-PKcs pS2056 rabbit antibody	Abcam	ab124918
Protein A/G magnetic beads	Fisher	88803
AMPure XP beads	Beckman Coulter	A63880
Nucleotide Removal Kit	Qiagen	28306
PowerUp SYBR green PCR mix	Fisher	A25742

3. Before you begin

1. Resuspend NU7441 at 5 mM in DMSO; store at 4 °C in 1 mL aliquots or at −20 °C for long-term storage
2. Resuspend 4OHT in methanol or ethanol and dilute to 600 μM for a 1000 × stock.
3. Prepare Phosphate-buffered saline (PBS) and cool it to 4 °C
4. 1.25 M glycine (dissolve in water)

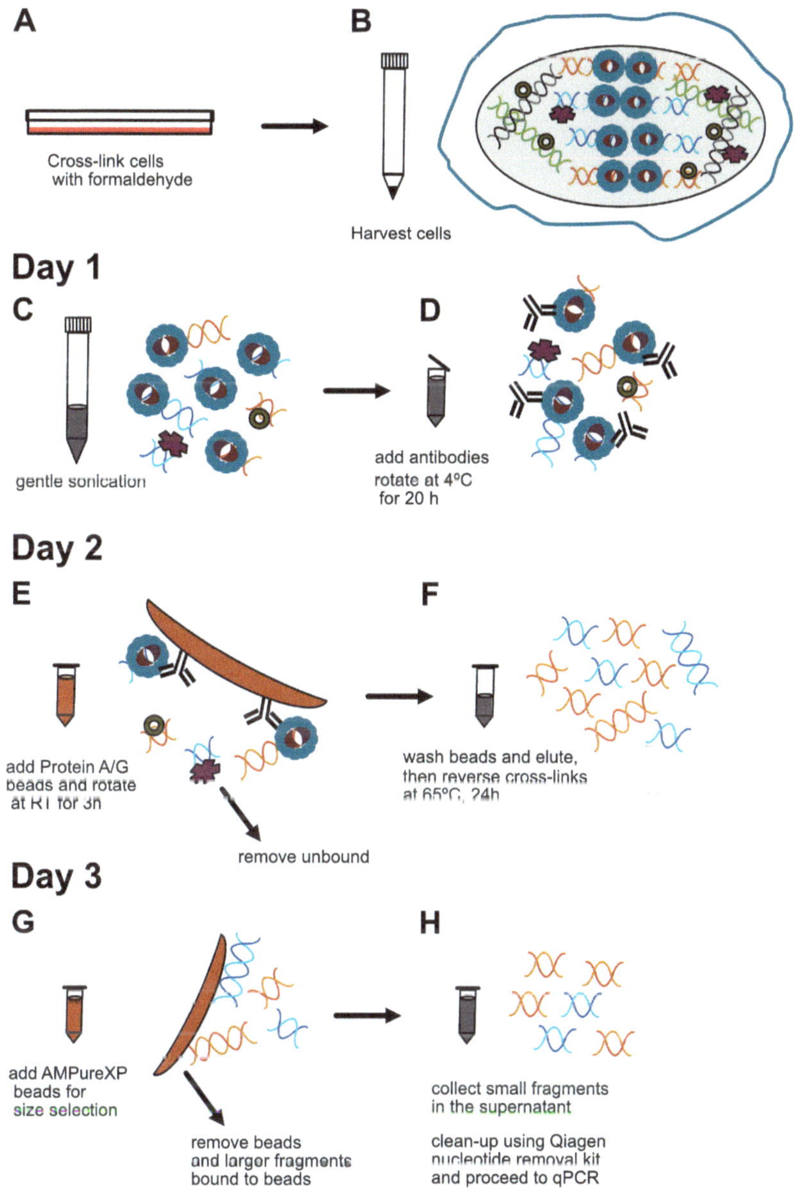

Fig. 2 Workflow of cell harvesting and GLASS-ChIP. (A) Dishes with cells to be used for GLASS-ChIP. Cells are crosslinked with formaldehyde. (B) Cells are rinsed with PBS after crosslinking and harvested in 15 mL conical tubes. At this stage the cells will have proteins and DNA crosslinked as shown on right, DNA-PK complex (blue and brown). (C–H) depict the steps in GLASS-ChIP. (C) On the first day, cell pellets are thawed and resuspended in RIPA buffer and sonicated gently. This will preferably release the

(Continued)

5. RIPA buffer: 50 mM Tris-HCl pH 8.0, 150 mM NaCl, 2 mM EDTA pH 8.0, 1% NP-40, 0.5% Sodium Deoxycholate, 0.1% SDS. Keep buffers cold until use
6. Low salt wash buffer (TSE150): 0.1% SDS, 1% Triton X-100, 2 mM EDTA, 20 mM Tris-HCl pH 8.0, 150 mM NaCl
7. High salt wash buffer (TSE500): 0.1% SDS, 1% Triton X-100, 2 mM EDTA, 20 mM Tris-HCl pH 8.0, 500 mM NaCl
8. LiCl wash buffer: 0.25 M LiCl, 1% NP-40, 1% Sodium Deoxycholate, 1 mM EDTA, 10 mM Tris-HCl pH 8.0
9. TE buffer: 10 mM Tris pH 8.0, 0.1 mM EDTA
10. qPCR primers specific for DSB site of interest (see Note 2)

4. Equipment required

Cell incubator
Centrifuge
Sonicator
Rotator for dishes
Rotator for tubes
30 °C shaker
Magnetic rack
65 °C incubator/oven
qPCR machine

Fig. 2—Cont'd DNA already fragmented. (D) Cell suspensions are transferred to microfuge tubes, centrifuged, and the supernatant used for immunoprecipitation with appropriate antibodies. We have used anti-DNA-PKcs antibodies (black Y shapes). Microfuge tubes are rotated 20 h at 4 °C. (E) On the second day, Protein A/G magnetic beads are added to the tubes and rotated at RT for 3 h. DNA fragments bound by DNA-PKcs complex are bound to the surface of magnetic beads (brown crescent) through the anti-DNA-PKcs antibodies (F) Separate beads using a magnetic rack and aspirate the supernatant thus removing the unbound material. Wash beads and elute the bound proteins. The elutions are then incubated at 65 °C for 24 h to reverse the cross-links. This will result in releasing the proteins cross-linked to DNA. (G) On Day 3, size selection is carried out by adding AMPure XP beads. Pull-down beads and longer DNA fragments bound to beads using a magnetic rack, leaving smaller DNA fragments in the supernatant. (H) The supernatant is collected after size selection and DNA is purified using a Qiagen nucleotide removal kit. This DNA is then quantified using qPCR.

5. Methods

5.1 Cell growth

(1) Grow target cells in appropriate media to 60%–70% confluency and perform any treatment required for the experiment. For example, use U2OS cells containing a virus encoding HA-*Asi*SI-ER to express the restriction enzyme AsiSI (DIvA system) (Iacovoni et al., 2010). The DNA-PKcs kinase inhibitor NU7441 can also be used to inhibit the NHEJ pathway and promote higher efficiency MRN cutting at DNA-PK-bound ends (Deshpande et al., 2020). To treat the cells with DNA-PKcs inhibitor, add NU7441 to 10 μM final concentration in a 150 mm dish in 15 mL, then incubate at 37 °C. After 1 h, add 4OHT (600 nM final concentration) and incubate at 37 °C for an additional 4 h.

5.2 Harvesting cells

Timing: 1–2 h

(1) Place dishes on a horizontal rotator (90–110 rpm) at RT. Add formaldehyde (37% stock) dropwise to 1% final concentration (400 μL for 15 mL media in a 150 mm dish) while shaking, continue incubation for 7 min.

(2) Add 1.25 M glycine to 0.139 M final concentration (1.7 mL for 15 mL media in a 150 mm dish) dropwise while shaking. Rotate for 5 min at RT.

(3) Remove the media, add 10 mL of cold PBS. Rinse the cells and discard the PBS. Add another 10 mL PBS. Leave cells in PBS until harvesting.

(4) To harvest cells, aspirate PBS and add 5 mL of fresh cold PBS. Scrape the cells from the plate using a cell scraper and move to a 15 mL conical tube on ice. Add another 5 mL PBS to the dish and scrape cells again to collect all cells. Combine with 5 mL in the tube. Keep the tubes cold during harvesting of all dishes.

(5) Centrifuge the harvested cells at 1865 × *g* for 15 min at 4 °C. Aspirate the supernatant as much possible without losing cells. Freeze pellets in liquid nitrogen and store at −70 °C. Cells from one 150 mm dish constitute one pellet in our experiment.

5.3 GLASS-ChIP

Timing: 3 days

Day 1

Timing: 2–3 h

1. Prepare RIPA buffer with protease inhibitors sufficient for the samples. Use one tablet of protease inhibitors for 10 mL of cold RIPA buffer. Dissolve the inhibitors completely by brief vortexing, crush the tablet for easier dissolving and keep buffers at 4 °C
2. Thaw cell pellets at RT for 5 min, vortex gently for 2–3 s, then transfer to ice bucket once thawed
3. Add 2.3 mL RIPA buffer with protease inhibitors to the pellet. Mix with pipetting to resuspend the pellet completely; keep on ice for 15 min
4. Keeping the tube on ice, sonicate at low power (Qsonica 55 with 1/8 in. microprobe at 15 setting) for 10 s, wait for 20 s and then sonicate with 10–12 pulses of 1 s each. (Note: This sonication primarily serves to dissociate cells and is mild compared to the standard ChIP sonication)
5. Transfer the lysate to two 1.5 mL microfuge tubes
6. Centrifuge the tubes at 845 × *g* (3000 RPM in microfuge) for 3 min. There will be a pellet at the bottom
7. Label 1.5 mL microfuge tubes, one each for input, minus antibody and plus antibody pull-down. Transfer 1 mL each of the supernatant to the tubes labeled minus and plus antibody. Transfer 50 μL of the supernatant to the tube labeled as input. Keep all tubes in an ice bucket during these transfers. Optional: analyze pellet fraction similar to standard ChIP-seq protocols to analyze protein bound in the chromatin fraction
8. Add 4 μL (1.6 μg) anti-DNA-PKcs-pS2056 antibody (Abcam, ab124918) to the plus antibody sample. The minus antibody tube serves as a control; mock antibody can be added to this sample
9. Incubate all (input, minus and plus antibody) tubes at 4 °C with rotation for 20 h (overnight).

Day 2

Timing: 5–7 h (excluding time required for reversal of cross-links)

1. Spin tubes briefly at 845 × *g* for 5 s. Keep the input tubes separately at 4 °C
2. Mix Protein A/G magnetic beads (Pierce) thoroughly and add 25 μL beads to the tubes labeled minus and plus antibody. Incubate these samples at RT for 3 h, with rotation
3. Keep the buffers to be used for washing beads in the following steps at RT for warming. The SDS in elution buffer precipitates in cold

temperatures, keep the elution buffer in 37 °C incubator till the precipitate dissolves completely, then keep it at RT

4. Centrifuge the tubes briefly at 845 × *g* for 5 s and then isolate the beads using a magnetic rack
5. Keeping the tubes on the magnetic rack, transfer supernatant to a new microfuge tube. Reserve for optional later use
6. Add 1 mL of TSE150 buffer to the tubes with magnetic beads. Mix the beads by tapping the tubes and rotate at RT for 15–20 min
7. Transfer the tubes to the magnetic rack. Aspirate the supernatant and add 1 mL of buffer TSE500. Mix the beads by tapping the tubes and rotate at RT for 15–20 min
8. Transfer the tubes to the magnetic rack. Aspirate the supernatant and add 1 mL of LiCl buffer. Mix the beads by tapping the tubes and rotate at RT for 15–20 min
9. Transfer the tubes to the magnetic rack. Aspirate the supernatant and add 1 mL of TE buffer. Mix the beads by pipetting and transfer to a new microfuge tube and rotate at RT for 15–20 min
10. Transfer the tubes to the magnetic rack. Aspirate the supernatant and add 0.1 mL of elution buffer. Mix the beads by pipetting and incubate at 30 °C for 30 min with shaking at 200 rpm (see Note 3)
11. Transfer the tubes to the magnetic rack. Aspirate the supernatant and transfer to a new tube (see Note 4)
12. Incubate the elutions and input tubes at 65 °C for 24 h (not less) to reverse the cross-links

5.4 Size selection

Timing: 2–3 h

Day 3

1. Cool down the sample tubes to room temperature for 30 min. At the same time, warm up AMPure XP beads to RT for 30 min. Keep the input tubes aside
2. Mix the AMPure XP beads thoroughly and add 65 μL beads to the minus antibody and plus antibody elutions. Mix well by pipetting up and down 10–15 times
3. Incubate at RT for exactly 10 min. Separate beads using a magnetic rack
4. Transfer the supernatant into a fresh PCR or microfuge tube (see Note 4)

5. Add 25 μL AMPure XP beads to this supernatant, mix well by pipetting up-down 10–15 times. Incubate at RT for exactly 10 min
6. Transfer the supernatant to a fresh 1.5 vial. This is the size selected sample where larger fragments are removed for our experiment (see Note 5)
7. Purify the size selected supernatant as well as the input using Qiagen Nucleotide removal kit. Elute in 60 μL elution buffer (Qiagen).

5.5 Quantitation of GLASS-ChIP DNA by qPCR

Timing: 1–2 h (setting up the qPCR plate)

1. Use 3 μL input or size selected minus antibody or plus antibody samples as template in a 20 μL qPCR reaction containing 10 μL 2× PowerUp SYBR green PCR Master Mix and 0.5 μM of each primer. Each 96 well contains: (a) 0.4 μL of 25 μM forward primer (b) 0.4 μL of 25 μM reverse primer (c) 10 μL of 2× PowerUp SYBR green PCR Master Mix (ABI) (e) 6.2 μL ddH_2O (f) 3 μL sample. Use PCR cycling as recommended by Thermo or other master mix suppliers. See Table 1 for examples of primer sets to use for *Asi*SI sites.
2. Calculate the percentage of DNA at selected sites. For each sample, a ΔCt is calculated by subtracting the Ct value of the sample from the Ct value of the input. The percentage of DNA in the pull-down is calculated with the following equation: $2^{(Ct(input)-Ct(test))} \times 100\%$. To compensate for the difference in volume used for input (50 μL) and pull-down (1000 μL), the values are divided by 20. DNA values obtained for minus antibody samples are subtracted from the values obtained in presence of antibody.

5.6 Expected outcome

When using the DIvA system (Iacovoni et al., 2010), DNA fragments near the *Asi*SI site can be identified in DNA-PK GLASS-ChIP products when measured by qPCR (Fig. 3). Recovery of these DNA fragments is dependent on 4-OHT exposure and is promoted by NU7441, the DNA-PKcs catalytic inhibitor. Recovery of fragments close to the DSB site (40–50 bp) is significantly more efficient compared to recovery of fragments farther from the DSB site (300 bp).

Table 1 Primers used for qPCR in *AsiS*I-expressing U2OS cells (Deshpande et al., 2020).

AsiSI site	Location as in hg18	Primer pair	Forward primer	Reverse primer	Amplicon size (bp)
DSB1	chr18:7556705	U1	TCGGGGCCAGCGGCGTGTA	CGCCAGCCCGCTCCC	52
		D1	CGCGGGGCTCGGCGC	GGGAGATGGCGCGGGAGC	40
		U2	GTGCTGGCTCAATGTGCTTATT	ACGATTTTGGGTCTGAGTGAA	132
		D2	CGCAGCCTCTTCCACAGTCA	GCCACTACCGCCGCCGAA	139
DSB2	chr21: 32167332	U1	GGGAGCGGCCGCCCAG	GCTCCTAGCCGTGCGCT	40
		D1	CGGGAGCCCGACCCAA	CGCCGTCTGGCCCGCA	40
		U2	CGAAAGGTCCAGAAAACCCAA	GAAGCCACCTGAGCGCCAGA	132
		D2	TTGTCTACGCGCCTCGCT	CGGCTTCCCCGGCTTCT	119
DSB3	chr9:129732985	U1	GACTGCGGCTGCATCCAA	CGCCAGCGCCTCCCGC	41
		D1	CGCCTGCGGGTCCCGC	CTGAAGGATGCTGCAGCCGT	40
		U2	CCGCACTGGATGAGAGCTT	CCTGGCGGATATCCCTCAA	112
		D2	GGACATCCATTCATTGAACACA	GATCACGCGGGCAGCTGA	113
DSB4	chr22:37194040	U1	CCCGGCCAAGAGTGCGT	CGCACCCGCGCGCCG	40
		D1	CGCGGAGCTGTGAGGC	GTCTCTAGGTGCCCCAGA	45
		U2	AAGATGAGGACAATAGCAGGAA	AAGCCCCAATCTCTGCCTCA	125
		D2	CAGGGCGCTCCAGGTGT	GGGTCCTCCTCCTCTGAT	118

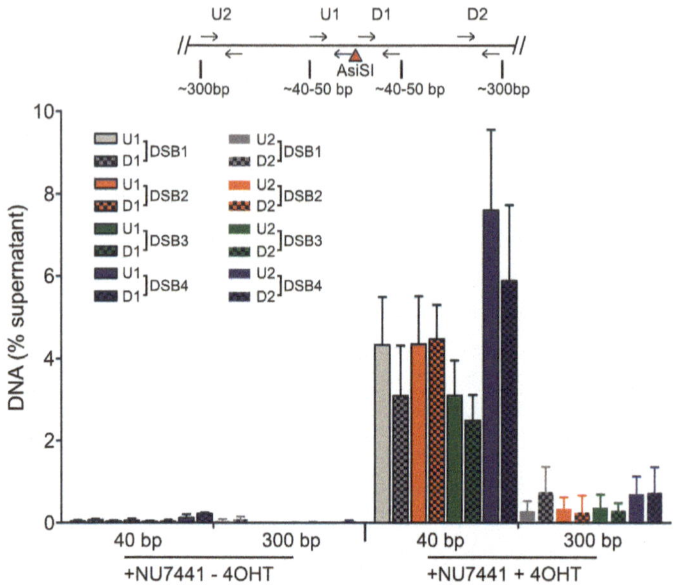

Fig. 3 Example of the expected outcome from GLASS-ChIP qPCR monitoring four *AsiS*I sites. Small dsDNA products resulting from nucleolytic cleavage of DNA-PK bound AsiSI-generated DNA ends were isolated using the GLASS-ChIP protocol from U2OS cells treated with 4-OHT or vehicle for 4 h as indicated. These cells were also treated with 10 μM NU7441, the DNA-PKcs inhibitor, as indicated for 5 h starting at 1 h prior to 4-OHT addition. qPCR was performed to quantify the isolated DNA by using primers located ~40 to 50 bp (primer sets U1 and D1) or ~300 bp (primer sets U2 and D2) from the AsiSI cut site. Primer sets U1 and U2 (solid) are upstream whereas D1 and D2 (checkered) are downstream of the AsiSI cut sites. At all four sites, DNA close to the AsiSI cut was detected with U1 and D1 primer sets upon induction of DSB with 4-OHT addition. In comparison, the yield of DNA with U2 and D2 primer sets ~300 bp away from the AsiSI cut was low.

6. Notes

Note 1: Sometimes formaldehyde forms crystals at bottom so it is best to use a fresh stock of formaldehyde or make sure the stock is clear.

Note 2: qPCR primers: Use previously published primer sets close to AsiSI DSBs (Table 1) (Deshpande et al., 2020) or any primer sets within approximately 100 bp of DSB location.

Note 3: Since the elution volume is low, a shaker with horizontal circular motion was used. Tubes were kept at a 45° angle.

Note 4: Optional: use DNA lo-bind tubes (Eppendorf) for elution.

Note 5: The amounts of AMPure beads used will remove DNA larger than 300 bp in size.

References

Blackford, A. N., & Jackson, S. P. (2017). ATM, ATR, and DNA-PK: The trinity at the heart of the DNA damage response. *Molecular Cell*, *66*, 801–817.

Deshpande, R. A., Myler, L. R., Soniat, M. M., Makharashvili, N., Lee, L., Lees-Miller, S. P., et al. (2020). DNA-dependent protein kinase promotes DNA end processing by MRN and CtIP. *Science Advances*, *6*, eaay0922.

Iacovoni, J. S., Caron, P., Lassadi, I., Nicolas, E., Massip, L., Trouche, D., et al. (2010). High-resolution profiling of gammaH2AX around DNA double strand breaks in the mammalian genome. *The EMBO Journal*, *29*, 1446–1457.

Jasin, M., & Rothstein, R. (2013). Repair of strand breaks by homologous recombination. *Cold Spring Harbor Perspectives in Biology*, *5*, a012740.

Jones, C. E., & Forsburg, S. L. (2021). Monitoring *Schizosaccharomyces pombe* genome stress by visualizing end-binding protein Ku. *Biology Open*, *10*. bio054346.

Kim, J.-S., Krasieva, T. B., Kurumizaka, H., Chen, D. J., Taylor, A. M. R., & Yokomori, K. (2005). Independent and sequential recruitment of NHEJ and HR factors to DNA damage sites in mammalian cells. *The Journal of Cell Biology*, *170*, 341–347.

Kochan, J. A., Desclos, E. C. B., Bosch, R., Meister, L., Vriend, L. E. M., van Attikum, H., et al. (2017). Meta-analysis of DNA double-strand break response kinetics. *Nucleic Acids Research*, *45*, 12625–12637.

Lobrich, M., Shibata, A., Beucher, A., Fisher, A., Ensminger, M., Goodarzi, A. A., et al. (2010). gammaH2AX foci analysis for monitoring DNA double-strand break repair: Strengths, limitations and optimization. *Cell Cycle*, *9*, 662–669.

Mao, Z., Bozzella, M., Seluanov, A., & Gorbunova, V. (2008). Comparison of non-homologous end joining and homologous recombination in human cells. *DNA Repair*, *7*, 1765–1771.

Pannunzio, N. R., Watanabe, G., & Lieber, M. R. (2018). Nonhomologous DNA end-joining for repair of DNA double-strand breaks. *The Journal of Biological Chemistry*, *293*, 10512–10523.

Paull, T. T. (2018). 20 years of Mre11 biology: No end in sight. *Molecular Cell*, *71*, 419–427.

Symington, L. S. (2016). Mechanism and regulation of DNA end resection in eukaryotes. *Critical Reviews in Biochemistry and Molecular Biology*, *51*, 195–212.

Vítor, A. C., Huertas, P., Legube, G., & de Almeida, S. F. (2020). Studying DNA double-strand break repair: An ever-growing toolbox. *Frontiers in Molecular Biosciences*, *7*, 24.

Whelan, D. R., & Rothenberg, E. (2021). Super-resolution mapping of cellular double-strand break resection complexes during homologous recombination. *Proceedings of the National Academy of Sciences*, *118*, e2021963118.

Wright, W. D., Shah, S. S., & Heyer, W.-D. (2018). Homologous recombination and the repair of DNA double-strand breaks. *The Journal of Biological Chemistry*, *293*, 10524–10535.

Wu, D., Topper, L. M., & Wilson, T. E. (2008). Recruitment and dissociation of non-homologous end joining proteins at a DNA double-strand break in Saccharomyces cerevisiae. *Genetics*, *178*, 1237–1249.

Yang, G., Liu, C., Chen, S.-H., Kassab, M. A., Hoff, J. D., Walter, N. G., et al. (2018). Super-resolution imaging identifies PARP1 and the Ku complex acting as DNA double-strand break sensors. *Nucleic Acids Research*, *46*, 3446–3457.

Zhou, Y., & Paull, T. T. (2021). Quantifying DNA end resection in human cells. *Methods in Molecular Biology (Clifton, N.J.)*, *2153*, 59–69.

CHAPTER ELEVEN

Light activation and deactivation of Cas9 for DNA repair studies

Roger S. Zou[a] **and Taekjip Ha**[a,b,c,d,*]
[a]Department of Biomedical Engineering, Johns Hopkins University, Baltimore, MA, United States
[b]Department of Biophysics and Biophysical Chemistry, Johns Hopkins University, Baltimore, MA, United States
[c]Department of Biophysics, Johns Hopkins University, Baltimore, MA, United States
[d]Howard Hughes Medical Institute, Baltimore, MA, United States
*Corresponding author: e-mail address: tjha@jhu.edu

Contents

Abstract

DNA double-strand breaks in DNA (DSBs) are common yet highly detrimental events in living organisms. To repair the damage, each cell requires a coordinated set of DNA damage response (DDR) proteins that can respond quickly, effectively, and precisely. Better understanding of these processes is therefore essential and would require an effective means of inducing targeted DSBs on demand, but previous methods are hampered by limited control over genomic location, timing, or lesion types. Tight spatiotemporal control of CRISPR-Cas9 activity has potential to overcome these limitations, which led to the development of two methods for rapid activation or deactivation of Cas9

Methods in Enzymology, Volume 661
ISSN 0076-6879
https://doi.org/10.1016/bs.mie.2021.08.007

using light. In this chapter, we discuss how control of Cas9 can advance DDR studies, describe protocols to control Cas9 activation and deactivation using this new technology, and finally outline three compatible readouts of DNA damage and the cellular response: DSB levels using droplet digital PCR, repair factor localization using ChIP-seq, and insertion-deletion (indel) repair outcomes using Sanger sequencing.

1. Introduction

1.1 DNA double-strand breaks (DSBs) and their repair in biology and medicine

Double-strand breaks (DSBs) to genomic DNA are highly deleterious events that can compromise genetic integrity. Each cell must have robust biochemical machinery involving a plethora of DNA damage response (DDR) proteins to faithfully repair the lesion. Inherited mutations in even a single DDR protein can lead to severe phenotypes and increased risk for cancer (Her & Bunting, 2018; Lieber, 2010). Intentional DNA damage is also used therapeutically; many cancer therapies, from chemotherapy to radiotherapy, rely on selective killing of cancer cells by inducing toxic DSBs (Cheung-Ong, Giaever, & Nislow, 2013). Many cancers also arise from somatically defective DDR pathways, such that pharmacologic inhibition of additional pathways leads to selective cancer cell death, a phenomenon termed synthetic lethality (Lord, Tutt, & Ashworth, 2015). Finally, genome editing with systems such as CRISPR-Cas9 involves intentionally generating targeted DSBs and relying on endogenous DDR pathways to catalyze the desired genomic modifications (Pickar-Oliver & Gersbach, 2019).

The cellular pathways for recognition and repair of DSBs are complex, comprised over 100 proteins operating in concert to maintain genomic integrity. Initial break recognition is believed to be primarily mediated by Ku70/80 heterodimers, which eventually leads to repair by classical non-homologous end joining (c-NHEJ) (Aleksandrov et al., 2018; Bunting et al., 2010; Caron et al., 2015; Gupta et al., 2018; Natale et al., 2017; Noordermeer et al., 2018; Ochs et al., 2019; Yang et al., 2018; Zhou et al., 2017). Ku70/80 binding competes with the MRN (MRE11-Rad50-Nbs1) complex, which leads to DNA resection that promotes repair by homologous recombination (HR) (Aymard et al., 2017; Pickar-Oliver & Gersbach, 2019; Scully, Panday, Elango, & Willis, 2019). MRE11-enabled resection is also essential for an alternate pathway termed microhomology mediated end joining (MMEJ or alt-NHEJ). Many large-scale changes also occur during DSB repair (Clouaire & Legube, 2019). Chromatin is decompacted (Gursoy-Yuzugullu, Ayrapetov, & Price, 2015; Hauer & Gasser, 2017; Price & D'Andrea, 2013; Xu et al., 2012) and the nucleosome variant H2AX is phosphorylated (γH2AX) megabases around the cut site (Aymard et al., 2017; Clouaire et al., 2018; Turinetto & Giachino, 2015). Transcription of coding genes by RNA polymerase II (RNAPII) is also repressed if the DSB is in coding regions (Shanbhag, Rafalska-Metcalf, Balane-Bolivar, Janicki, & Greenberg, 2010; Vitor et al., 2019), but transcription of damage-induced long non-coding RNAs (dilncRNAs) are initiated from the cut site, synergizing with 53BP1 to form a large focus that stabilizes the region undergoing active repair (Michelini et al., 2017; Noordermeer et al., 2018; Ochs et al., 2019; Pessina et al., 2019).

Given its complexity and importance to biological systems, study of DNA damage and repair is therefore crucial to fundamental understanding of this essential biological process, to medicine due to diseases involving defective DNA repair processes, and to biotechnology as an integral component of genome editing and other emerging applications.

1.2 Advantages and disadvantages of previous methods for inducing DSBs

The presence of multiple DDR pathways and network of interacting proteins have made studies of DSB repair extremely challenging. Previous studies often rely on measurements of unsynchronized repair progression or at unknown genomic locations (Clouaire & Legube, 2019). Given that repair proteins are recruited within seconds and the process can finish within an hour, capturing both temporal and sequence information in a single experiment is desired but challenging. The challenges stem from the lack of a technique for creating sequence specific DSBs at high spatiotemporal resolutions (Clouaire & Legube, 2019). The common laser microirradiation method has precise temporal control but generates complex DNA lesions and suffers from lack of sequence specificity (Aleksandrov et al., 2018; Blackford & Jackson, 2017). Chemical mutagens also have good temporal control but lack sequence specificity (Blackford & Jackson, 2017). Endonucleases such as meganucleases, transcription activator-like effector nucleases (TALENs), zinc finger nucleases (ZFNs), or CRISPR-Cas9 exhibit sequence specificity but lack desired temporal control (Pickar-Oliver & Gersbach, 2019). An ideal system would combine sequence specificity with precise temporal control to generate synchronized DSBs that enable multiple methods for readout of the DSB response. We hypothesized that such a system can be developed by engineering ways to effectively control the activation and deactivation of CRISPR-Cas9 on demand.

1.3 Very fast activation and deactivation of Cas9 using light

To engineer rapid control of Cas9 activity, we leveraged mechanistic understandings of Cas9 binding and cleavage. Cas9 first interrogates potential binding sequences by evaluating its protein binding affinity to the trinucleotide protospacer adjacent motif (PAM) (Jinek et al., 2012). After PAM recognition, Cas9 begins unwinding double-stranded DNA to allow guide RNA (gRNA)-to-DNA heteroduplex formation, starting from PAM (PAM-proximal) and proceeding distally (PAM-distal) (Sternberg, LaFrance, Kaplan, & Doudna, 2015). Only 12 complementary base pairs at the PAM-proximal side are needed for stable Cas9 binding. In contrast, gRNA-to-DNA heteroduplex formation up to 20 base pairs is needed to convert Cas9 to a cleavage-competent conformation (Chen et al., 2017; Singh, Sternberg, Fei, Doudna, & Ha, 2016).

We hypothesized that introducing a few bulky photocaged oligonucleotides (6-nitropiperonyloxymethyl deoxythymidine, or NPOM-dT) to the PAM-distal section of gRNA would still allow heteroduplex formation at the PAM-proximal side and facilitate stable Cas9 binding, but block cleavage until the bulky light-sensitive moieties are removed with light exposure. Cas9 with this special caged gRNA (cgRNA) would natively bind to the target site, but not cleave until exposed to light. This forms the basis for a new method for Cas9 activation termed very fast CRISPR (vfCRISPR) (Liu et al., 2020). Utilizing these same properties, we hypothesized that having an unobtrusive photocleavable moiety (Photocleavable Linker, or PC-Linker) in the PAM-distal region of gRNA would retain Cas9 cleavage capability, but exposure to light would abolish cleavage by separating away the last few nucleotides of the gRNA. Cas9 with this special photocleavable gRNA (pcRNA) would natively be cleavage-competent and deactivate upon light exposure (Zou, Liu, Wu, & Ha, 2021). We demonstrated that these developments enabled the fastest and most complete activation and deactivation functionality for CRISPR-Cas9 systems, with activation to full Cas9 activity or deactivation to no Cas9 activity, all within seconds after light exposure (Liu et al., 2020; Zou et al., 2021). We demonstrated diverse application of these technologies to improve the specificity of genome editing, characterize the kinetics of Cas9 DNA damage and repair, and further understand the cellular response to Cas9-mediated DNA damage.

1.4 Readouts to characterize DSB repair

After control of Cas9 activation or deactivation, convenient readouts are needed to characterize Cas9-mediated DNA damage and its induced cellular responses as a function of time. First, Cas9 induces DSBs at its intended target position, so measurement of DSB levels at that specific position would directly reflect the cleavage activity of Cas9. We adapted a recently developed method using droplet digital polymerase chain reaction (PCR) called DSB-ddPCR (Rose et al., 2017) to measure the amount of DSBs at the Cas9 cut site as a function of time after Cas9 activation or deactivation. After DNA damage is induced using Cas9, the cell recruits DNA repair factors around the break site. To measure these binding dynamics as a function of time, we adapted chromatin immunoprecipitation with sequencing (ChIP-seq), which measures the level of binding by a protein of interest, in this case, DNA repair factors, to DNA genome-wide. After the DNA repair factors guide the progression and completion of repair, artifacts of imperfect repair at the cut site may appear as short insertion or deletion mutations (indels). To measure mutagenesis as a function of time, we adapted PCR followed by Sanger sequencing to estimate the percentage of these mutations (Brinkman, Chen, Amendola, & van Steensel, 2014).

1.5 Overview of this chapter

In summary, in this chapter we present protocols for use of Cas9 activation or deactivation to study DNA damage and repair. First, we detail the protocol for delivering Cas9 protein in complex with either photo-activatable or photo-deactivatable Cas9 into cells using electroporation. This is followed by protocols for light exposure to activate or deactivate Cas9. Finally, we outline protocols for three readouts of DNA damage and its response: (1) DSB-ddPCR to measure the proportion of transient DSBs in a cell population at a given time (Rose et al., 2017), (2) Sanger sequencing to measure repair outcomes as insertion-deletion mutations (indels) (Brinkman et al., 2014), and (3) chromatin immunoprecipitation (ChIP) (Park, 2009) to measure localization of specific repair factors to sites of genomic DNA damage.

2. Methods

2.1 Overview of setup and required reagents

The protocols require a basic laboratory facility for mammalian cell culture and molecular biology. The cell culture facility should contain equipment such as laminar flow cell culture hoods, light microscopes, and cell incubators. A method for Cas9 delivery should also be available; our protocol outlines nucleofection using the Lonza electroporator system, which allows efficient delivery to numerous cell types including primary cells. First, we deliver Cas9 with either the (light-activatable) cgRNA or (light-deactivatable) pcRNA into HEK293T cells. For Cas9/cgRNA, we wait 12h to allow stable binding to the target sequence without cleavage, followed by light-based activation and time-resolved sample collection. For Cas9/pcRNA, we wait 2h to allow cleavage of DNA targets, followed by light-based deactivation and time-resolved sample collection. Afterwards, we outline assays to measure DSB levels, DNA damage response, and repair outcomes across all time points. For the purposes of this chapter, all protocols will assume Cas9 targeting of a sequence in *ACTB* in HEK293T cells.

2.2 Delivery of Cas9 ribonucleoprotein (RNP) complex into cells

First, Cas9 in complex with gRNA is delivered into HEK293T cells using electroporation with the Lonza 4D nucleofector system. We have also successfully performed Cas9 electroporation into two other cell types, K562 cells and WTC-11 iPSCs, with minor modifications to the protocol. The Cas9 delivery protocol is identical between standard crRNA, (light-activatable) cgRNA, or (light-deactivatable) pcRNAs, so for the remainder of this protocol, the word crRNA is interchangeable with cgRNA or pcRNA. We will electroporate Cas9 into 24 million HEK293T cells using two cuvettes from the SF Cell Line 4D-Nucleofector™ X Kit L (Lonza).

2.2.1 Reagents

- SF Cell Line 4D-Nucleofector™ X Kit L (Lonza V4XC-2012)
- Cas9 (10mg/mL) (Integrated DNA Technologies)
- crRNA, tracrRNA (Integrated DNA Technologies), cgRNA (Bio-Synthesis, Inc.), pcRNA (Integrated DNA Technologies) (sequences in Supplementary Table S1 in the online version at https://doi.org/10.1016/bs.mie.2021.08.007)
- Alt-R® Cas9 Electroporation Enhancer, 10nmol (Integrated DNA Technologies 1075916)
- HEK293T cells (ATCC)
- Trypsin-EDTA (0.05%), phenol red (Thermo Fisher 25300062)
- Corning™ DMEM with L-Glutamine, Glucose, Sodium Pyruvate (Fisher Scientific MT10013CV)
- Corning™ Regular Fetal Bovine Serum (Fisher Scientific MT35011CV)

- Corning™ Cell Culture Phosphate Buffered Saline (1 ×) (Fisher Scientific MT21040CV)
- Penicillin-Streptomycin (Thermo Fisher 15070063)
- Nalgene™ Rapid-Flow™ Sterile Single Use Vacuum Filter Units (Fisher Scientific 09-741-02)
- Collagen I, rat tail (Thermo Fisher A1048301)
- 12-well tissue culture plates (Corning)

2.2.2 Specialized equipment

- 4D-Nucleofector™ Core Unit and 4D-Nucleofector™ X Unit (Lonza AAF-1002B & AAF-1002 ×)
- Hemocytometer

2.2.3 Protocol

1. Before starting the experiment, prepare Dialysis buffer (20 mM HEPES pH 7.5, 500 mM KCl, 20% glycerol) that we use to dilute Cas9.
2. Before starting the experiment, prepare over 24 million HEK293T cells for electroporation. We grow HEK293T cells in "DMEM complete" media, composed of 450 mL of DMEM mixed with 50 mL fetal bovine serum and 5 mL of penicillin-streptomycin that is sterile filtered (using Nalgene™ Rapid-Flow™ Sterile Single Use Vacuum Filter Units).
3. The crRNA and tracrRNA are first annealed in a PCR tube. Because cgRNAs and pcRNAs are light-sensitive, make sure to avoid direct sunlight and cover with aluminum foil whenever possible. 4 μL of crRNA is mixed with 4 μL of tracrRNA to form 8 μL of 50 μM cr/trRNA. This PCR tube is placed in a thermocycler set to a constant 95 °C for 3 min, followed by removal and cooling on a benchtop (at room temperature) for 5 min
4. To form the full Cas9-cr/tracrRNA ribonucleoprotein (RNP) complex, first mix the 8 μL cr/tracrRNA solution with 6 μL of 10 mg/mL (~60 μM) Cas9. Then, mix in 16 μL of Dialysis Buffer to form a 30 μL "RNP mixture."
5. To make the "Lonza SF mixture" in a separate tube, mix 147.6 μL of SF nucleofection reagent with 32.4 μL of Supplement 1 to form 180 μL of mixed nucleofection reagent. Both the SF nucleofection reagent and Supplement 1 are supplied with the SF Cell Line 4D-Nucleofector™ X Kit L.
6. Proceed to prepare cells for electroporation, leaving the RNP complex covered with foil at benchtop (room temperature).

7. At the tissue culture room, take out four 12-well plates (A, B, C, D), oriented horizontally. First, prepare the wells intended to hold cells by coating with collagen. This step is critical to promote cell adherence and health after electroporation. For 12-well plates, we add 500 μL of 1:100 collagen in PBS into each of the middle 6 wells (middle two columns). Place in the 37 °C CO_2 incubator.
8. Take out 24+ million HEK293T cells in culture, aspirate the cell culture media, then gently rinse cells with prewarmed PBS. Then, add enough pre-warmed 0.05% Trypsin-EDTA into the culture flask to completely cover the cells, then incubate for 5 min in the incubator. Waiting the full 5 min is important to promote better dissociation of the cells from the culture plate. Afterwards, add equal volumes of "DMEM complete" to inactivate the trypsin, then gently mix and collect the cell suspension into a 15 mL or 50 mL centrifuge tube.
9. Count the cells using a hemocytometer. Add two volumes of cell suspension that each contain 12 million cells into separate 15 mL tubes, and centrifuge at 200 × *g* for 3 min to pellet the cells. Then, carefully aspirate the supernatant and resuspend in 2 mL PBS to wash the cells. Pellet the cells again with 200 × *g* for 3 min in the centrifuge, then remove as much supernatant as possible using a pipette without disturbing the pellet.
10. We will now mix in the electroporation solution and Cas9 RNP. First, add all 180 μL of the "Lonza SF mixture" into the cell pellet of one 15 mL tube and mix well. Transfer all to other 15 mL tube and mix well. Now, all 24 million cells are suspended in 180 μL of "Lonza SF mixture," resulting in an approximately 215 μL solution. Make sure that the cell culture hood lights are turned OFF. Move the 30 μL "RNP mixture" into the culture hood and remove the aluminum foil covering. Transfer all cells (215 μL solution) into the 30 μL RNP mixture. Add 4 μL Cas9 EP enhancer. Mix by pipetting up and down 10 times, then split the total volume equally into two 100 μL Lonza electroporation cuvettes.
11. Before going to the 4D-Nucleofector for electroporation, be sure to take out the four 12-well plates in the 37 °C CO_2 incubator. Completely remove the collagen solution and let it dry inside the cell culture hood. At the 4D-Nucleofector, enter "Cell Line SF" and pulse code CA-189 (Fig. 1A) into the machine's graphical user interface (GUI). Then, accept the settings by pressing "OK" on the GUI. This will open the machine door to allow placement of the cuvettes.

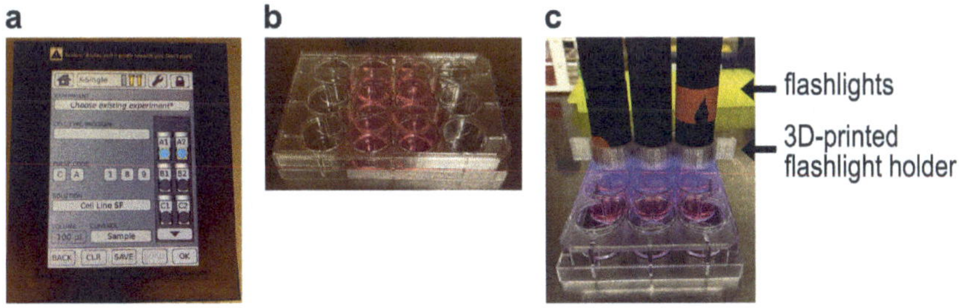

Fig. 1 Electroporation, cell plating, and light activation setup. (A) Screen with graphical user interface (GUI) on the Lonza 4D-Nucleofector. The code (CA-189) and the solution (Cell Line SF) are entered. Two wells are selected for electroporation (represented in the GUI as two circles filled with blue color and labeled A1 and A2). (B) The 12-well culture plate after transfer of electroporated cells in media to the middle six wells. (C) Process of light exposure onto HEK293T cells in a 12-well plate for Cas9 activation. A 3D-printed flashlight holder holds multiple LED flashlights to enable simultaneous illumination of multiple wells.

After placing the cuvettes into the available holder, press "START" on the GUI to begin the electroporation protocol, which only takes a few seconds. Afterwards, the machine door automatically opens to allow retrieval of the cuvettes. There may be some white precipitate visible after electroporation, but it does not appear to impact the electroporation efficiency or cell health in any way.

12. After electroporation, go back to the culture hood with lights still OFF. Add 200 μL of "DMEM complete" into each cuvette and transfer all the cells from the two cuvettes into 24 mL of "DMEM complete" in a 50 mL falcon tube. Therefore, the cell concentration is 1 million cells per mL. Aliquot 1 mL of electroporated cells in "DMEM complete" to each well of 12-well plates with collagen coating (24 wells total) (Fig. 1B). Cover plates with aluminum foil then transfer to the 37 °C CO_2 incubator.
13. For experiments with light-deactivating pcRNA, we complete this electroporation protocol in the morning (around 9 am), followed by light-exposure 2 h later. For experiments with light-activating cgRNA, we complete the electroporation in the evening (around 8 pm), followed by light-exposure approximately 12 h later the next morning at 8 am.

2.3 Light exposure for Cas9 activation with caged guide RNAs (cgRNAs)

Cas9 in complex cgRNAs are naturally cleavage-deficient because caging molecules in the PAM-distal region of gRNA allow stable binding but not cleavage. After exposure to 365 nm wavelength light, the caging molecules are liberated and allow Cas9 activity. Experimentally, we wait approximately 12 h for Cas9 inside cells to approach steady-state binding with its genomic target. Next, we expose cells to 1 min of light using a commercially available LED flashlight rated for 365 nm. This synchronizes the activation of Cas9 inside cells to facilitate time-resolved measurements of response phenotypes.

2.3.1 Reagents

- None

2.3.2 Specialized equipment

- JAXMAN 365 nm LED flashlights (amazon.com https://www.amazon.com/JAXMAN-ultraviolet-365nm-detector-flashlight/dp/B06XW7S1CS/)
- (Optional) 3D printer design for LED flashlight holder (https://github.com/rogerzou/chipseq_pcRNA/blob/master/Jaxman_LED_flashlight_holder_design/files/8zeFECPViSo.stl)
- (Optional) MakerBot Replicator + 3D printer

2.3.3 Protocol

1. After performing Cas9 delivery via electroporation at night, we resume the next day approximately 12 h later. Check that the cells are well-adherent using light microscopy. For the purposes of this protocol, each 12-well plate (A, B, C, D) containing 6 wells of cells will correspond to different time points—A: no light exposure, B: 10 min after light exposure, C: 30 min after light exposure, D: 1 h after light exposure. The flashlight we use is a 365 nm LED light (JAXMAN) purchased from amazon.com. We activate Cas9 in the plates labeled B, C, and D.
2. We can shine light to one well at once using a single flashlight, or up to 3 wells and 6 wells at once in a 12-well plate using a 3D-printed flashlight holder design. To shine light, remove the plate cover from the 12-well plate containing cells, turn on the flashlight(s), then place the flashlight(s) on top of the wells (Fig. 1C). Slowly move the flashlight around in a circle such that the brightest region illuminated by the flashlight will reach all parts of the well. After 1 min of illumination, replace the plate cover and return the cells to the incubator.

2.4 Light exposure for Cas9 deactivation with photocleavable guide RNAs (pcRNAs)

Cas9 in complex with pcRNAs are naturally cleavage-competent despite the photocleavable group in the PAM-distal region of gRNA. However, after exposure to 365 nm wavelength light, the pcRNA becomes cleaved at the photocleavable group. Experimentally, we first allow 2 h after Cas9/pcRNA delivery for Cas9 to induce genomic DNA cleavage inside cells. Next, we expose cells to 1 min of light using the same commercially available LED flashlight as the previous Cas9 cgRNA activation protocol. This synchronizes Cas9 deactivation inside cells to facilitate time-resolved measurements of response phenotypes. One difference from the Cas9 activation protocol is that with pcRNAs, we first remove the majority of the media from the well before light exposure. Cell media normally absorbs part of the light before it reaches cells, reducing the final delivered dose. We found that Cas9 activation using cgRNAs is successfully accomplished with the weaker dose of light without media removal. However, for deactivation, each cell needs to be properly deactivated to avoid leakage of Cas9 cleavage, so we remove the media before light delivery to maximize the light dose.

2.4.1 Reagents

– "DMEM complete" (to add back to the wells after light illumination)

2.4.2 Specialized equipment

– JAXMAN 365 nm LED flashlights (Amazon https://www.amazon.com/JAXMAN-ultraviolet-365nm-detector-flashlight/dp/B06XW7S1CS/)
– (Optional) 3D printer design for LED flashlight holder (https://github.com/rogerzou/chipseq_pcRNA/blob/master/Jaxman_LED_flashlight_holder_design/files/8zeFECPViSo.stl)
– (Optional) MakerBot Replicator + 3D printer

2.4.3 Protocol

1. By 2 h after electroporation, the cells should be adherent if collagen is coated properly. Check that is the case using light microscopy. We deactivate the 12-well plates labeled B, C, and D.
2. To deactivate, first remove most of the media, because "DMEM complete" can partially block the light which greatly reduces deactivation efficiency. Make sure the plate cover is removed. Turn on and place the 365 nm LED flashlight resting above the well for 1 min. Just like with activation, slowly move the flashlight around in a circle such that the

brightest region illuminated by the flashlight will reach all parts of the well. After 1 min of illumination, add back 1 mL fresh "DMEM complete" media, place the plate cover back on the 12-well plate, and return the plate back to the incubator.

3. For deactivating more than 1 well simultaneously, use the 3D printed flashlight holder design, which is designed for use with 12-well plates and can support up 3 or 6 flashlights at once.

2.5 Quantification of DSBs and insertion-deletion mutations (indels)

At various timepoints after activation or deactivation outlined in the previous section, we terminate the Cas9 reaction inside cells by cellular lysis and extraction of its genomic DNA. The extracted genomic DNA can be used for downstream assays to measure DSBs and indels as a function of time after Cas9 activation or deactivation.

2.5.1 Reagents

- Single-stranded oligonucleotides (Integrated DNA Technologies, sequences in Supplementary Table S1 in the online version at https://doi.org/10.1016/bs.mie.2021.08.007)
- DNeasy Blood & Tissue Kit (Qiagen 69504)
- QIAquick PCR Purification Kit (Qiagen 28104)
- Q5® High-Fidelity 2× Master Mix (New England BioLabs M0492)
- BamHI-HF (New England BioLabs)
- ddPCR Supermix for Probes (no dUTP) (Bio-Rad)
- Droplet Generation Oil for Probes (Bio-Rad)

2.5.2 Specialized equipment

- DG8 Gaskets, DG8 Cartridges, and other consumables for the QX200 Droplet Digital system (Bio-Rad)
- QX200 Droplet Generator and QX200 Droplet Reader (Bio-Rad)
- Thermocycler supporting 96-well plates (Bio-Rad)
- Fisherbrand™ Cell Scraper (Fisher Scientific 08–100-241)
- (micro) Centrifuge 5424 R (Eppendorf)

2.5.3 Protocol for genomic DNA extraction

1. First, remove the media and add 250 μL PBS. Use a cell scraper to dissociate the cells from the well. Transfer the entire cell-containing solution to a 1.5 mL Eppendorf tube. Proceed with gDNA extraction using

the DNeasy Blood & Tissue Kit (Qiagen) following manufacturer's protocol. Elute in 100 μL of AE buffer supplied by the kit. This gDNA can be stored long-term in −20 °C.

2.5.4 Protocol for DSB readout

1. The ddPCR protocol is adapted from previous literature (Rose et al., 2017). Two amplicons were designed for the locus of interest. One amplicon (target1) includes the Cas9 cleavage site; the other (target2) is a nearby sequence that is not cleaved. One dual-quenched qPCR probe (IDT) was designed for each amplicon. The probe with HEX fluorophore corresponds to target1 while the probe with FAM fluorophore corresponds to target2 (Fig. 2A). Primers and probes were designed using guidelines by

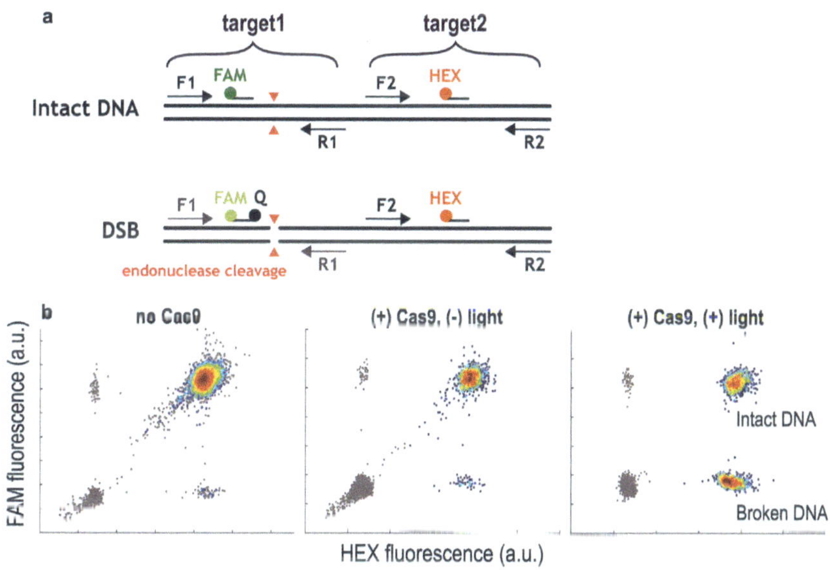

Fig. 2 Concept and results of DSB-ddPCR. (A) Schematic of DSB-ddPCR. For droplets that contain cleaved DNA at the expected position, target1 will not amplify, resulting in low FAM fluorescence because the probes will not be de-quenched. In contrast, intact DNA will amplify, digesting the probes to liberate the FAM fluorophore from its quencher and yielding high fluorescence. From 40 cycles total, the reaction in each droplet is "all or none". (B) Plots of FAM and HEX fluorescence for each detected droplet after PCR, either for samples without Cas9 exposure, Cas9 exposure but no Cas9 activation, or Cas9 exposure and 30 s after Cas9 activation. There are four clusters in each plot, of which the right two are important. The top right cluster corresponds to droplets with intact DNA at target1 and target2. The bottom right corresponds to droplets with intact DNA at target2, but broken DNA at target1. Note that the number of droplets in the bottom right cluster greatly increases upon Cas9 activation, suggesting more DNA is cleaved within target1.

Bio-Rad (http://www.bio-rad.com/en-us/applications-technologies/planning-droplet-digital-pcr-experiments ?ID=MDV33OKG4). The primer (ddPCR_ACTB_*) and probe (ddPCR_ACTB_probe*) sequences are in Supplementary Table S1 in the online version at https://doi.org/10.1016/bs.mie.2021.08.007.

2. For the ddPCR reaction, initially place each reaction in a PCR tube. Use enough volume of genomic DNA for 10–50 ng per reaction. For the purposes of this protocol, suppose X μL is used. Besides for genomic DNA, each ddPCR reaction is composed of 10 μL ddPCR Supermix for Probes (no dUTP), 1.8 μL of 10 μM primer mix, 1 μL of 5 μM probe mix, (8.2—X) μL of nuclease free water, and 1 μL of BamHI-HF (for the *ACTB* locus, 1 μL BamHI-HF was added to the reaction mixture to ensure better separation of signals during amplification. BamHI-HF was chosen because it is not expected to cleavage genomic DNA within target1 or target2). After genomic DNA is included, the total reaction volume is 22 μL, of which 20 μL will be used to generate droplets.
3. To expedite this process across multiple reactions, a master mix composed of all reagents except genomic DNA can be created. After each genomic DNA sample is aliquoted into a different PCR tube, add the master mix to each tube to make 22 μL total. Make sure that the number of reactions used is a multiple of 8. If not, use 10 μL ddPCR Supermix for Probes (no dUTP) with 12 μL water for the remaining reactions.
4. Create ddPCR droplets by first placing a DG8 disposable cartridge into the DG8 gasket. There are three rows of 8-wells in the cartridge. Add 70 μL of Droplet Generation Oil for Probes to the bottom 8-well row. Add 20 μL of samples to the middle 8-well row composed of smaller wells meant to hold samples. Place a red rubber cover (a consumable of the QX200 Droplet Digital system) over the gasket. Open the door of the QX200 Droplet Generator (Bio-Rad) by pressing the green arrow button. The door should slide open. Then place the entire container into the droplet generator, and close the door by pressing the green arrow button again. After closing the door, the machine will automatically begin droplet creation.
5. After droplet creation, 40 μL of generated droplets will be found in the previously empty top 8-well row of the DG8 gasket. Transfer them to a 96-well PCR plate. Repeat the droplet creation process, transferring each set of 8 samples into a different set of wells in the 96-well PCR plate. Once that is complete, place an aluminum plate cover (a consumable of the QX200 Droplet Digital system) and heat-seal the plate cover onto the PCR plate using the PX1 PCR Plate Sealer (Bio-Rad).

6. Next, transfer the PCR plate into the thermocycler. Perform PCR amplification with the following thermocycler conditions: 95 °C for 10 min, 40 cycles of {94 °C for 30 s, variable annealing temperatures for 30 s, 72 °C for 2 min}, 98 °C for 10 min, 12 °C hold. The annealing temperature for our *ACTB* primer/probe set is 60 °C, but should be experimentally determined for each primer/probe set.
7. Following PCR, which takes approximately 3 h, transfer the PCR plate into the QX200 Droplet Reader (Bio-Rad), where each droplet of each sample is individually scanned to measure its fluorescence signal. Droplets plotted in each fluorescent channel (FAM/HEX) should be well separated into four distinct clusters (Fig. 2B), and thresholds should be set to distinguish between droplets that are either positive or negative for FAM (target1) or HEX (target2).
8. Calculate the double-strand break (DSB) frequency using the following formula: [target1-,target2+]/([target1-,target2+]+[target1+,target2+]), where target1- indicates negative droplets for the target1 amplicon, target2+ indicates positive droplets for the target2 amplicon. Note that only droplets with positive target2 are considered.

2.5.5 Protocol for insertion-deletion (indel) readout

1. To measure insertion deletion mutations at the Cas9 target site, first perform genomic PCR to amplify the specific region of interest. Prepare the PCR reaction by mixing together the following components. The genomic DNA corresponds to the samples obtained previously using the DNeasy Blood & Tissue Kit (Qiagen). The primer set sequences (Sanger_ACTB_*) are found in Supplementary Table S1 in the online version at https://doi.org/10.1016/bs.mie.2021.08.007.

Component	Volume (μL)
Nuclease Free Water (NFW)	3
Genomic DNA	1
Fwd/Rev. primer set (10 μM)	1
Q5® High-Fidelity 2× Master Mix	5
Total	10

2. Start the thermocycling protocol. The 71 °C annealing temperature is specific to this *ACTB* primer set—use of other primer sets require empirical determination of the best annealing temperature.

Step	Temp. (°C)	Time
Initial Denaturation	98	30 s
35 cycles	98	10 s
	71 (*ACTB*)	10 s
	72	20 s
Final extension	72	2 min
Hold	4	Inf

3. Extract PCR-amplified target DNA with the QIAquick PCR Purification Kit following manufacturer's protocol. Elute with 30 μL of EB supplied with the kit.
4. For Sanger Sequencing using GeneWiz, first add 10 μL of the eluted PCR amplicon to a PCR tube. Add in 5 μL of 5 μM sequencing primer, which can be either the Fwd or Rev. PCR primer. Also sequence wild type HEK293T cells for a "negative control" sample with no genome editing.
5. Use TIDE to estimate % of indels (http://shinyapps.datacurators.nl/tide/). The software requires (1) input of the Cas9 gRNA sequence, (2) the .ab sequencing file for the sample of interest, and (3) the .ab sequencing file for the wild type "negative control" for reference (Fig. 3).

2.6 Measurement of localized repair factor recruitment using ChIP-seq

ChIP-seq is a protocol with numerous steps and some variations in protocols between different research groups (Park, 2009; Wienert et al., 2019). The protocol we use is derived from previous work that used ChIP-seq for detection of MRE11 binding after Cas9 exposure as a means of off-target detection (Wienert et al., 2019). For the purposes of this chapter, we will detect MRE11 recruitment to sites of Cas9 DSBs at numerous time points after light exposure (no light, 10 min, 30 min, and 1 h). We have successfully detected MRE11 as well as other proteins such as Cas9, 53BP1, and

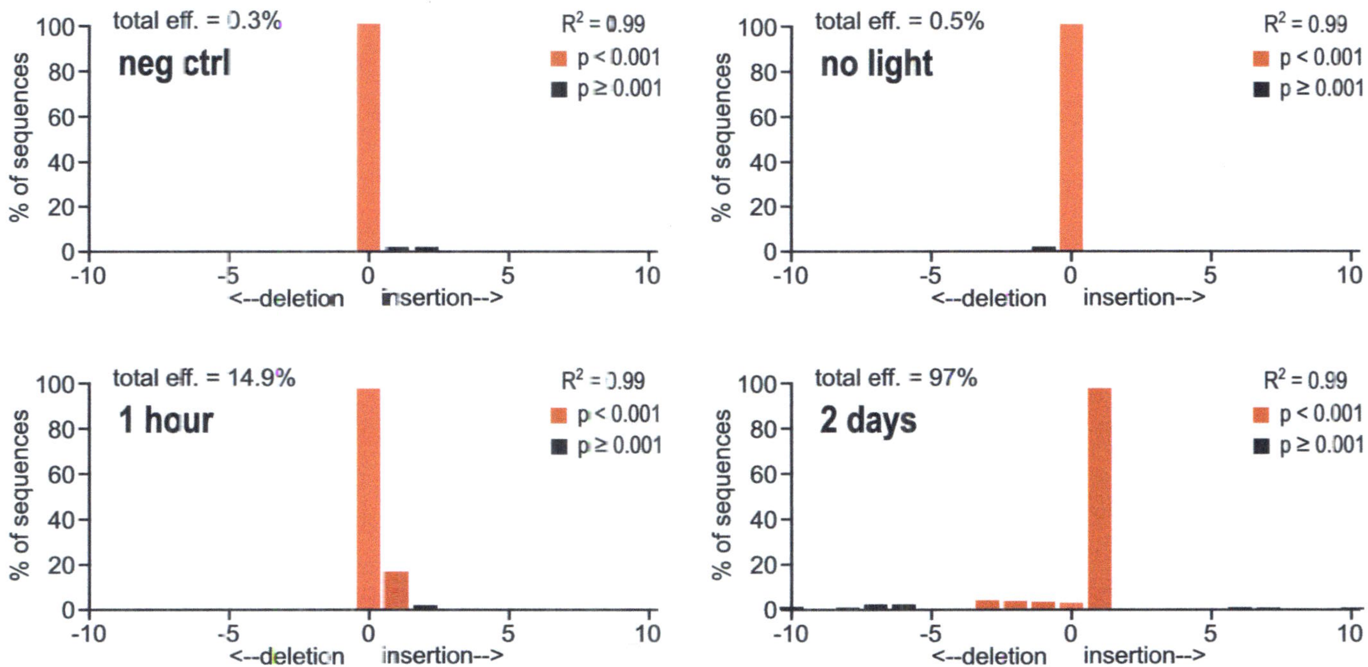

Fig. 3 Results of indel estimation from Sanger Sequencing. The output from TIDE for 4 samples—cells without Cas9 exposure (neg ctrl), cells exposed to Cas9 with no activation (no light), 1 h after activation (1 h), or 2 days after activation (2 days). "total eff." on the top left indicates the indel percentage, and the bars indicate the type of indel product. At the *ACTB* target site, the majority of indel products are +1 insertions.

gH2AX using this exact protocol but simply replacing the antibody for immunoprecipitation (Liu et al., 2020; Zou et al., 2021).

2.6.1 Reagents

- All oligo sequences are in Supplementary Table S1 in the online version at https://doi.org/10.1016/bs.mie.2021.08.007 (MNase_F, MNase_R, PE_i5, PE_i7XX for XX in {01–24})
- Corning™ DMEM with L-Glutamine, Glucose, Sodium Pyruvate (Fisher Scientific MT10013CV)
- Pierce™ 16% Formaldehyde (w/v), Methanol-free (Thermo Fisher 28908)
- Dynabeads™ Protein A for Immunoprecipitation (Thermo Fisher 10002D)
- Halt™ Protease Inhibitor Cocktail (100×) (Thermo Fisher 78438)
- MRE11 antibody (Novus NB100-142)
- IDTE pH 8.0 (Integrated DNA Technologies 11-05-01-09)
- AMPure XP for PCR Purification (Beckman Coulter A63881)
- NEBNext® Ultra™ II End Repair/dA-Tailing Module (New England BioLabs E7546S)
- T4 DNA Ligase (New England BioLabs M0202S)
- Phusion® High-Fidelity PCR Master Mix with HF Buffer (New England BioLabs M0531S)

2.6.2 Specialized equipment

- Argos Technologies™ RotoFlex™ Tube Rotator (Fisher Scientific 22-505-001)
- (micro) Centrifuge 5424 R (Eppendorf)
- Centrifuge 5810 R (Eppendorf)
- Q125 sonicator (Qsonica Q125-110)
- Some magnetic rack to manipulate magnetic beads. We use the items listed below as Optional
- (Optional) 96 well Magnet Plate (Alpaqua A001322)
- (Optional) 3D printed magnetic rack (https://www.thingiverse.com/thing:319772 or https://www.thingiverse.com/thing:79424)
- (Optional) MakerBot Replicator+3D printer
- (Optional) Whole Housewares 4" Square Glass Vase, Candle Holder, 6 Pack Clear Cube Centerpiece (Amazon https://www.amazon.com/whole-housewares-square-candle-centerpiece/dp/B07MT8TDJ8/)

– (Optional) 3D printed tube holder for sonication (https://github.com/rogerzou/chipseq_pcRNA/blob/master/2mL_tube_holder_for_sonication/files/k77AEoSKdcg.stl)
– Fisherbrand™ Cell Scraper (Fisher Scientific 08-100-241)

2.6.3 Protocol for buffer preparation

1. Before starting the ChIP protocol, first prepare some general stock buffers, whose recipes are below. Store all in 4 °C.

Reagent	Final concentration	Volume
5 M NaCl solution (50 mL)	5 M NaCl	58.44 g/mol * 5 mol/L * 0.05 L = 14.61 g NaCl
5 M LiCl solution (50 mL)	5 M LiCl	42.39 g/mol * 5 mol/L * 0.05 L = 10.60 g LiCl
10% Triton X-100 solution (50 mL)	10% Triton X-100	5 mL of 100% Triton X-100 in 45 mL NFW
10% Igepal CA-630 (50 mL)	10% Igepal CA-630	5 mL of 100% Igepal CA-630 in 45 mL NFW
10% Na-Deoxycholate (50 mL) *Protect from light	10% Na-Deoxycholate	5 g Na-Deoxycholate in 45 mL NFW
2 M glycine in PBS (50 mL)	2 M glycine in PBS	75.07 g/mol * 2 mol/L * 0.05 L = 7.507 g glycine
		Up to 50 mL with PBS
0.5% BSA in PBS (50 mL)	0.5% BSA in PBS	2.5 mL of 10% BSA in 47.5 mL PBS
PEG/NaCl solution for SPRI cleanup (10 mL)	2.5 M NaCl	2.5 mol/L * 58.44 g/mol * 0.01 L = 1.461 g NaCl
	20% PEG8000	20% PEG8000 = 2 g PEG8000
	In nuclease free water	Fill with water to 10 mL. it is viscous and slow to dissolve, so place on rotator overnight. 2000 × g centrifugation removed bubbles. Invert to mix, store in 4 °C

2. Next, before starting the ChIP protocol, prepare the ChIP buffers, whose recipes are below. Store all in 4 °C except LB3 and elution buffer (EB), which can be left at room temperature because those buffers can have precipitation due to presence of ionic detergents if stored in 4 °C long-term.

Reagent	Final concentration	Volume
LB1 (500 mL) Non-ionic detergents to lyse only the cell membrane	50 mM HEPES 140 mM NaCl 1 mM EDTA 10% glycerol 0.5% Igepal CA-630 0.25% Triton X-100 KOH pH to 7.5 1% of 100 × PI right before	238.3 g/mol * 0.05 mol/L * 0.5 L = 5.96 g HEPES powder 58.44 g/mol * 0.14 mol/L * 0.5 L = 4.1 g NaCl pellets 1 mL of 0.5 M EDTA 50 mL of 100% glycerol 2.5 mL of Igepal CA-630 1.25 mL of 100% Triton X-100 2 M KOH to pH 7.5 (starts around pH 5.5) 1% of 100 × PI right before
LB2 (500 mL) Wash step to remove free cytoplasmic components, but keeping nuclei intact	10 mM Tris–HCl pH 8 200 mM NaCl 1 mM EDTA 0.5 mM EGTA HCl pH to 8.0 1% of 100 × PI right before	5 mL of 1 M Tris–HCl pH 8.0 58.44 g/mol * 0.2 mol/L * 0.5 L = 5.844 g NaCl pellets 1 mL of 0.5 M EDTA 568 μL of 0.44 M EGTA 1 N HCl to pH 8.0 (starts around pH 8.25) 1% of 100 × PI right before
LB3 (500 mL) Ionic detergents to lyse nuclear membrane	10 mM Tris-HCl pH 8 100 mM NaCl 1 mM EDTA 0.5 mM EGTA 0.1% Na-deoxycholate 0.5% N-lauroylsarcosine HCl pH to 8.0 1% of 100 × PI right before	5 mL of 1 M Tris–HCl pH 8.0 58.44 g/mol * 0.1 mol/L * 0.5 L = 2.922 g NaCl pellets 1 mL of 0.5 M EDTA 568 μL of 0.44 M EGTA 500 mg Na-deoxycholate powder 2.5 g of N-lauroylsarcosine powder 1 N HCl to pH 8.0 (starts around pH 8.25) 1% of 100 × PI right before

—cont'd

Reagent	Final concentration	Volume
RIPA WB1 (500 mL)	50 mM HEPES 500 mM LiCl 1 mM EDTA 1% Igepal CA-630 0.7% Na-deoxycholate KOH pH to 7.5	238.3 g/mol * 0.05 mol/L * 0.5 L = 5.96 g HEPES powder 42.39 g/mol * 0.5 mol/L * 0.5 L = 10.6 g LiCl powder 1 mL of 0.5 M EDTA 5 mL of Igepal CA-630 3.5 g of Na-deoxycholate powder 2 M KOH to pH 7.5 (starts around pH 6.5)
TBS WB2 (50 mL)	20 mM Tris–HCl pH 7.5 150 mM NaCl	1 mL of 1 M Tris–HCl pH 7.5 3.75 mL of 2 M NaCl
ChIP-EB (50 mL)	50 mM Tris–HCl pH 8.0 10 mM EDTA 1% SDS	2.5 mL of 1 M Tris–HCl pH 8.0 1 mL of 0.5 M EDTA 2.5 mL of 20% SDS

3. We have a specific protocol for washing magnetic Protein A beads, which we call the *Magnetic Beads Wash Protocol*. Assume the beads are in a 1.5 mL or 2 mL Eppendorf tube in a magnetic rack. The protocol is: add washing liquid to tube, close cap, invert the magnetic rack twice with tubes inside, set down, lift tube up from magnetic rack, turn tube clockwise 180 degrees in the air, set down on magnetic rack—magnetic beads should flow from one side of tube to other side —repeat this with another 180-degree rotation after lifting tube away from rack, decant by pipetting the liquid away.

2.6.4 Protocol for cell harvesting and fixation

1. At any time point after activation or deactivation, we harvest cells by fixation using formaldehyde, followed by flash-freezing for storage until the time of ChIP. Generally, more than 3 million cells are required for each ChIP sample. For the purposes of this protocol, one well of the six 12-wells per timepoint was previously used for ddPCR/indel analysis, so the remaining five wells will be used for ChIP.
2. With the 12-well plates out, first remove the "DMEM complete" media, then replace with 0.5 mL DMEM. Use a cell scraper to dissociate the

cells from the well. Combine all cells from the same time point into a 15 mL falcon tube. Ensure that the final cell suspension in DMEM is 10 mL.

3. For fixation, add 721 μL of 16% formaldehyde to a final 1.077% concentration, then incubate at room temperature for 15 min on a rotator. Only mix by inversion or rotation, and never mix the liquid using a pipette from this step to before addition of lysis buffer at a later step—the cells will readily stick to the pipette tip during these steps, resulting in significant cell loss. Afterwards, add 750 μL of 2 M glycine in PBS to 130 mM final concentration to quench the formaldehyde.
4. Centrifuge to pellet the cells at 1200 × g for 3 min at 4 °C. Decant by pouring. Then add 10 mL of ice-cold PBS, invert tube multiple times to rinse, then centrifuge at 1000 × g for 3 min at 4 °C. Repeat this step, then decant by pouring. Remember to never use pipettes to mix.
5. Cap the 15 mL falcon tubes tightly and place into liquid nitrogen to flash freeze. Store pellets in −80 °C until the next step. The fixed, frozen cell pellets are stable for at least a few months.

2.6.5 Protocol for prebinding of antibody to magnetic beads

1. We first add antibody to magnetic beads and allow room temperature incubation under rotation for 1–2 h for the antibody to fully adhere to magnetic beads. This conjugates the antibodies to magnetic beads.
2. For each ChIP sample, use 50 μL of Protein A with 3 μL of antibody. In this protocol, there are four samples corresponding to different time points, so use 200 μL of Protein A. Therefore add 200 μL Protein A to a 2 mL tube. Collect beads on the magnetic rack, then remove the supernatant. Wash with 1 mL 0.5% BSA in PBS two times following *Magnetic Beads Wash Protocol.*
3. Resuspend Protein A beads in 400 μL 0.5% BSA in PBS (100 μL for each sample). Add 12 μL of MRE11 antibody (3 μL for each sample), then incubate 1–2 h on the rotating platform at room temperature. Immediately move on to cell lysis during incubation.

2.6.6 Protocol for cell lysis

1. For each ChIP sample, we use 4 mL LB1, 4 mL LB2, and 1.5 mL LB3. Therefore, for the four ChIP samples, prepare stock solutions of 16 mL LB1 (add 160 μL of 100 × proteinase inhibitor), 16 mL LB2 (add 160 μL of 100 × proteinase inhibitor), and 6 mL LB3 (add 60 μL of 100 × proteinase inhibitor).

2. For each cell sample, resuspend pellet of crosslinked cells in 4 mL of LB1 (w/1 × protease inhibitor). Pipettes can now be used again, because the cells in lysis buffer do not stick to pipette tips anymore. Use a 1 mL pipette to gently break up the cell pellet, then place on the rotator in 4 °C for 10 min. Pellet nuclei at 2000 × *g* for 3 min at 4 °C and decant by pouring. Then, for each cell sample, resuspend pellet in 4 mL of LB2 (w/1 × protease inhibitor). Use a 1 mL pipette to gently break up the cell pellet, then place on rotator in 4 °C for 5 min. Then, pellet nuclei at 2000 × *g* for 3 min at 4 °C and decant by pouring. Finally, resuspend pellet in 1.5 mL LB3 (w/1 × protease inhibitor), use 1 mL pipette to thoroughly mix, and transfer to a 2 mL Eppendorf tube.

2.6.7 Protocol for sonication

1. We use a probe sonicator (QSonica Q125 sonicator) with a very convenient setup where a custom 3D-printed Eppendorf tube holder is secured in a glass vase. The vase is filled with ice water. This allows the Eppendorf tube holding the cell nuclei to be cooled by the ice water during sonication. Alternative setups likely work, but the key is to reduce excessive heating of the sample due to sonication using ice water. Alternative sonicators will likely work as well, but likely have differences in power that require calibration for the optimal sonication protocol.
2. We clean the sonicator probe using a wet kimwipe, then dry it with a dry kimwipe. We then place the tube into the Eppendorf holder, open the tube cap, then insert the sonicator tip halfway into the tube liquid (Fig. 4A). We set the sonication protocol to be 30s ON, 30s OFF, 12 min total time, 55% amplitude. Be sure to wear earplugs because the sonication can be loud. After the sonication is complete, remove and cap the sample tube, then thoroughly clean the sonicator tip using a wet kimwipe then a dry kimwipe to avoid sample cross-contamination. Then, sonicate the next sample using the same protocol. Be sure to keep samples on ice before and after sonication.
3. After sonication of all samples, spin each sample at 20,000 × *g* for 10 min at 4 °C to pellet debris. There may be a small dark pellet after centrifugation. This is normal and the amount should look similar for all samples. A sample without a small dark pellet may indicate insufficient sonication.

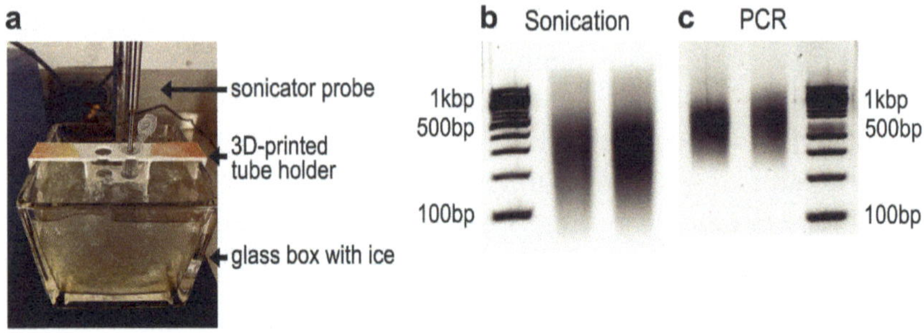

Fig. 4 Electrophoresis gel after sonication and PCR amplification for ChIP-seq. (A) Image of probe-based sonication setup. A 3D-printed Eppendorf tube holder secures a 2mL Eppendorf tube in a glass container with ice water. The sonicator probe is able to insert directly into the Eppendorf tube without touching the sides. (B) Gel of fixed chromatin after probe-based sonication. The majority of DNA is between 200 and 500bp. The two lanes correspond to two biological replicates. (C) Gel of final product for sequencing. The majority of DNA is between 300 and 800bp. The ligated adapter sequences and additional bases added from PCR accounts for the increase in length. The two lanes correspond to two biological replicates.

2.6.8 Protocol for ChIP antibody binding

1. Transfer 1.2mL of sample supernatant to a new 5mL Eppendorf tube. Add an additional 1.5mL LB3 (no protease inhibitor). Then, add 300μL of 10% Triton X-100 to yield ~3mL total per sample.
2. For each sample, first take a small amount for quality control (QC) samples by pipetting 15μL of the final 3mL mixture into a PCR tube. To each tube, add 40μL of ChIP-EB, mix, then incubate at 65°C overnight for reverse crosslinking of QC samples.
3. Retrieve the antibody/magnetic bead tubes currently rotating at room temperature and move onto magnetic racks. Wash beads three times each using 1mL 0.5% BSA in PBS, following *Magnetic Beads Wash Protocol*. Then, resuspend in 200μL 0.5% BSA in PBS (50μL per IP).
4. Add 50μL of antibody/bead to each of four 3mL sonicated cell lysate mixtures. Mix on a rotator at 4°C for at least 7h or overnight.

2.6.9 Protocol for ChIP wash and reverse crosslinking setup

1. Transfer the 3mL liquid from the ChIP tubes into 2mL Eppendorf tubes on a magnetic stand. First, transfer 1.5mL, wait 2min to allow magnetic beads to settle, then remove supernatant. Repeat until all the beads are transferred to the Eppendorf tube on the magnetic stand.

2. For each tube, add 1 mL RIPA WB, wash following *Magnetic Beads Wash Protocol.* Repeat 5 more times. Then, wash once with 1 mL TBS following *Magnetic Beads Wash Protocol.*
3. After decanting all liquid, add 50 μL ChIP-EB to suspend the decanted bead-antibody-chromatin samples. Transfer liquid to PCR tubes, then incubate at 65 °C for over 7 h or overnight, ideally with shaking for reverse crosslinking of these ChIP samples.

2.6.10 Protocol for ChIP DNA extraction

1. For both the QC samples and ChIP samples, add 40 μL IDTE to each tube, mix, add 2 μL RNase A, mix, then incubate at 37 °C for 15 min. Then, add 4 μL Proteinase K, mix, and incubate at 55 °C for 30 min.
2. For ChIP samples in PCR tubes which still contain magnetic beads, first use the magnetic plate (Alpaqua) to separate supernatant from magnetic beads, moving the separate supernatant to another set of tubes. Discard the beads.
3. Use the MinElute Reaction Cleanup Kit (Qiagen) to extract DNA from both the ChIP and QC samples, following manufacturer protocols. Elute in 34 μL EB supplied with the kit. Run 10 μL of only QC samples on a 2% agarose gel to verify appropriately sheared DNA between 300 and 1000 bp (Fig. 4B). Store all extracted DNA (both ChIP and QC samples) in −20 °C.

2.6.11 Protocol for sequencing library preparation

1. We now have 34 μL of DNA that were previously bound to MRE11 inside cells. To determine the sequence identity of this DNA population, we will prepare this DNA for high-throughput sequencing using the Illumina platform. As an overview, the DNA will first be blunted and appended an adenine at the 3′ end. Then, prepared adapters that contain additional sequences necessary for sequencing are ligated on both ends. Finally, PCR is performed using primers that amplify from the previously ligated adapters, producing DNA whose ends permit initiation of Illumina sequencing and whose middle, sequenced region contains the genomic sequence of interest. After sequencing, the genomic region is identified, and bioinformatic analysis can determine the genome-wide position of each sequenced piece of DNA.
2. First, prepare the double-stranded DNA adapter from single stranded oligonucleotides (MNase_F and MNase_R), whose sequences are listed in Supplementary Table S1 in the online version at https://doi.org/

10.1016/bs.mie.2021.08.007. Dilute MNase_F and MNase_R in 100 μM TE buffer, then mix 5 μL MNase_F, 5 μL MNase_R, and 40 μL duplex buffer (IDT) in a PCR tube. Next, anneal the oligonucleotides by placing the PCR tube in a thermocycler that first heats the sample at 95 °C for 5 min, then lowers the temperature at 1 °C/min until reaching 4 °C. Store this annealed adapter DNA in −20 °C.

3. The first step to preparing the ChIP-ed DNA for sequencing is end repair, which blunts the DNA, then A-tailing, which adds a 3′ adenine. These can be combined into a single reaction using the NEBNext® Ultra™ II End Repair/dA-Tailing Module (New England BioLabs). To prepare this reaction, mix the following in a PCR tube for each sample:

Reagent	Volume (μL)
ChIPed DNA	16.67
NEBNext Ultra II end prep reaction buffer	2.33
NEBNext Ultra II end prep enzyme mix	1
Total	20

4. Run the following program on a thermocycler: 20 °C 30 min ->65 °C 30 min ->4 °C hold. Then, clean up the reaction using 1.2× (24 μL volume) of AMPure beads (Beckman Coulter) following manufacturer's instructions. Elute in 16.5 μL nuclease free water with beads still in solution.
5. Next, ligate the MNase_F/R adapter onto the end-repaired/A-tailed DNA using T4 Ligase (New England BioLabs). Mix in a PCR tube:

Reagent	Volume (μL)
DNA in water mixed with beads	16.5
MNase_F/R annealed adapter (10 μM)	0.5
T4 DNA ligase buffer	2
T4 DNA ligase	1
Total	20

6. Run the following program on a thermocycler: 16 °C 30 min ->22 °C 30 min ->4 °C. After PCR is complete, add 24 μL of PEG/NaCl solution to the ligation reaction which already contains AMPure beads, resulting in a 1.2× SPRI cleanup with reused beads; elute in 20 μL TE buffer (pH 8).
7. We now have adapter-ligated ChIP-ed DNA. Next, perform low cycle number PCR to amplify enough DNA for sequencing, while at the same time incorporating the complete DNA ends from the primer sequences that allow library binding to the Illumina flow cell. The primers for PCR are listed in Supplementary Table S1 in the online version at https://doi.org/10.1016/bs.mie.2021.08.007. For each sample that is expected to be sequenced together, be sure that the PE_i7XX primer is different. Mix the following together in a PCR tube.

Reagent	Volume (μL)
DNA	9.4
10 μM PE_i5 primer	0.3
10 μM PE_i7XX primer	0.3
Phusion 2× master mix	10
Total	20

8. Next, run the following program on a thermocycler: 98 °C for 3 min, then 13 cycles of {10 s at 98 °C, 30 s at 68 °C, 30 s at 72 °C}, then hold indefinitely at 4 °C. After PCR is complete, clean up the reaction using 1× volume of AMPure beads (Beckman Coulter) following manufacturer's instructions. Elute in 40 μL TE buffer (pH 8).
9. Run on a 2% agarose gel to verify the correct library size (~300 bp–1 kb) (Fig. 4C). This DNA is now ready for sequencing. Store in −20 °C until right before sequencing. We use a sequencing core facility affiliated with the university, but many other options exist. We usually perform paired-end sequencing on a NextSeq 500 (Illumina), sequencing 36 to 75 base pairs on either end of the DNA. Other platforms such as the HiSeq or NovaSeq (Illumina) can also be compatible. The bottom line is to ensure at least 15 million high-quality sequencing reads per sample

2.7 Design principles for photosensitive cgRNAs and pcRNAs

Lastly, we outline design principles for creating new photosensitive cgRNAs and pcRNAs for Cas9 activation and deactivation, respectively, which promote identification of new target positions that leverage light control of CRISPR for novel studies of DNA damage and repair. For both cgRNAs and pcRNAs, be sure to target regions that previously have high editing efficiencies, which often correspond to promoter and/or chromatin accessible regions (Wienert et al., 2019; Wu et al., 2014), and result in higher synchronization after light exposure for more robust readout signal.

For cgRNAs, first identify candidate 20 bp target sequences with 2–3 well-spaced uracils in the PAM-distal region of crRNA (12 bp or more away from PAM). cgRNAs will replace these uracils with the bulky NPOM-dT moiety, which still allow Cas9 binding but not cleavage (Liu et al., 2020); after light exposure, the NPOM components will be liberated and leave behind deoxythymidines at these positions in the crRNA, which now confers efficient Cas9 activity (Yin et al., 2018). After identifying candidate target sequences with uracils at the desired positions, these candidates can be screened for promising function by purchasing or synthesizing crRNAs with (1) the original sequences or (2) with the 2–3 PAM-distal uracils replaced by adenine or guanine to mimic the steric (mismatch) disruption of NPOM-dTs. Using protocols from Sections 2.1 to 2.5, evaluate for high Cas9 editing efficiency (at the indel or DSB level) inside cells using the crRNA without mismatches, and complete abrogation of Cas9 activity using the mismatched crRNA. The target sequences corresponding to no residual activity with the 2–3 PAM-distal mismatches in the gRNA and high activity without mismatches are the best choices. Proceed with this target sequence for cgRNA synthesis. Provide the exact RNA sequence with NPOM-dT replacements of the 2–3 PAM-distal uracils to Bio-Synthesis, LLC. for custom oligonucleotide synthesis.

For pcRNAs, replace the 15th nucleotide from PAM (also in the PAM-distal region) with the PC-Linker moiety (Zou et al., 2021). Nucleotide replacement in the PAM-distal (and not PAM-proximal) region is crucial to minimize native disruption of Cas9 binding and/or cleavage (Singh et al., 2016). We provide the exact RNA sequence with the PC-Linker replacement of the 15th nucleotide from PAM to Integrated DNA Technologies for custom oligonucleotide synthesis.

3. Summary

Cellular responses to DNA damage are complex processes important to biology and medicine that require extensive investigation. Investigators studying these processes would benefit from versatile methods of targeted DNA damage that is sequence-specific, fast, and has tight control over activation or deactivation. This chapter presents protocols for two such methods that leverage newly developed controls over CRISPR-Cas9 activity using light. Activation or deactivation of DNA damage is followed by three time-resolved readouts that evaluate the entire DNA repair process from DSB induction, repair protein recruitment, to repair completion and mutagenesis. We anticipate that adoption of these methods will open the horizon to new studies of DNA damage and repair.

Acknowledgments

This work was supported by the grants from the National Institutes of Health (R35 GM 122569 and U01 DK 127432 to T.H.; T32 GM 73009 and F30 CA 254160 to R.S.Z.) and the National Science Foundation (PHY 1430124 to T.H.). T.H. is an investigator of the Howard Hughes Medical Institute.

References

Aleksandrov, R., Dotchev, A., Poser, I., Krastev, D., Georgiev, G., Panova, G., et al. (2018). Protein dynamics in complex DNA lesions. *Molecular Cell*, *69*, 1046–1061.

Aymard, F., Aguirrebengoa, M., Guillou, E., Javierre, B. M., Bugler, B., Arnould, C., et al. (2017). Genome-wide mapping of long-range contacts unveils clustering of DNA double-strand breaks at damaged active genes. *Nature Structural & Molecular Biology*, *24*, 353–361.

Blackford, A. N., & Jackson, S. P. (2017). ATM, ATR, and DNA-PK: The trinity at the heart of the DNA damage response. *Molecular Cell*, *66*, 801–817.

Brinkman, E. K., Chen, T., Amendola, M., & van Steensel, B. (2014). Easy quantitative assessment of genome editing by sequence trace decomposition. *Nucleic Acids Research*, *42*, 168.

Bunting, S. F., Callén, E., Wong, N., Chen, H. T., Polato, F., Gunn, A., et al. (2010). 53BP1 inhibits homologous recombination in Brca1-deficient cells by blocking resection of DNA breaks. *Cell*, *141*, 243–254.

Caron, P., Choudjaye, J., Clouaire, T., Bugler, B., Daburon, V., Aguirrebengoa, M., et al. (2015). Non-redundant functions of ATM and DNA-PKcs in response to DNA double-strand breaks. *Cell Reports*, *13*, 1598–1609.

Chen, J. S., Dagdas, Y. S., Kleinstiver, B. P., Welch, M. M., Sousa, A. A., Harrington, L. B., et al. (2017). Enhanced proofreading governs CRISPR–Cas9 targeting accuracy. *Nature*, *550*, 407–410.

Cheung-Ong, K., Giaever, G., & Nislow, C. (2013). DNA-damaging agents in cancer chemotherapy: Serendipity and chemical biology. *Chemistry & Biology*, *20*, 648–659.

Clouaire, T., & Legube, G. (2019). A snapshot on the cis chromatin response to DNA double-strand breaks. *Trends in Genetics*, *35*, 330–345.

Clouaire, T., Rocher, V., Lashgari, A., Arnould, C., Aguirrebengoa, M., Biernacka, A., et al. (2018). Comprehensive mapping of histone modifications at DNA double-strand breaks deciphers repair pathway chromatin signatures. *Molecular Cell*, *72*, 250–262.

Gupta, R., Somyajit, K., Narita, T., Maskey, E., Stanlie, A., Kremer, M., et al. (2018). DNA repair network analysis reveals shieldin as a key regulator of NHEJ and PARP inhibitor sensitivity. *Cell*, *173*, 972–988.

Gursoy-Yuzugullu, O., Ayrapetov, M. K., & Price, B. D. (2015). Histone chaperone Anp32e removes H2A. Z from DNA double-strand breaks and promotes nucleosome reorganization and DNA repair. *Proceedings of the National Academy of Sciences of the United States of America*, *112*, 7507–7512.

Hauer, M. H., & Gasser, S. M. (2017). Chromatin and nucleosome dynamics in DNA damage and repair. *Genes & Development*, *31*, 2204–2221.

Her, J., & Bunting, S. F. (2018). How cells ensure correct repair of DNA double-strand breaks. *Journal of Biological Chemistry*, *293*, 10502–10511.

Jinek, M., Chylinski, K., Fonfara, I., Hauer, M., Doudna, J. A., & Charpentier, E. (2012). A programmable dual-RNA-guided DNA endonuclease in adaptive bacterial immunity. *Science*, *337*, 816–821.

Lieber, M. R. (2010). The mechanism of double-strand DNA break repair by the non-homologous DNA end-joining pathway. *Annual Review of Biochemistry*, *79*, 181–211.

Liu, Y., Zou, R. S., He, S., Nihongaki, Y., Li, X., Razavi, S., et al. (2020). Very fast CRISPR on demand. *Science*, *368*, 1265–1269.

Lord, C. J., Tutt, A. N., & Ashworth, A. (2015). Synthetic lethality and cancer therapy: Lessons learned from the development of PARP inhibitors. *Annual Review of Medicine*, *66*, 455–470.

Michelini, F., Pitchiaya, S., Vitelli, V., Sharma, S., Gioia, U., Pessina, F., et al. (2017). Damage-induced lncRNAs control the DNA damage response through interaction with DDRNAs at individual double-strand breaks. *Nature Cell Biology*, *19*, 1400–1411.

Natale, F., Rapp, A., Yu, W., Maiser, A., Harz, H., Scholl, A., et al. (2017). Identification of the elementary structural units of the DNA damage response. *Nature Communications*, *8*, 1–18.

Noordermeer, S. M., Adam, S., Setiaputra, D., Barazas, M., Pettitt, S. J., Ling, A. K., et al. (2018). The shield in complex mediates 53BP1-dependent DNA repair. *Nature*, *560*, 117–121.

Ochs, F., Karemore, G., Miron, E., Brown, J., Sedlackova, H., Rask, M. B., et al. (2019). Stabilization of chromatin topology safeguards genome integrity. *Nature*, *574*, 571–574.

Park, P. J. (2009). ChIP–seq: Advantages and challenges of a maturing technology. *Nature Reviews Genetics*, *10*, 669–680.

Pessina, F., Giavazzi, F., Yin, Y., Gioia, U., Vitelli, V., Galbiati, A., et al. (2019). Functional transcription promoters at DNA double-strand breaks mediate RNA-driven phase separation of damage-response factors. *Nature Cell Biology*, *21*, 1286–1299.

Pickar-Oliver, A., & Gersbach, C. A. (2019). The next generation of CRISPR–Cas technologies and applications. *Nature Reviews Molecular Cell Biology*, *20*, 490–507.

Price, B. D., & D'Andrea, A. D. (2013). Chromatin remodeling at DNA double-strand breaks. *Cell*, *152*, 1344–1354.

Rose, J. C., Stephany, J. J., Valente, W. J., Trevillian, B. M., Dang, H. V., Bielas, J. H., et al. (2017). Rapidly inducible Cas9 and DSB-ddPCR to probe editing kinetics. *Nature Methods*, *14*, 891–896.

Scully, R., Panday, A., Elango, R., & Willis, N. A. (2019). DNA double-strand break repair-pathway choice in somatic mammalian cells. *Nature Reviews Molecular Cell Biology*, *20*, 698–714.

Shanbhag, N. M., Rafalska-Metcalf, I. U., Balane-Bolivar, C., Janicki, S. M., & Greenberg, R. A. (2010). ATM-dependent chromatin changes silence transcription in cis to DNA double-strand breaks. *Cell*, *141*, 970–981.

Singh, D., Sternberg, S. H., Fei, J., Doudna, J. A., & Ha, T. (2016). Real-time observation of DNA recognition and rejection by the RNA-guided endonuclease Cas9. *Nature Communications*, 7, 1–8.

Sternberg, S. H., LaFrance, B., Kaplan, M., & Doudna, J. A. (2015). Conformational control of DNA target cleavage by CRISPR–Cas9. *Nature*, *527*, 110–113.

Turinetto, V., & Giachino, C. (2015). Multiple facets of histone variant H2AX: A DNA double-strand-break marker with several biological functions. *Nucleic Acids Research*, *43*, 2489–2498.

Vitor, A. C., Sridhara, S. C., Sabino, J. C., Afonso, A. I., Grosso, A. R., Martin, R. M., et al. (2019). Single-molecule imaging of transcription at damaged chromatin. *Science Advances*, *5*, 1249.

Wienert, B., Wyman, S. K., Richardson, C. D., Yeh, C. D., Akcakaya, P., Porritt, M. J., et al. (2019). Unbiased detection of CRISPR off-targets in vivo using DISCOVER-Seq. *Science*, *364*, 286–289.

Wu, X., Scott, D. A., Kriz, A. J., Chiu, A. C., Hsu, P. D., Dadon, D. B., et al. (2014). Genome-wide binding of the CRISPR endonuclease Cas9 in mammalian cells. *Nature Biotechnology*, *32*, 670–676.

Xu, Y., Ayrapetov, M. K., Xu, C., Gursoy-Yuzugullu, O., Hu, Y., & Price, B. D. (2012). Histone H2A. Z controls a critical chromatin remodeling step required for DNA double-strand break repair. *Molecular Cell*, *48*, 723–733.

Yang, G., Liu, C., Chen, S. H., Kassab, M. A., Hoff, J. D., Walter, N. G., et al. (2018). Super-resolution imaging identifies PARP1 and the Ku complex acting as DNA double-strand break sensors. *Nucleic Acids Research*, *46*, 3446–3457.

Yin, H., Song, C. Q., Suresh, S., Kwan, S. Y., Wu, Q., Walsh, S., et al. (2018). Partial DNA-guided Cas9 enables genome editing with reduced off-target activity. *Nature Chemical Biology*, *14*, 311–316.

Zhou, Y., Lee, J. H., Jiang, W., Crowe, J. L., Zha, S., & Paull, T. T. (2017). Regulation of the DNA damage response by DNA-PKcs inhibitory phosphorylation of ATM. *Molecular Cell*, *65*, 91–104.

Zou, R. S., Liu, Y., Wu, B., & Ha, T. (2021). Cas9 deactivation with photocleavable guide RNAs. *Molecular Cell*, *81*, 1553–1565.

CHAPTER TWELVE

A detection method for the capture of genomic signatures: From disease diagnosis to genome editing

Orléna Benamozig[a,b,c], **Lou Baudrier**[a,b,c], **and Pierre Billon**[a,b,c,*]
[a]The University of Calgary, Cumming School of Medicine, Department of Biochemistry and Molecular Biology, Calgary, AB, Canada
[b]Robson DNA Science Centre, Calgary, AB, Canada
[c]Arnie Charbonneau Cancer Institute, Calgary, AB, Canada
[*]Corresponding author: e-mail address: pierre.billon@ucalgary.ca

Contents

Methods in Enzymology, Volume 661
ISSN 0076-6879
https://doi.org/10.1016/bs.mie.2021.08.012

Abstract

Variations in the genetic information originate from errors during DNA replication, error-prone repair of DNA damages, or genome editing. The most common approach to detect changes in DNA sequences employs sequencing technologies. However, they remain expensive and time-consuming, limiting their utility for routine laboratory experiments. We recently developed DinucleoTidE Signature CapTure (DTECT). DTECT is a marker-free and versatile detection method that captures targeted dinucleotide signatures resulting from the digestion of genomic amplicons by the type IIS restriction enzyme AcuI. Here, we describe the DTECT protocol to identify mutations introduced by CRISPR-based precision genome editing technologies or resulting from genetic variation. DTECT enables accurate detection of mutations using basic laboratory equipment and off-the-shelf reagents with qualitative or quantitative capture of signatures.

1. Introduction

Nucleic acids encode critical information for human physiology and medicine (Miga et al., 2020). A spectrum of variations contribute to the evolution and adaptation of species, underlies, or predisposes to disorders, and impacts disease burden (Ellegren & Galtier, 2016; Genomes Project et al., 2015; The International HapMap Consortium, 2015; Karczewski et al., 2020; Landrum et al., 2020; Lek et al., 2016). Mutations originate from error-prone repair of DNA damage induced by cellular processes or environmental mutagens and from errors introduced during replication but unresolved by the mismatch repair pathway (Alexandrov et al., 2013; Helleday, Eshtad, & Nik-Zainal, 2014; Zou et al., 2021). Although decoding genetic variants with a role in human disease is crucial, the large fraction of documented variants is uncharacterized. Connecting variants to phenotypes requires installing desired mutations with genome editing technologies, and methods to detect nucleic acid changes.

CRISPR-based precision genome editing technologies, such as CRISPR-mediated homology-directed repair, base editing, or prime editing, introduce desired modifications by directing the activity of nucleases and mutagenic enzymes at desired genomic locations (Anzalone, Koblan, & Liu, 2020; Billon et al., 2017). For instance, CRISPR-prime editing utilizes an engineered reverse transcriptase that copies the information encoded in an

RNA template directly in the genome, thereby enabling the insertion of any desired genomic changes. These technologies introduce site-specific DNA lesions, such as double-strand breaks (DSBs), resolved by various cellular DNA repair mechanisms (Nambiar et al., 2019; Yeh, Richardson, & Corn, 2019). Precision genome editing enables manipulating genetic information in cellular and animal models for functional interrogation of coding sequences and non-coding regulatory elements (Pickar-Oliver & Gersbach, 2019). Our ability to rapidly and accurately model or correct disease-associated variants in model systems is crucial to accelerate basic and translational research.

Typical approaches used to detect nucleic acid sequences are mostly limited to sequencing technologies, primer- or probe-specific mutations, or approaches that require specific instrumentation (Brinkman, Chen, Amendola, & van Steensel, 2014; Clement et al., 2019; Findlay, Vincent, Berman, & Postovit, 2016; Guell, Yang, & Church, 2014; Lindsay et al., 2016; Qiu et al., 2004). However, these methods require sophisticated detection strategies with weak sensitivity, precision, and accuracy, limiting their broad adoption (Germini et al., 2018). Also, despite recent spectacular advances, sequencing technologies remain expensive and time-consuming for basic laboratory experiments. Therefore, there is a need for rapid, versatile, and easy to implement laboratory detection assays.

Restriction endonucleases and ligases have played a pivotal role in the DNA recombinant technology revolution by enabling the assembly of selected DNA molecules (Khan et al., 2016). Restriction endonucleases are members of restriction-modification systems, a primitive bacterial innate immune system that protects hosts by destroying mobile genetic elements (Labrie, Samson, & Moineau, 2010). Endonucleases have revolutionized the DNA repair field by developing unique tools that simplify the efficient insertion of site-specific DSBs (Berkovich, Monnat, & Kastan, 2007; Chailleux et al., 2014; Rouet, Smih, & Jasin, 1994; Shanbhag, Rafalska-Metcalf, Balane-Bolivar, Janicki, & Greenberg, 2010). DNA ligases exhibit critical enzymatic activities during DNA replication and repair (Pascal, 2008). Ligases seal two DNA ends by forming phosphodiester bonds between 3′-hydroxyl and 5′-phosphoryl termini (Green & Sambrook, 2019). Endonucleases and ligases are ubiquitous tools in modern molecular biology for laboratory DNA manipulation and genetic studies. In particular, restriction enzymes from the type IIS family ("S" for "shifted," because DNA cleavage is shifted to one side of the recognition motif) have been utilized in molecular biology for their properties to cleave unknown sequences (Drmanac et al., 2010; Klompe, Vo, Halpin-Healy, & Sternberg, 2019; Pingoud, Wilson, & Wende, 2014; Ran et al., 2013; Shendure et al., 2005).

Here, we describe a rapid-to-execute and easy-to-implement protocol that utilizes AcuI, a type IIS restriction endonuclease discovered in *Acinetobacter calcoaceticus*, and a DNA ligase to identify nucleic acid signatures (Billon et al., 2020). AcuI is programmed to expose desired nucleic acid signatures that are subsequently captured by the ligation of standard DNA adaptors. This protocol allows the quantitative and qualitative identification of genetic variation induced by precision genome editing or natural variation. We illustrate the protocol by detecting two mutations introduced with CRISPR technologies in the *DPY30* gene, which encodes for a core subunit of the SET1/MLL histone methyltransferase complex (Xue et al., 2019), and the *SAGE1* (sarcoma antigen 1) gene, which encodes for a cancer biomarker and target for immunotherapy (Djureinovic et al., 2016). We introduce these transversion mutations in human cells using CRISPR-mediated prime editing (Anzalone et al., 2019) and use DTECT to determine editing efficiency (Billon et al., 2020). We also describe how positive and negative controls from non-edited samples are used to generate high confidence results. Finally, we discuss the strengths and limitations of DTECT compared to other detection methods.

2. Overview of the procedure

DTECT is a capture-based detection method that requires the digestion of a particular amplicon and ligation of specific adaptors (Billon et al., 2020). The experimental workflow of DTECT comprises six steps (Fig. 1). Briefly, DTECT requires the selection of the desired dinucleotide that includes the variant base(s). The selection of a "*dinucleotide signature*" of interest enables the design of a particular primer named "AcuI-*tagging primer*" (Fig. 1, Step I). This primer is adjacent to the targeted dinucleotide (Fig. 1, Step I dinucleotide in red) and amplifies the DNA sequence by PCR (Fig. 1, Step II). The AcuI-tagging PCR allows the insertion of an AcuI motif (5′-CTGAAG-3′) 14 bp upstream of the dinucleotide of interest (Fig. 1, Step II in green) and the addition of a 5′ handle used to detect the ligated product in step VI (Fig. 1, Step II in blue). The resulting amplicon is then digested with AcuI to generate two products of digestion, a small fragment (60 bp) and a long DNA fragment (>100 bp) (Fig. 1, Step III). Next, the small DNA fragment (60 bp), which contains the 3′ overhang dinucleotide signature, is isolated using solid-phase reversible immobilization (SPRI) magnetic beads (Fig. 1, Step IV). Then, the dinucleotide signature is captured by ligating complementary or non-complementary DNA adaptors (Fig. 1, step V). The adaptors are selected from a library of 16 standard adaptors that covers all the possible dinucleotides (Fig. 2A).

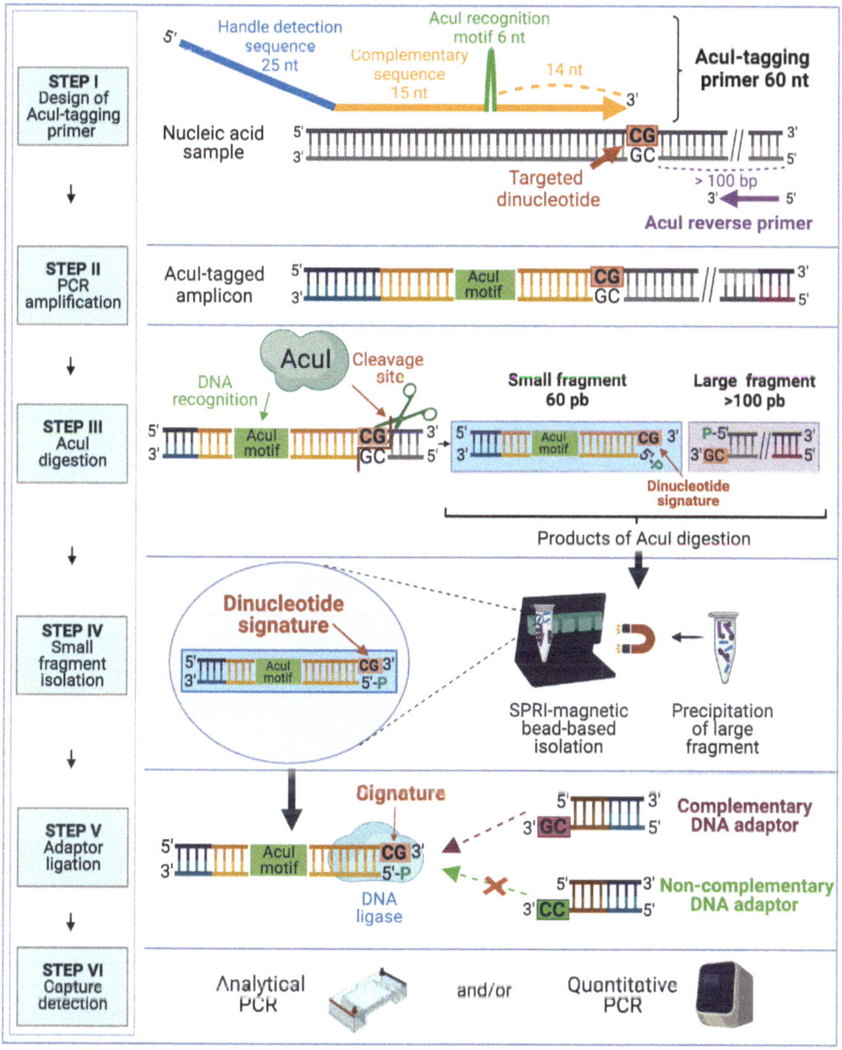

Fig. 1 Schematic representation of DTECT. DTECT is composed of six steps (Step I–VI). Step I: The targeted genomic locus contains a dinucleotide of interest (CG, in red). The AcuI-tagging oligonucleotide contains a detection handle (blue), two complementary sequences to the genomic locus (yellow) and an AcuI hairpin (green). The AcuI-tagging oligonucleotide is adjacent to the targeted dinucleotide. The reverse primer is positioned at >100 bp from the targeted dinucleotide (purple). Step II: The selected dinucleotide signature is amplified. Step III: AcuI digestion of the AcuI-tagged amplicon generates a 3′ dinucleotide overhang at the dinucleotide of interest (CG) and 5′ phosphates at the cleavage site. Two products of digestion are generated, a small fragment of 60 bp and a large fragment of >100 bp. Step IV: The small fragment is isolated using SPRI-magnetic beads. Step V: The signature is captured using complementary (CG) and non-complementary (CC) adaptors. Step VI: The captured material is amplified and revealed by analytical gel or quantitative PCR.

The ligated product serves as a template for detection by analytical or quantitative PCR (Fig. 1, Step VI) using unique primers that anneal to the 5′ end of the AcuI-tagging oligo and 3′ end of the adaptors (Fig. 1, Step V in blue). These PCRs quantify and reveal the presence of the genomic signatures in the original nucleic acid samples.

3. Materials and equipment

Here is the list of materials, reagents, and equipment required for DTECT. All reagents are commercially available and can be purchased from different sources. Moreover, the materials required for DTECT constitute standard equipment for a basic biology laboratory.

3.1 Key resources table

Reagent or resource	Source	Identifier
Chemicals, peptides, and recombinant proteins		
AcuI	NEB	R0641
T4 DNA Ligase	Invitrogen	15224025
Q5 High-Fidelity DNA Polymerase	NEB	M0491
dNTPs	NEB	N0447
Loading Dye	NEB	B7024
1 kb Plus DNA Ladder	NEB	N3200
Critical commercial assays		
Agencourt AMPure XP magnetic beads	Beckman Coulter	A63881
Power SYBR Green PCR Master Mix	Applied Biosystems	4367659
SYBR Gold Nucleic Acid Gel Stain	Invitrogen	S11494
Zymoclean gel DNA recovery Kit	Zymo Research	D4008
Quick Extract DNA Extraction Solution	Lucigen	QE09050
Deposited data		
DTECT—Plasmid for standard curve	Addgene	139333
Experimental models: Cell lines		
HEK293T cells	ATCC	CRL-11268

Oligonucleotides		
Primers for PCR	This paper	Table 1
Software and algorithms		
Excel		Microsoft
Other		
QuantStudio 6 Flex Real-Time PCR System	Applied Biosystems	4484642
ProFlex 3 × 32-well PCR System	Applied Biosystems	4484073
ChemiDoc Touch Gel Imaging System	Bio-Rad	1708370
12-Tube Magnetic Separation Rack	NEB	S1509

3.2 Reagents

- Agarose (Fisher BioReagents, #BP160-500)
- Tris Base (Fisher BioReagents, #BP152-5)
- Ethylenediaminetetraacetic Acid (Fisher BioReagents, #BP120-1)
- Sodium Hydroxide (Fisher BioReagents, #BP359-212)
- Acetic Acid, Glacial (Fisher BioReagents, #BP2401-212)
- Sodium Chloride (Fisher BioReagents, #BP358-10)
- Dimethyl Sulfoxide (Fisher BioReagents, #BP231-1)
- Phosphate buffered saline (Sigma-Aldrich, #P4417-100TAB)
- Molecular Biology Grade Water (Fisher BioReagents, #BP281920)

3.3 Equipment

- NanoDrop 2000c Spectrophotometer (Thermo Scientific, #ND-2000C)
- Mini Centrifuge (DLAB Scientific, #9031001012)
- Vortex Mixer (DLAB Scientific, #8031102000)
- Centrifuge 5425 (Eppendorf, #5405000042)
- Centrifuge 5810R (Eppendorf, #022628258)
- Dry Block Heaters (VWR, #12621-104)
- PowerPac Basic Power Supply (Bio-Rad, #1645050)
- Blue Light Transilluminator (VWR, #76151-834)
- Horizontal Mini S Gel Electrophoresis System (VWR, #76314-718)

3.4 Consumables

- MicroAmp Optical 384-Well Reaction Plate (Applied Biosystems, #4309849)
- MicroAmp Optical Adhesive Film (Applied Biosystems, #4311971)
- PCR tubes (Sarstedt, #72.991.002)
- Microtubes 1.5 mL (Sarstedt, #72.690.301)
- Tips, 10 μL, Eppendorf/Gilson type (Sarstedt, #70.1116.210)
- Filter Pipette tip 1–20 μL (Globe Scientific, #150910)
- Filter Pipette tip 1–200 μL (Globe Scientific, #150920)
- Filter Pipette tip 1–1000 μL (Globe Scientific, #150935)
- Razor blades (VWR, #55411-050)

3.5 Reagent setup and stock solutions

Timing: 1 h

- Preparation of the 500 mM EDTA stock solution: 186 g EDTA is dissolved into 800 mL of distilled H_2O. The pH is adjusted to 8.0 with NaOH. Finally, the final volume is adjusted to 1 L with distilled H_2O.
- Preparation of the 1 M Tris HCl pH 8.0 stock solution: 121.1 g of tris base is dissolved into 800 mL of distilled H_2O. HCl is slowly added until the pH reaches 8.0. Then, the final volume of the solution is adjusted to 1 L with distilled H_2O.
- Preparation of TAE 50 ×: The 50 × concentrated stock solution is prepared by dissolving 242 g of tris base into ~700 mL distilled H_2O. Then, 57.1 mL of glacial acetic acid and 100 mL of 500 mM EDTA (pH 8.0) are added. Finally, the final volume is adjusted to 1 L with distilled H_2O.
- The TE (10 mM Tris HCl pH 8.0, 1 mM EDTA pH 8.0) solution is prepared from the 1 M Tris HCl pH 8.0 and 500 mM EDTA pH 8.0 stock solutions. 1 mL of 1 M Tris HCl and 200 μL of 500 mM EDTA are mixed into 98.8 mL distilled H_2O.
- Dilution of the SYBR Gold solution: 10 μL of the SYBR Gold stock solution is diluted in 990 μL DMSO.

3.6 Software for analysis

- Spreadsheet software such as Microsoft Excel, Apple Numbers, or Google Sheets is needed for statistical analysis and quantification.

3.7 Preparation of the adaptor library

Timing: ~30 min

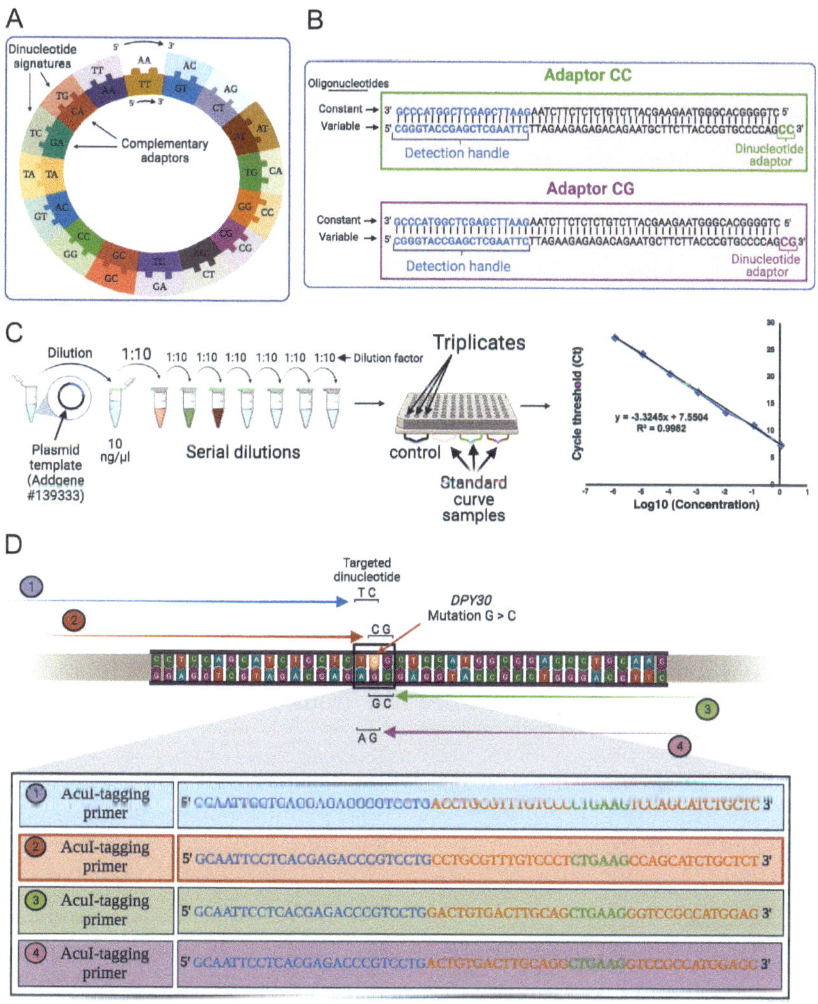

Fig. 2 Preparation of the library of adaptors and standard curve, and illustration of the flexibility of DTECT. (A) Illustration of the library of 16 adaptors and their respective complementary dinucleotide signatures. (B) Sequences of the complementary constant and variable oligonucleotides to generate the dsDNA adaptors. The common detection handle (blue) and the dinucleotides of the CC and CG adaptors (green and purple) are represented. (C) Steps required for the generation of the standard curve. Serial dilutions of a plasmid template are prepared and tested in triplicate to generate a unique standard curve common to all DTECT experiments. (D) Schematic representation of the targeted *DPY30* sequence surrounding the edited base (C > G). Four independent dinucleotides from the two strands can be generated to detect the base of interest. The sequences for the four respective AcuI-tagging primers used to capture the respective signatures are indicated.

A unique library of 16 double-stranded DNA adaptors is used to capture each possible dinucleotide signature. Fig. 2A illustrates the signature/adaptor combinations. This library is prepared from 17 individual oligonucleotides, consisting of 16 variable oligonucleotides of 60 nt and one constant oligonucleotide of 58 nt (oligonucleotide sequences are in Table 1). Each variable oligonucleotide contains a 58 nt sequence complementary to the constant oligonucleotide and has one of the 16 dinucleotides at the 3′ end (Fig. 2B, green and purple). In addition, adaptors contain a handle sequence (Fig. 2B, in blue) to analyze the captured products (detailed in part 5).

The double-stranded DNA library is prepared by combining the constant oligonucleotide with each of the 16 variable oligonucleotides as described below.

a. Adaptor oligonucleotides are regular oligos resuspended at a concentration of 100 μM in TE buffer.
b. Oligonucleotides are assembled in a 20 μL reaction. 2.5 μL of the constant oligonucleotide (100 μM) and 2.5 μL of each variable oligonucleotide (100 μM) are mixed with 4 μL of 1 × ligase buffer (5 ×) and 11 μL H_2O.
c. The mixture containing the constant oligonucleotide and one of the variable oligonucleotides is annealed by incubating for 5 min at 95 °C, followed by a temperature decrease from 95 to 15 °C at a ramp of 0.5 °C/s.
d. After annealing, 100 μL of molecular biology grade H_2O is added to generate a stock of double-stranded DNA adaptors at 5 μM.
e. Adaptors are stored at −20 °C. Each adaptor preparation captures up to 240 reactions.
f. (*Optional*) The adaptors can be visualized on agarose or polyacrylamide gels to confirm successful annealing.

3.8 Generation of the standard curve

Timing: 120 min

The detection of the ligated product requires a single pair of primers that anneal to the 5′ handle of the AcuI-tagging oligos and 3′ end of the adaptors (Fig. 1, step V in blue). Consequently, a unique standard curve is necessary to detect the captured material in all experiments, irrespective of the mutation type or genomic locus.

Here is the protocol to generate the unique standard curve (Fig. 2C).

a. The plasmid containing a product of DTECT ligation (Addgene #139333) is diluted at a concentration of approximately 10 ng/μL.

Table 1 Oligonucleotides used in this study.

Name	Sequence (5′ → 3′)	Description
OB1	ctggggcacgggtaagaagcattctgtctctcttctaagaattcgagctcggtacccg	Constant adaptor
OB2	cgggtaccgagctcgaattcttagaagagagacagaatgcttcttacccgtgccccag**aa**	AA-Variable adaptor
OB3	cgggtaccgagctcgaattcttagaagagagacagaatgcttcttacccgtgccccag**ac**	AC-Variable adaptor
OB4	cgggtaccgagctcgaattcttagaagagagacagaatgcttcttacccgtgccccag**ag**	AG-Variable adaptor
OB5	cgggtaccgagctcgaattcttagaagagagacagaatgcttcttacccgtgccccag**at**	AT-Variable adaptor
OB6	cgggtaccgagctcgaattcttagaagagagacagaatgcttcttacccgtgccccag**ca**	CA-Variable adaptor
OB7	cgggtaccgagctcgaattcttagaagagagacagaatgcttcttacccgtgccccag**cc**	CC-Variable adaptor
OB8	cgggtaccgagctcgaattcttagaagagagacagaatgcttcttacccgtgccccag**cg**	CG-Variable adaptor
OB9	cgggtaccgagctcgaattcttagaagagagacagaatgcttcttacccgtgccccag**ct**	CT-Variable adaptor
OB10	cgggtaccgagctcgaattcttagaagagagacagaatgcttcttacccgtgccccag**ga**	GA-Variable adaptor
OB11	cgggtaccgagctcgaattcttagaagagagacagaatgcttcttacccgtgccccag**gc**	GC-Variable adaptor
OB12	cgggtaccgagctcgaattcttagaagagagacagaatgcttcttacccgtgccccag**gg**	GG-Variable adaptor
OB13	cgggtaccgagctcgaattcttagaagagagacagaatgcttcttacccgtgccccag**gt**	GT-Variable adaptor
OB14	cgggtaccgagctcgaattcttagaagagagacagaatgcttcttacccgtgccccag**ta**	TA-Variable adaptor
OB15	cgggtaccgagctcgaattcttagaagagagacagaatgcttcttacccgtgccccag**tc**	TC-Variable adaptor
OB16	cgggtaccgagctcgaattcttagaagagagacagaatgcttcttacccgtgccccag**tg**	TG-Variable adaptor

Continued

Table 1 Oligonucleotides used in this study.—cont'd

Name	Sequence (5′ → 3′)	Description
OB17	cgggtaccgagctcgaattcttagaagagagacagaatgcttcttacccgtgccccag**<u>tt</u>**	TT-Variable adaptor
OB18	gcaattcctcacgagacccgtcctg	DTECT qPCR for
OB19	cgggtaccgagctcgaattcttagaag	DTECT qPCR rev
OB210	gcaattcctcacgagacccgtcctgttgtgaagagaggatctgaagaaataacggccaac	AcuI tagging PCR *SAGE1*
OB211	taagttgaggtcacgaaagataattca	PCR *SAGE1* rev
OB212	gcaattcctcacgagacccgtcctgcctgcgtttgtccctctgaagccagcatctgctct	AcuI tagging PCR *DPY30*
OB213	gatgaaggaagattgtagctcgtag	PCR *DPY30* rev

List of oligonucleotide sequences used to generate the library of adaptors, detect the captured material, and detect *SAGE1* and *DPY30* mutations. OB1-OB17 are used to construct the library of adaptors (3.7). OB1 corresponds to the constant oligonucleotide. OB2-OB17 correspond to the variable oligonucleotides. The dinucleotide specific to each adaptor is underlined. OB18 and OB19 are primers used to detect the captured material by analytical or quantitative PCR (3.7, 5.1, and 5.2). OB210-OB213 are DTECT oligos to capture *SAGE1* and *DPY30* mutations.

The exact concentration of the dilution must be accurately determined with a spectrophotometer (e.g., nanodrop). In our example, the measured concentration is 11.4 ng/μL.

b. Several 10-fold serial dilutions of the plasmid are prepared by adding 1 μL of the plasmid in 9 μL of H_2O. This dilution corresponds to "sample 1" at approximatively 1 ng/μL (1.14 ng/μL in our example).

c. 1 μL of "sample 1" is added into 9 μL of H_2O to obtain "sample 2" at approximatively ~0.1 ng/μL (0.114 ng/μL in our example).

d. The successive 10-fold serial dilutions are repeated to obtain a total of 7 samples.

e. The qPCR master mix containing the DTECT primers, named OB18 and OB19 (sequences listed in Table 1), is prepared. Each sample is tested in technical triplicates. In addition, a "negative control" sample that consists of no DNA template is also tested.

Master mix component	Final concentration	Volume (μL)—10 μL/reaction	Volume (μL)—Master mix for 26 reactions
H_2O	–	3.8	98.8
SYBR Green (2 ×)	1 ×	5	130
OB18, 100 μM	1 μM	0.1	2.6
OB19, 100 μM	1 μM	0.1	2.6
Diluted samples	–	1	–

f. 9 μL of the master mix is loaded into 24 qPCR wells.

g. 1 μL of the respective sample dilutions is added to the wells in triplicate.

h. 1 μL of H_2O is added in the last three qPCR wells as negative controls.

i. The qPCR machine is set up with the following program:

Temperature (°C)	Time (s)	PCR phase	Number of cycles
50	120	Hold stage	1
95	600		
95	10	PCR stage	40
60	30		
95	15	Melting	1
60	60		
95	15		

j. The cycle threshold (Ct) values are retrieved for each triplicate sample (columns 1–3 in the table below).

k. The average Ct values obtained from the triplicates are calculated using the formula "=AVERAGE(A2:C2)" (column 4 in the table below). Next, the concentrations of the 10-fold dilutions are determined using the formula "=LOG10(E2)" (columns 5 and 6).

Triplicate 1 (Ct)	Triplicate 2 (Ct)	Triplicate 3 (Ct)	Average Ct	Concentration (ng)	Log10 (concentration)
7.173	7.690	7.527	7.464	1.14	0.057
11.195	11.025	10.958	11.059	0.114	−0.943
13.671	13.412	13.285	13.456	0.0114	−1.943
17.009	17.143	17.482	17.211	0.00114	−2.943
20.427	20.540	20.742	20.570	0.000114	−3.943
24.120	24.351	24.333	24.268	0.0000114	−4.943
27.567	27.171	27.208	27.315	0.00000114	−5.943
30.759	30.506	31.126	30.797	H_2O (control)	

l. The threshold cycle as a function of the dilution factor is plotted in a graph representing the linear regression of the "Average Ct" as a function of "log10 (concentration)."

Each experimental point of the standard curve should be aligned to confirm the inverse correlation between the log of the quantity of DNA and the Ct value. In addition, the efficiency of OB18 and OB19 primers can be calculated from the slope using the formula: Efficiency = $10^{(-1/\text{slope})}$. Primer efficiency should be close to 100%, and the coefficient correlation (R^2) should be close to 1. In our example, the following standard curve formula determines the amount of captured material: $y = -3.3245 \times + 7.5504$. Part 5.2 describes how to measure the relative frequency of genomic signatures using the standard curve.

4. Step-by-step dinucleotide signature capture

The DTECT protocol contains five steps for the signature capture (4.1–4.5) and a detection analysis step (5.1 and/or 5.2). The signature capture consists of (4.1) the design of two primers for the AcuI-tagging PCR,

(4.2) AcuI-tagging PCR to tag the desired signature of interest with the AcuI motif and insert the detection handle sequence, (4.3) digestion of the amplicon with AcuI, (4.4) isolation of the small digestion fragment and (4.5) DNA adaptor ligation to capture desired signatures. The DNA ligated product is subsequently analyzed by quantitative (5.1) and/or qualitative PCR (5.2).

4.1 Design of primers for the AcuI-tagging PCR

a. Selection of the dinucleotide of interest

The first objective is to select a dinucleotide signature of interest that must include the base (or bases) probed for modification (Fig. 1, step I in red). Notably, a selected genomic base can be included in four independent dinucleotides (Fig. 2D). Precisely, two dinucleotides can be designed using the upstream or downstream nucleotide of the targeted base from each DNA strand, as illustrated in Fig. 2D for the *DPY30* target. The selection of the dinucleotide provides substantial flexibility in the design of the AcuI-tagging primer. The generation of diverse dinucleotide signatures to capture the same genomic base is a striking advantage of DTECT (advantages are described in Section 6.2), which helps bypass certain limitations of the method (limitations are described in Section 6.1).

b. General design guidelines for AcuI-tagging primers

Timing: ~5 min

The AcuI-tagging primer has a total length of 60 nt and includes three essential parts. First, the 5′ end contains a 25 nt handle detection sequence (5′-GCAATTCCTCACGAGACCCGTCCTG-3′) to detect the captured material in step 5. Second, it includes 29 nt of the targeted nucleic acid sequence adjacent to the targeted genomic sequence. Third, an AcuI recognition motif (5′-CTGAAG-3′) is inserted strictly at 14 nt from the 3′ end (Fig. 1, step I in green). Consequently, the targeted sequence complementary to the nucleic acid sequence is interspaced by the AcuI motif in two independent parts of 15 nt and 14 nt, respectively (Fig. 1, step I in yellow).

c. Design of PCR reverse primers

Timing: ~5 min

In addition to the AcuI-tagging primer, a reverse primer is designed to amplify the targeted dinucleotide during the AcuI-tagging PCR (Fig. 1, step I). The reverse primer is a locus-specific primer located at a distance greater than 100 bp from the selected dinucleotide. The selection of a primer sequence at a distance > 100 bp is important for the rapid and efficient separation of the two DNA fragments generated during AcuI digestion, as detailed in Section 4.4 and illustrated in Fig. 1, step IV.

For the design of the reverse primer, we recommend using "*Primer 3*" (https://bioinfo.ut.ee/primer3/) (Koressaar et al., 2018) with the following parameters: The primer length is selected between 25 and 30 nt, and the optimal melting temperature (Tm) should be between 57 and 63 °C. We also recommend selecting the "Mispriming library" with the species of interest to design a highly specific primer. For example, to amplify human genomic DNA, "HUMAN" should be selected at "Mispriming library =."

d. Primer design to capture the DPY30 and SAGE1 mutations introduced with prime editing

To illustrate this protocol, we introduce two transversion mutations into *DPY30* and *SAGE1* genes using prime editing (Fig. 3). We design AcuI-tagging primers and reverse primers to amplify the signatures and distinguish the edits from the reference signatures (Fig. 3).

Design of the AcuI-tagging oligonucleotide for the detection of the *DPY30* mutation (G > C):

First, we collect the 29 nt genomic sequence upstream of the nucleotide targeted for modification: "5′-CCTGCGTTTGTCCCTCCAGCATC TGCTCT-3.′" Next, we insert the AcuI motif sequence "5′-CTGAAG-3′" 14 nt upstream of the 3′ end from the targeted nucleotide "5′-CCTGCGTTTGTCCCT-*CTGAAG*-CCAGCATCTGCTCT-3′." Finally, we introduce the handle detection sequence (5′-GCAATT CCTCACGAGACCCGTCCTG-3′) at the 5′ end of the oligonucleotide: "5′-*GCAATTCCTCACGAGACCCGTCCTG*CCTGCGTTTGTCCC TCTGAAGCCAGCATCT
GCTCT-3′" (Oligonucleotide named OB212). Oligonucleotide sequences are available in Table 1.

Design of the AcuI-tagging oligonucleotide for the detection of the *SAGE1* mutation (C > G):

First, we collect the 29 nt genomic sequence that is upstream of the nucleotide targeted for modification: "5′-TTGTGAAGAGAGGATAAA TAACGGCCAAC-3′." Next, we insert the AcuI motif sequence "5′-CTGAAG-3′" 14 nt upstream of the 3′ end from the targeted nucleotide "5′-TTGTGAAGAGAGGAT-*CTGAAG*-AAATAACGGCCAAC-3′". Finally, we introduce the handle detection sequence (5′-GCAATTCC TCACGAGACCCGTCCTG-3′) at the 5′ end of the oligonucleotide: "5′-*GCAATTCCTCACGAGACCCGTCCTG*TTGTGAAGAGAGGA TCTGAAGAAATAACGGCCAAC-3′" (Oligonucleotide named OB210). Oligonucleotide sequences are available in Table 1.

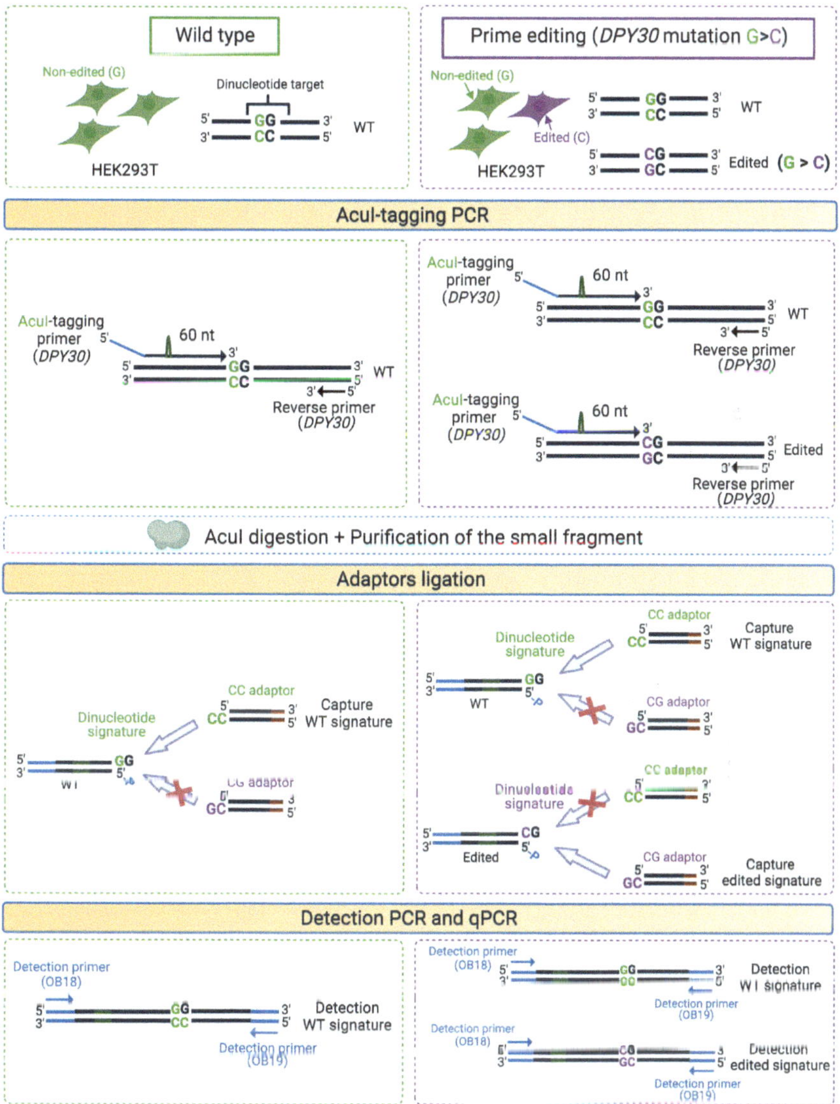

Fig. 3 DTECT capture of newly created signatures by CRISPR-mediated prime editing at the *DPY30* and *SAGE1* loci. Illustration of DTECT to capture the WT and edited signatures on the control (left) and edited samples (right) at the *DPY30* locus. An AcuI-tagging PCR is performed to amplify the signatures of interest in the respective samples. Adaptors complementary to the WT and edited signatures are utilized to capture signatures and are detected by PCR and qPCR using oligos OB18 and OB19.

The reverse primers are designed computationally using "*primer 3*," as described in 4.1.c (sequences available in Table 1). Reverse oligonucleotide sequences are named OB211 and OB213 for *SAGE1* and *DPY30*, respectively. Oligonucleotide sequences are available in Table 1.

4.2 Amplification of the targeted dinucleotide using the AcuI-tagging primer and PCR purification

Timing: ~60–90 min

The second step of the protocol consists of amplifying the locus of interest with the desired targeted dinucleotide by PCR using primers designed in Section 4.1. This protocol focuses on DTECT, so we refer the readers to excellent protocols for detailed experimental design for genome editing (Anzalone et al., 2019; Bak, Dever, & Porteus, 2018; Huang, Newby, & Liu, 2021; Ran et al., 2013; Santos, Kiskinis, Eggan, & Merkle, 2016). A critical negative control consists of the transfection of a non-targeting guide RNA to obtain an unedited control genomic sample.

a. Cells edited at the *SAGE1* and *DPY30* loci and unedited (control) cells are harvested by centrifugation at 300 × *g* for 3 min. Next, the medium is removed, and the cell pellet is washed with PBS. Then, cells are centrifuged at 300 × *g* for 3 min, and the PBS is removed.

b. The genomic DNA (gDNA) is extracted with the Quick Extract solution. First, the cell pellet is resuspended in 50 μL Quick Extract solution. Next, the solution is mixed by vortexing for 15 s and incubated at 65 °C for 6 min. Once the incubation is completed, the solution is mixed by vortexing for 15 s.

c. The tube is transferred to a heat block for 2 min at 98 °C.

d. H_2O is added to dilute the gDNA at approximatively 200 ng/μL as determined by nanodrop quantification. The gDNA preparation can be stored in 50–100 μL aliquots at −80 °C or immediately used for the AcuI-tagging PCR. If the genomic DNA samples are stored, the sample should be heated for 2 min at 98 °C before adding samples in the PCR reactions.

Note: The DTECT protocol can start here if DNA has been extracted with different approaches.

e. The AcuI-tagging PCR is conducted in a 25 μL reaction. First, a PCR master mix is prepared to amplify the loci of interest (*SAGE1* and *DPY30*) in the edited and unedited samples.

Master mix component	Final concentration	Volume (μL)—25 μL/reaction	Volume (μL)—Master mix for 2.5 reactions
H_2O	–	18.05	45.125
Q5 buffer, 5 ×	1 ×	5	12.5
dNTP, 10 mM	0.1 μM	0.2	0.5
Q5 polymerase, 2 units/μL	0.5 units	0.25	0.625
AcuI-tagging primer, 100 μM	1 μM	0.25	0.625
Reverse primer, 100 μM	1 μM	0.25	0.625
gDNA, 200 ng/μL	–	1	–

f. One AcuI-tagging PCR is performed per edited and unedited genomic DNA samples. 24 μL of the master mix is loaded in two PCR tubes, and 1 μL (~200 ng/μL) of the respective genomic DNA is added. The AcuI-tagging of the *SAGE1* targeted dinucleotide is conducted with primers OB210 and OB211. The AcuI-tagging of the *DPY30* targeted dinucleotide is conducted with primers OB212 and OB213.

g. The PCR machine is set up with the following conditions.

Temperature	Time	PCR phase	Cycle number
95 °C	30 s	Denaturation	1
95 °C 58 °C 72 °C	10s 10 s 45 s	Denaturation Annealing Extension	40
72 °C	60 s	Extension	1

h. A 2% (wt/vol) agarose gel is prepared by combining 1 g of agarose in 50 mL 1 × TAE. The mixture is melted by heating until agarose has dissolved (approximatively 1–2 min). The solution is swirled carefully to ensure complete solubilization of the agarose and prevent excess evaporation of the buffer.

Note: Appropriate gloves must be worn to protect from burns.

i. The agarose is carefully poured into a gel cassette, bubbles are removed with a pipette tip, and a gel comb is added. The gel is left to cool down at room temperature.
j. Once the agarose gel has solidified, the molecular weight marker is prepared by mixing 5 μL DNA ladder, 1 μL loading dye, 0.5 μL diluted SYBR gold. Then, the samples are prepared by adding 1 μL diluted SYBR gold and 5 μL loading dye to the PCR tubes.
k. The AcuI-tagging PCRs are loaded on the gel to verify the successful amplification of the loci and extract the amplicons.

Expected result: A successful PCR should be a single band at the size of the expected amplicon. In the *DPY30* and *SAGE1* examples, the amplicons have an expected length of 582 and 646 bp, respectively.

l. The four AcuI-tagged amplicons are excised from the gel with a clean razor blade and purified using a Zymoclean Gel DNA Recovery Kit. To avoid contamination between samples, the razor blade is washed with ethanol and H_2O between samples. Alternatively, the AcuI-tagging PCR can be purified on a column.
m. The purified AcuI tagged amplicons are quantified using a nanodrop and diluted at ~20 ng/μL in the elution buffer.

4.3 Digestion of the AcuI-tagged genomic amplicon with AcuI

Timing: 80 min

The third step of the DTECT protocol requires digestion of the AcuI-tagged amplicons with AcuI (Fig. 3). AcuI cleaves at a fixed and predictable position 14/16 bp downstream of its recognition motif, generating a 3′ dinucleotide signature and leaving 5′ phosphates (Fig. 3).

a. The AcuI digestion master mix is prepared as indicated below.

Master mix component	Final concentration	Volume (μL)—20 μL/reaction	Volume (μL)—Master mix for 4.5 reactions
H_2O	–	15.75	70.875
CutSmart Buffer, 10 ×	1 ×	2	9
AcuI, 5000 units/mL	1.25 units	0.25	1.125
AcuI-tagged PCR, 20 ng/μL	40 ng	2	–

b. 18 µL of the master mix is loaded into tubes, and 2 µL of the purified AcuI-tagged amplicon (40 ng/µL) is added.

c. The reactions are incubated at 37 °C for 60 min and heated at 65 °C for 20 min to inactivate AcuI.

Pause point: The AcuI-digested amplicons can be stored at −20 °C or immediately used for the isolation step (step 4.4).

4.4 Isolation of the small AcuI-digested fragments by solid-phase reversible immobilization (SPRI)

Timing: ~10 min

DTECT requires the isolation of the small fragments (60 bp) to facilitate the ligation of DNA adaptors. The isolation of the small fragment is achieved by the precipitation of the large fragment by SPRI using magnetic beads. The stock of beads must be resuspended by mixing the stock bottle. Once well resuspended, a small aliquot (~800 µL) is placed in a 1.5 mL microcentrifuge tube and stored at 4 °C. The aliquot of beads should be placed at room temperature approximately 1 h before step 4.4.

a. 10 µL of the digestion reactions are transferred in new microcentrifuge tubes, and 18 µL of AMPure XP magnetic beads are added (volume ratio of DNA:beads = 1:1.8). The solutions are mixed by pipetting up and down 10 times and are incubated at room temperature for 5 min.

b. After incubation, the tubes are placed on a magnetic rack until the beads are pulled to the side of the tube by the magnetic field, which takes approximately 3–5 min. The small fragment is solubilized in the supernatant, and the large fragment is bound to the magnetic beads.

c. 20 µL of the supernatant is transferred into a new 1.5 mL microcentrifuge tube, and 40 µL H_2O is added to dilute the DNA.

Pause point: The isolated fragments can be stored at −20 °C or immediately used for the ligation.

4.5 The capture of dinucleotide signatures

Timing: 70 min

The final step of the capture is the ligation of DNA adaptors that are complementary or non-complementary to the reference signatures and edited signatures (Fig. 3). In the examples presented here, the selected dinucleotide signature is 5′-GG-3′ for *DPY30* and 5′-CA-3′ for *SAGE1*; and the expected edited signatures is 5′-CG-3′ for *DPY30* and 5′-GA-3′ for *SAGE1*. For simplification, only the *DPY30* target is illustrated in Fig. 3.

a. Adaptors are selected for their complementarity to the reference and edited dinucleotide signatures. We, therefore, use the complementary adaptors 5′-CC-3′ and 5′-TG-3′ to capture the reference dinucleotide signatures; and the complementary adaptors 5′-CG-3′ and 5′-TC-3′ for the mutations inserted with prime editing into *DPY30* and *SAGE1*, respectively.

b. Independent ligation master mixes for each adaptor are prepared.

Master mix component	Final concentration	Volume (μL)—10 μL/reaction	Volume (μL)—Master mix for 2.5 reactions
H_2O	–	6.5	16.25
T4 ligase buffer, 5 ×	1 ×	2	5
T4 ligase, 1 unit	0.5 units/μL	0.5	1.25
Adaptor, 20 ×	1 ×	0.5	1.25
Isolated material (Step IV)	–	0.5	–

c. 9.5 μL of the master mix is loaded into PCR tubes, and 0.5 μL of the isolated material is added to the ligation reactions.

d. The reaction is incubated at 25 °C for 60 min followed by 65 °C for 10 min to heat inactivate the DNA ligase.

Pause point: The ligation tubes can be stored at −20 °C or immediately used to analyze the captured material (steps 5.1 or/and 5.2).

5. Capture detection and quantification

The ligation reactions can either be analyzed by analytical or quantitative PCR (Fig. 3). While analytical PCRs determine the presence of a particular signature, quantitative PCRs determine the relative quantity of the respective signatures in the nucleic acid samples.

5.1 Analytical detection

Timing: 45 min

The analytical PCR detects the ligation product (Fig. 4A–B). The detection of the ligated product reveals the presence of the particular signature in the original nucleic acid sample.

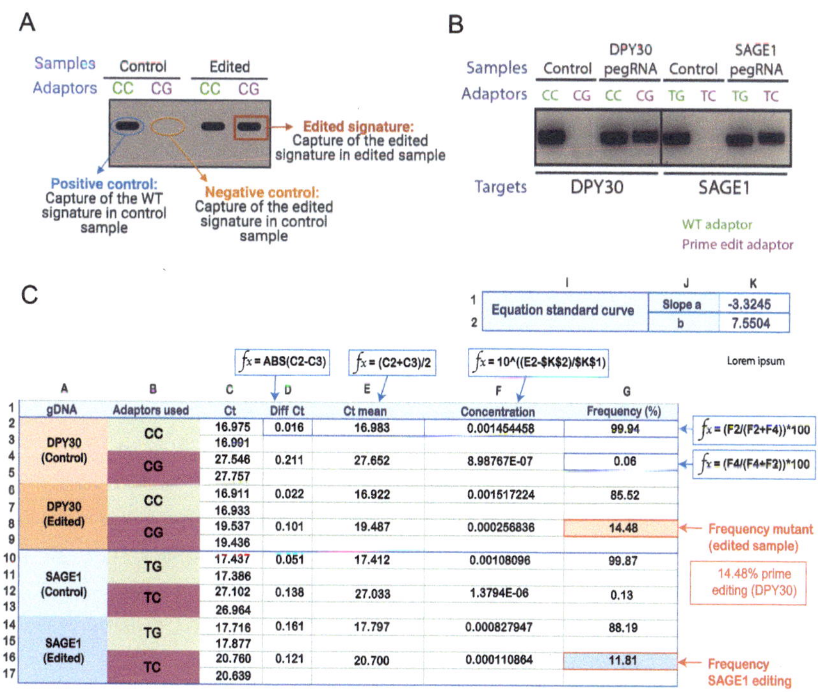

Fig. 4 Example of analytical and quantitative analysis of *DPY30* and *SAGE1* edited sites with prime editing. (A) Illustration of expected results from an analytical gel. WT signature (positive control) is specifically detected in the control sample but not edited signature (negative control). The detection of a captured signature in the edited sample indicates successful editing. (B) Agarose gel of DTECT for the capture of prime editing at the *DPY30* and *SAGE1* loci. (C) Mathematical formulas and quantifications of the newly created signatures with prime editing using qPCR.

a. A PCR master mix is prepared to amplify the captured material

Master mix component	Final concentration	Volume (μL) 12.5 μL/reaction	Volume (μL)—Master mix for 4.5 reactions
H_2O	–	9.075	40.84
Q5 buffer, 5×	1×	2.5	11.25
dNTP, 10 mM	0.05 mM	0.2	0.9
OB18, 100 μM	0.5 μM	0.0625	0.281
OB19, 100 μM	0.5 μM	0.0625	0.281
Q5 polymerase, 2 units/μL	1 unit	0.1	0.45
Ligation products (Step V)	–	0.5	–

b. 12 μL of the PCR master mix is loaded into PCR tubes, and 0.5 μL of the respective ligations is added.

c. The analytical PCR is conducted under the following conditions:

Temperature (°C)	Time (s)	PCR phase	Cycle number
95	60	Denaturation	1
95	10	Denaturation	
65	5	Annealing	15–25 (see below)
72	7	Extension	
72	60	Extension	

The number of PCR cycles is adjusted based on the expected efficiency. For example, a low abundance of genomic variants (≤1% frequency) is detected with 23–25 PCR cycles. On the other hand, the detection of greater amounts of edited products can be achieved between 15 and 22 PCR cycles. For instance, given that we did not know the expected editing level of *DPY30* and *SAGE1*, we conducted 22 PCR cycles (Fig. 4B).

d. Before starting the PCR, a 2% agarose gel is prepared, as described in 4.2f–i.

e. 0.5 μL of diluted SYBR gold and 2.5 μL loading dye are added to the reaction, and the PCR and DNA ladder is loaded on the agarose gel. Then, the gel is run in 1 × TAE until the DNA has resolved.

f. Gels are developed using a gel imaging system (e.g., ChemiDoc or LI-COR Odyssey).

Expected result: If the experiment is successful, it is expected that the capture of the reference signature in the non-edited sample (positive control) generates an unambiguous amplicon that migrates at 120 bp (Fig. 4A, in blue control sample). As expected, this amplicon corresponds to the detection handle, the genomic DNA sequence interspaced with the AcuI motif, the dinucleotide signature, and the adaptor (Billon et al., 2020). In addition, the capture using the variant adaptor in the non-edited sample (negative control) generates no amplification (or background) (Fig. 4A, in orange control sample). A signal in the edited sample captured with the adaptor corresponding to the variant confirms the successful incorporation of the mutation (Fig. 4A, edited signature in red edited sample). The development of the analytical gels for the *SAGE1* and *DPY30* prime editing experiment is shown in Fig. 4B. As expected, a strong capture was observed in the positive control and no detection in the negative control. However, by capturing the edited signature in the samples that were edited with prime editing, the newly created signature was unambiguously detected with DTECT (Fig. 4B, CG adaptor for *DPY30* and TC adaptor for *SAGE1* in pegRNA samples).

5.2 Quantitative detection

Timing: 90 min

Quantitative detection enables the determination of the relative amount of captured material by the respective adaptors (Fig. 4C).

a. The amount of captured material is quantified by qPCR in technical duplicates.

b. A qPCR master mix is prepared as indicated below.

Master mix component	Final concentration	Volume (μL)—10 μL/reaction	Volume (μL)—Master mix for 18 reactions
H_2O	–	3.8	68.4
SYBR green, 2 ×	1 ×	5	90
OB18, 100 μM	1 μM	0.1	1.8
OB19, 100 μM	1 μM	0.1	1.8
Ligated products	–	1	–

c. 9 μL of the master mix is loaded into a qPCR plate, and 1 μL of the respective ligation reactions is added in duplicates.

d. The qPCR plate is sealed with a transparent adhesive film to prevent sample evaporation during PCR heating and loss during transportation. The plate is centrifuged to ensure the samples are at the bottom of the plate.

e. The qPCR machine is set with the following program.

Temperature (°C)	Time (s)	PCR phase	Number of cycles
50	120	Hold stage	1
95	600		
95	10	PCR stage	40
60	30		
95	15	Melting	1
60	60		
95	15		

5.2.1 Quantification and statistical analysis

Downstream analysis of the experiment is shown in Fig. 4C. The standard curve equation (determined in Section 3.8) and experimental Ct scores (Fig. 4C, column C) determine the relative frequency of each genomic signature captured by the respective adaptors (Fig. 4C, column G).

First, the difference in Ct between technical duplicates is calculated (Fig. 4C, column D). Usually, when the difference between technical duplicates is >0.5 Ct, the qPCR samples are repeated. Then, the average cycle threshold (Ct mean) is calculated from the technical duplicates of each sample (Fig. 4C, column E) using the formula "=(C2 + C3)/2." Next, the standard curve generated in step 3.7 is used to determine the relative quantity of each signature captured using the formula: "Concentration = 10^((Mean Ct − 7.5504)/−3.3245)" (Fig. 4C, column F).

The relative abundance between the wild type (WT) and the variant is determined as follows: "Frequency Mutant = 10^(Concentration Mutant/(Concentration Mutant + Concentration WT)) x 100" and "Frequency WT = 10^(Concentration WT/(Concentration Mutant + Concentration WT)) x 100" (Fig. 4C, column G).

Expected result: If successful, the relative abundance associated with the negative control should be close to 0%, representing the background capture. Here, the DPY30 background capture is at 0.06% (Fig. 4C, G4) and the SAGE1 capture background is at 0.13% (Fig. 4C, G12). Interestingly, the prime editing frequency at these loci is 14.48% (Fig. 4C, G8) and 11.81% (Fig. 4C, G16) for *DPY30* and *SAGE1*, respectively.

6. Limitations and advantages of DTECT

6.1 Limitations of DTECT

DTECT has three limitations caused by the sequences surrounding the targeted dinucleotide, the specificity of AcuI, and the error-prone repair of lesions introduced by genome editing technologies.

The first limitation occurs when one or multiple AcuI recognition motifs (5′-CTGAAG-3′) are present in the targeted nucleic acid sequence near the dinucleotide of interest. If genomic AcuI motif(s) are included in the AcuI-tagged amplicon, the AcuI digestion generates additional undesired signature(s), thereby potentially affecting targeted capture. However, given that AcuI-tagging primers can be designed from both strands (Fig. 2D), only when two AcuI motifs are positioned on both sides of the targeted dinucleotide DTECT is affected. Consequently, genomic AcuI motifs present near the target site are problematic only if two AcuI motifs are close (<30 nt) to the targeted dinucleotide both upstream and downstream. Nevertheless, a 6-nucleotide motif is approximately present only every 4 kb, limiting the probability of two interfering AcuI motifs.

Although type IIS enzymes cleave far from their recognition motif, their helpful activity can come at the expense of their precision. The slippage is an unspecific activity that occurs when cleavage is shifted by one nucleotide. Slippage activity of certain type IIS enzymes is more frequent than the canonical cleavage (Lundin et al., 2015). However, AcuI is not prone to frequent slippage, estimated at only 1.1% (Lundin et al., 2015), which explains the high precision of DTECT. Moreover, the dinucleotide signatures induced by AcuI slippage are predictable based on the sequence flanking the targeted dinucleotide. Therefore, the dinucleotide induced by the slippage can be avoided by the appropriate selection of the adaptors (Billon et al., 2020).

Finally, DTECT can only capture the identity of a two-nucleotide window. Certain byproducts of the repair of lesions introduced by genome editing technologies can occur in a larger window. For instance, DSB repair by end-joining leads to the generation of predictable insertion-deletion (indel) mutations. Consequently, the indels generated near the targeted dinucleotide signature are susceptible to disrupt the proximal sequence of the targeted signature, thereby inducing erroneous AcuI-tagging PCR amplification. This limitation results in an overestimation of the frequency of precision genome editing by DTECT, as we revealed previously (Billon et al., 2020). However, we also demonstrated that it is possible to position the targeted dinucleotide at a location where no indels are induced (generally ~10 bp from the DSB). However, indels are rare byproducts of CRISPR-mediated base editing, prime editing, and naturally occurring genetic variants, suggesting that DTECT is less affected in these contexts.

6.2 Strengths and advantages of DTECT

DTECT is a simple detection method that relies on standard molecular biology techniques (PCR, digestion, and ligation), off-the-shelf reagents, and enzymes (AcuI and T4 ligase) and requires minimal equipment (thermocycler and qPCR). In addition, DTECT has a short turnaround time (~4–5 h) with limited hands on time.

Notably, one of the most significant strengths of DTECT is its versatility and flexibility. A unique library of 16 standard adaptors is sufficient to identify all dinucleotide signature types (Fig. 2A), and each genomic base can be detected using 4 distinct signatures (Fig. 2D). Importantly, DTECT is sensitive as it can distinguish signatures present at low frequency (<1%) (Fig. 4C) in complex samples, including precious patient specimens (Billon et al., 2020). The high sensitivity of DTECT is mediated by the complementarity between the signature and the specific adaptor. In addition, the ligation creates a stable ligation product by covalently linking the adaptors and the signature. Finally, the detection is compatible with analytical and quantitative detection modalities, which provide complementary advantages. Indeed, it has the dual benefit of quantifying the relative frequency of signatures for precision genome editing experiments (Fig. 4C) and detecting the presence of a specific signature for the rapid genotyping of cellular and animal models, like we previously demonstrated (Billon et al., 2020).

Strikingly, DTECT has limited technical variabilities across distinct experimental conditions, samples, and mutation types and is conducted in a highly controlled experimental setup. Indeed, a unique pair of detection primers are used in all experiments to quantify the amount of captured material, regardless of the mutation or sample type. This is a significant advantage because a unique standard curve is sufficient (see part 3.7) to quantify the capture of all signatures. In contrast, methods using mutation-specific primers or probes require separate evaluation of the primer efficiencies and do not provide positive controls to ensure experiment quality. This is a problem because, unlike with DTECT, primers/probes-specific mutations have variable and unpredictable efficiencies due to the competition between reference and variant alleles. Another remarkable advantage of DTECT is the inherent controls to determine the efficiency and specificity of signature capture, as shown in Fig. 4 for the prime editing experiments. For instance, negative controls provide the experimental background as measured by capturing the variant signature in the control sample (Fig. 4C). Positive controls ensure that DTECT assays have been successfully conducted by capturing the WT signature in reference nucleic acid samples (Fig. 4C). Positive controls are, therefore, crucial to help new users to implement DTECT. The ease of implementing DTECT and the presence of controls are substantial advantages for its adoption.

DTECT is a robust assay with several advantages over alternative detection methods for the rapid and accurate identification of nucleic acid sequences. We previously compared the performance of DTECT to sequencing technologies to detect precision genome editing and genetic variants (Billon et al., 2020). Strikingly, DTECT shows comparable performance to Next-Generation Sequencing (NGS) for its precision to detect desired signatures (Billon et al., 2020). However, NGS has a larger window of detection and provides highly precise quantification of the mixture of modified alleles. Unlike DTECT, which takes a few hours to perform, sequencing technologies are time-consuming (several days to weeks) and expensive. DTECT and NGS are complementary as DTECT can validate the presence of genetic signatures before a deeper analysis of low-frequency mutations/editing byproducts by NGS. We also compared DTECT with Sanger sequencing-based detection methods (Billon et al., 2020). We observed that a significant fraction of reactions does not generate the high-quality profiles required for accurate quantification, making DTECT more reliable than Sanger sequencing.

DTECT can be applied to identify changes introduced by modern precision genome editing technologies and natural variants in virtually any biological system. DTECT requires a small amount of nucleic acid as starting material, only sufficient to conduct the AcuI-tagging PCR. This is advantageous for clinical applications in which samples are limited. For example, we previously demonstrated the utility of DTECT to identify oncogenic signatures in the bone marrow specimens from pediatric cancer patients (Billon et al., 2020). Consequently, DTECT will be helpful for the identification of pathogenic variants in cellular and animal models, human biopsies, and fluid sampling.

6.3 Optimizations and troubleshooting

With the presence of positive and negative controls and the use of commercial reagents, no optimizations are required to implement DTECT. Nevertheless, in rare cases (<5%), we observed that the AcuI-tagging PCR is inefficient at hard-to-amplify loci. One solution is to amplify the targeted genomic locus using standard primers and conduct the AcuI-tagging PCR using the locus amplicon as template DNA.

7. Concluding remarks

This step-by-step protocol describes how to identify nucleic acid signatures and associated genetic variation introduced by modern precision genome editing technologies, including CRISPR-mediated base editing and prime editing or naturally occurring mutations. The DTECT protocol is easy to conduct. It only requires oligo design, PCR amplification, restriction digestion, DNA ligation, and analytical/quantitative PCR. Furthermore, DTECT only requires commercially available reagents, is rapid to execute (~4–5 h), providing same-day quantitative determination of editing frequency. The development of novel detection methods for genetic signatures will improve the characterization of variations of uncertain significance and accelerate clinical diagnostics.

References

Genomes Project, C., Auton, A., Brooks, L. D., Durbin, R. M., Garrison, E. P., Kang, H. M., et al. (2015). A global reference for human genetic variation. *Nature, 526*(7571), 68–74. https://doi.org/10.1038/nature15393.

Alexandrov, L. B., Nik-Zainal, S., Wedge, D. C., Aparicio, S. A., Behjati, S., Biankin, A. V., et al. (2013). Signatures of mutational processes in human cancer. *Nature, 500*(7463), 415–421. https://doi.org/10.1038/nature12477.

Anzalone, A. V., Koblan, L. W., & Liu, D. R. (2020). Genome editing with CRISPR-Cas nucleases, base editors, transposases and prime editors. *Nature Biotechnology, 38*(7), 824–844. https://doi.org/10.1038/s41587-020-0561-9.

Anzalone, A. V., Randolph, P. B., Davis, J. R., Sousa, A. A., Koblan, L. W., Levy, J. M., et al. (2019). Search-and-replace genome editing without double-strand breaks or donor DNA. *Nature, 576*(7785), 149–157. https://doi.org/10.1038/s41586-019-1711-4.

Bak, R. O., Dever, D. P., & Porteus, M. H. (2018). CRISPR/Cas9 genome editing in human hematopoietic stem cells. *Nature Protocols, 13*(2), 358–376. https://doi.org/10.1038/nprot.2017.143.

Berkovich, E., Monnat, R. J., Jr., & Kastan, M. B. (2007). Roles of ATM and NBS1 in chromatin structure modulation and DNA double-strand break repair. *Nature Cell Biology, 9*(6), 683–690. https://doi.org/10.1038/ncb1599.

Billon, P., Bryant, E. E., Joseph, S. A., Nambiar, T. S., Hayward, S. B., Rothstein, R., et al. (2017). CRISPR-mediated base editing enables efficient disruption of eukaryotic genes through induction of STOP codons. *Molecular Cell, 67*(6), e1064. https://doi.org/10.1016/j.molcel.2017.08.008. 1068-1079.

Billon, P., Nambiar, T. S., Hayward, S. B., Zafra, M. P., Schatoff, E. M., Oshima, K., et al. (2020). Detection of marker-free precision genome editing and genetic variation through the capture of genomic signatures. *Cell Reports, 30*(10), e3286. https://doi.org/10.1016/j.celrep.2020.02.068. 3280–3295.

Brinkman, E. K., Chen, T., Amendola, M., & van Steensel, B. (2014). Easy quantitative assessment of genome editing by sequence trace decomposition. *Nucleic Acids Research, 42*(22), e168. https://doi.org/10.1093/nar/gku936.

Chailleux, C., Aymard, F., Caron, P., Daburon, V., Courilleau, C., Canitrot, Y., et al. (2014). Quantifying DNA double-strand breaks induced by site-specific endonucleases in living cells by ligation-mediated purification. *Nature Protocols, 9*(3), 517–528. https://doi.org/10.1038/nprot.2014.031.

Clement, K., Rees, H., Canver, M. C., Gehrke, J. M., Farouni, R., Hsu, J. Y., et al. (2019). CRISPResso2 provides accurate and rapid genome editing sequence analysis. *Nature Biotechnology, 37*(3), 224–226. https://doi.org/10.1038/s41587-019-0032-3.

Djureinovic, D., Hallstrom, B. M., Horie, M., Mattsson, J. S. M., La Fleur, L., Fagerberg, L., et al. (2016). Profiling cancer testis antigens in non-small-cell lung cancer. *JCI Insight, 1*(10), e86837. https://doi.org/10.1172/jci.insight.86837.

Drmanac, R., Sparks, A. B., Callow, M. J., Halpern, A. L., Burns, N. L., Kermani, B. G., et al. (2010). Human genome sequencing using unchained base reads on self-assembling DNA nanoarrays. *Science*, *327*(5961), 78–81. https://doi.org/10.1126/science.1181498.

Ellegren, H., & Galtier, N. (2016). Determinants of genetic diversity. *Nature Reviews. Genetics*, *17*(7), 422–433. https://doi.org/10.1038/nrg.2016.58.

Findlay, S. D., Vincent, K. M., Berman, J. R., & Postovit, L. M. (2016). A digital PCR-based method for efficient and highly specific screening of genome edited cells. *PLoS One*, *11*(4), e0153901. https://doi.org/10.1371/journal.pone.0153901.

Germini, D., Tsfasman, T., Zakharova, V. V., Sjakste, N., Lipinski, M., & Vassetzky, Y. (2018). A comparison of techniques to evaluate the effectiveness of genome editing. *Trends in Biotechnology*, *36*(2), 147–159. https://doi.org/10.1016/j.tibtech.2017.10.008.

Green, M. R., & Sambrook, J. (2019). Ligation and ligases. *Cold Spring Harbor Protocols*, *2019*-(8). https://doi.org/10.1101/pdb.top101352.

Guell, M., Yang, L., & Church, G. M. (2014). Genome editing assessment using CRISPR Genome Analyzer (CRISPR-GA). *Bioinformatics*, *30*(20), 2968–2970. https://doi.org/10.1093/bioinformatics/btu427.

Helleday, T., Eshtad, S., & Nik-Zainal, S. (2014). Mechanisms underlying mutational signatures in human cancers. *Nature Reviews. Genetics*, *15*(9), 585–598. https://doi.org/10.1038/nrg3729.

Huang, T. P., Newby, G. A., & Liu, D. R. (2021). Precision genome editing using cytosine and adenine base editors in mammalian cells. *Nature Protocols*, *16*(2), 1089–1128. https://doi.org/10.1038/s41596-020-00450-9.

The International HapMap Consortium. (2005). A haplotype map of the human genome. *Nature*, *437*(7063), 1299–1320. https://doi.org/10.1038/nature04226.

Karczewski, K. J., Francioli, L. C., Tiao, G., Cummings, B. B., Alfoldi, J., Wang, Q., et al. (2020). The mutational constraint spectrum quantified from variation in 141,456 humans. *Nature*, *581*(7809), 434–443. https://doi.org/10.1038/s41586-020-2308-7.

Khan, S., Ullah, M. W., Siddique, R., Nabi, G., Manan, S., Yousaf, M., et al. (2016). Role of recombinant DNA technology to improve life. *International Journal of Genomics*, *2016*, 2405954. https://doi.org/10.1155/2016/2405954.

Klompe, S. E., Vo, P. L. H., Halpin-Healy, T. S., & Sternberg, S. H. (2019). Transposon-encoded CRISPR-Cas systems direct RNA-guided DNA integration. *Nature*, *571*(7764), 219–225. https://doi.org/10.1038/s41586-019-1323-z.

Koressaar, T., Lepamets, M., Kaplinski, L., Raime, K., Andreson, R., & Remm, M. (2018). Primer3_masker: Integrating masking of template sequence with primer design software. *Bioinformatics*, *34*(11), 1937–1938. https://doi.org/10.1093/bioinformatics/bty036.

Labrie, S. J., Samson, J. E., & Moineau, S. (2010). Bacteriophage resistance mechanisms. *Nature Reviews. Microbiology*, *8*(5), 317–327. https://doi.org/10.1038/nrmicro2315.

Landrum, M. J., Chitipiralla, S., Brown, G. R., Chen, C., Gu, B., Hart, J., et al. (2020). ClinVar: Improvements to accessing data. *Nucleic Acids Research*, *48*(D1), D835–D844. https://doi.org/10.1093/nar/gkz972.

Lek, M., Karczewski, K. J., Minikel, E. V., Samocha, K. E., Banks, E., Fennell, T., et al. (2016). Analysis of protein-coding genetic variation in 60,706 humans. *Nature*, *536*(7616), 285–291. https://doi.org/10.1038/nature19057.

Lindsay, H., Burger, A., Biyong, B., Felker, A., Hess, C., Zaugg, J., et al. (2016). CrispRVariants charts the mutation spectrum of genome engineering experiments. *Nature Biotechnology*, *34*(7), 701–702. https://doi.org/10.1038/nbt.3628.

Lundin, S., Jemt, A., Terje-Hegge, F., Foam, N., Pettersson, E., Kaller, M., et al. (2015). Endonuclease specificity and sequence dependence of type IIS restriction enzymes. *PLoS One*, *10*(1), e0117059. https://doi.org/10.1371/journal.pone.0117059.

Miga, K. H., Koren, S., Rhie, A., Vollger, M. R., Gershman, A., Bzikadze, A., et al. (2020). Telomere-to-telomere assembly of a complete human X chromosome. *Nature*, *585*(7823), 79–84. https://doi.org/10.1038/s41586-020-2547-7.

Nambiar, T. S., Billon, P., Diedenhofen, G., Hayward, S. B., Taglialatela, A., Cai, K., et al. (2019). Stimulation of CRISPR-mediated homology-directed repair by an engineered RAD18 variant. *Nature Communications*, *10*(1), 3395. https://doi.org/10.1038/s41467-019-11105-z.

Pascal, J. M. (2008). DNA and RNA ligases: Structural variations and shared mechanisms. *Current Opinion in Structural Biology*, *18*(1), 96–105. https://doi.org/10.1016/j.sbi.2007.12.008.

Pickar-Oliver, A., & Gersbach, C. A. (2019). The next generation of CRISPR-Cas technologies and applications. *Nature Reviews. Molecular Cell Biology*, *20*(8), 490–507. https://doi.org/10.1038/s41580-019-0131-5.

Pingoud, A., Wilson, G. G., & Wende, W. (2014). Type II restriction endonucleases- -a historical perspective and more. *Nucleic Acids Research*, *42*(12), 7489–7527. https://doi.org/10.1093/nar/gku447.

Qiu, P., Shandilya, H., D'Alessio, J. M., O'Connor, K., Durocher, J., & Gerard, G. F. (2004). Mutation detection using surveyor nuclease. *BioTechniques*, *36*(4), 702–707. https://doi.org/10.2144/04364PF01.

Ran, F. A., Hsu, P. D., Wright, J., Agarwala, V., Scott, D. A., & Zhang, F. (2013). Genome engineering using the CRISPR-Cas9 system. *Nature Protocols*, *8*(11), 2281–2308. https://doi.org/10.1038/nprot.2013.143.

Rouet, P., Smih, F., & Jasin, M. (1994). Introduction of double-strand breaks into the genome of mouse cells by expression of a rare-cutting endonuclease. *Molecular and Cellular Biology*, *14*(12), 8096–8106. https://doi.org/10.1128/mcb.14.12.8096-8106.1994.

Santos, D. P., Kiskinis, E., Eggan, K., & Merkle, F. T. (2016). Comprehensive protocols for CRISPR/Cas9-based gene editing in human pluripotent stem cells. *Current Protocols in Stem Cell Biology*, *38*, 5B 6 1–5B 6 60. https://doi.org/10.1002/cpsc.15.

Shanbhag, N. M., Rafalska-Metcalf, I. U., Balane-Bolivar, C., Janicki, S. M., & Greenberg, R. A. (2010). ATM-dependent chromatin changes silence transcription in cis to DNA double-strand breaks. *Cell*, *141*(6), 970–981. https://doi.org/10.1016/j.cell.2010.04.038.

Shendure, J., Porreca, G. J., Reppas, N. B., Lin, X., McCutcheon, J. P., Rosenbaum, A. M., et al. (2005). Accurate multiplex polony sequencing of an evolved bacterial genome. *Science*, *309*(5741), 1728–1732. https://doi.org/10.1126/science.1117389.

Xue, H., Yao, T., Cao, M., Zhu, G., Li, Y., Yuan, G., et al. (2019). Structural basis of nucleosome recognition and modification by MLL methyltransferases. *Nature*, *573*(7774), 445–449. https://doi.org/10.1038/s41586-019-1528-1.

Yeh, C. D., Richardson, C. D., & Corn, J. E. (2019). Advances in genome editing through control of DNA repair pathways. *Nature Cell Biology*, *21*(12), 1468–1478. https://doi.org/10.1038/s41556-019-0425-z.

Zou, X., Koh, G. C. C., Nanda, A. S., Degasperi, A., Urgo, K., Roumeliotis, T. I., et al. (2021). A systematic CRISPR screen defines mutational mechanisms underpinning signatures caused by replication errors and endogenous DNA damage. *Nature Cancer*, *2*(6), 643–657. https://doi.org/10.1038/s43018-021-00200-0.

CHAPTER THIRTEEN

A method to sequence genomic sites of mitotic DNA synthesis in mammalian cells

Jonathan Mailler[†], Laura Padayachy[†], and Thanos D. Halazonetis*
Department of Molecular Biology, University of Geneva, Geneva, Switzerland
*Corresponding author: e-mail address: thanos.halazonetis@unige.ch

Contents

Abstract

Under normal conditions, the genome of eukaryotic cells is faithfully replicated during S phase. However, in cells exposed to DNA polymerase inhibitors, some regions of the genome may fail to be replicated prior to mitotic entry. To prevent chromosomal breakage and loss of genomic information, mitotic DNA synthesis (MiDAS) completes replication of the genome prior to the onset of anaphase. We have developed a

[†] These authors contributed equally.

Methods in Enzymology, Volume 661
ISSN 0076-6879
https://doi.org/10.1016/bs.mie.2021.08.013

protocol that allows one to map the genomic regions that are replicated by MiDAS in mammalian cells. The protocol involves incorporation of a thymidine analog in nascent DNA in mitotic cells and then capture and high throughput sequencing of the nascent DNA. With this approach, sites of MiDAS can be identified at high resolution.

1. Introduction

1.1 Mitotic DNA synthesis and common fragile sites

DNA replication is a crucial process whereby cells duplicate their genome prior to cell division. To maintain genomic integrity, it is important that this process is accurate and includes the entire genome. Interestingly, in the presence of the DNA polymerase inhibitor aphidicolin, certain genomic regions fail to be replicated before cells enter mitosis. These genomic regions are referred to as common fragile sites (CFSs), because they were originally described as sites of chromosomal gaps or breaks in aphidicolin-treated cells (Glover, Berger, Coyle, & Echo, 1984). It is now understood that these chromosomal gaps and breaks correspond to unreplicated genomic regions (Blin et al., 2019). The term "common" was used to indicate that these regions are recurrently unreplicated in a large variety of cells. Importantly, expression of CFSs, i.e., the presence of chromosomal breaks or gaps at CFSs, is not limited to cells treated with aphidicolin, but is also observed in cancer cells. In fact, genomic rearrangements, such as focal deletions, are very frequent at CFSs in human cancers, due to the presence of oncogene-induced DNA replication stress (Dereli-Öz, Versini, & Halazonetis, 2011).

The etiology of CFS expression is beginning to be understood. Recent studies have revealed that CFSs correspond to late-replicating genomic regions that are also devoid of DNA replication origins; as a result, CFSs are among the last genomic regions to be replicated, even under normal replication conditions (Letessier et al., 2011; Macheret et al., 2020; Sarni et al., 2019; Zhao, Sasaki, & Gilbert, 2020). Many CFSs map within genes, whose length often exceeds 800 kb and CFS expression correlates very well with the presence of transcriptional activity within these large genes (Helmrich, Ballarino, & Tora, 2011; Le Tallec et al., 2011; Macheret et al., 2020; Park et al., 2021; Wilson et al., 2015). The link of large transcription units to CFS expression may relate to the effect that transcription has on suppressing replication origin firing (Macheret & Halazonetis, 2018). Thus, large transcription units can create large origin-poor genomic regions that fail to be replicated before cells enter mitosis. The possibility that collisions between the replication and transcription machineries also contribute to CFS expression has not been excluded.

Recently, Mitotic DNA Synthesis (MiDAS) has been demonstrated to occur at CFSs (Minocherhomji et al., 2015). From the perspective of ensuring complete replication of the genome, the evolution of a mechanism by which unreplicated genomic regions can be replicated in mitosis, makes sense. If MiDAS ensures complete replication of the genome, then expression of CFSs could be prevented and it is reasonable to argue that MiDAS does, in fact, limit CFS expression. The existence of MiDAS can now be well documented using cells treated with aphidicolin, which slows down DNA replication. A CDK1 inhibitor is also added to keep the cells in interphase. Once the CDK1 inhibitor is washed away, a significant fraction of the cells enter mitosis and initiates MIDAS at the CFSs (Minocherhomji et al., 2015). On chromosome mitotic spreads, the sites of MiDAS can be visualized, as they incorporate thymidine analogs, and are seen to colocalize with the CFSs; high throughput sequencing of newly replicated DNA further confirms that the nascent DNA maps to the CFSs (Ji et al., 2020; Macheret et al., 2020).

Several studies suggest that MiDAS is a type of break-induced replication (BIR) (Bhowmick, Minocherhomji, & Hickson, 2016). BIR is a pathway for repair of single-end DNA double-strand breaks that can arise, for example, at stalled or collapsed DNA replication forks (Sotiriou et al., 2016). Proteins that have been implicated in BIR, such as PolD3, Rad52 and Mus81, are required for MiDAS and, as in BIR, DNA replication in MiDAS can be conservative, rather than semi-conservative (Bhowmick et al., 2016; Di Marco et al., 2017; Minocherhomji et al., 2015). Given the importance of CFSs in genome stability and cancer, methods to map sites of MiDAS can help improve our understanding of how cells maintain the stability of their genome.

1.2 Mapping of CFSs

CFSs were originally mapped as sites of chromosomal breaks on metaphase spreads (Glover et al., 1984). The mapping precision was limited by the size of the chromatin bands, which generally correspond to genomic regions that are several megabases long. Despite this low resolution, these efforts revealed the presence of chromosomal break hotspots and led to the first map of CFSs.

Later on, FISH, YAC and BAC improved the mapping of select CFSs (Mrasek et al., 2015; Paradee et al., 1995; Wilke et al., 1994). Condensed chromosomes were stained with specific probes, allowing a more precise mapping of the CFSs. These approaches and the evolution of sequencing techniques allowed researchers to progress in the identification and sequencing of CFS regions. The method, however, relies on the previous mapping of CFSs, since the design of novel probes for each CFS requires a known target region. One study, nevertheless, used random BAC probes to map known CFSs and to discover new ones, in an analysis that involved thousands of metaphase spreads (Weise, 2010).

Yet another method to map CFSs involves chromatin immunoprecipitation (ChIP). With the knowledge of proteins localizing at CFSs, it is possible to identify the position of these proteins on the genome and therefore to determine the CFS regions. FANCD2 is one such protein, as it has been demonstrated by immunofluorescence to localize at sites of chromosomal breaks (Howlett, Taniguchi, Durkin, D'Andrea, & Glover, 2005). This technique allows one to identify all the positions where FANCD2 is enriched on the genome, therefore promoting the discovery of new CFSs, while improving the level of resolution (Okamoto et al., 2018; Pentzold et al., 2018). However, the use of good antibodies is needed to produce accurate results. Moreover, the localization of most proteins is often not restricted to CFSs, which means that the results need to be manually filtered to eliminate the genomic sites that score positive, but are not CFSs.

Recently, Brison et al. developed a new method to map CFSs (Brison et al., 2020). By comparing Repli-Seq data obtained in the presence and absence of aphidicolin, they were able to identify regions that are replicated significantly later in cells exposed to aphidicolin. Combining their significantly delayed replication (SDR) index with Okazaki fragment sequencing data (Ok-seq) and known FISH data of CFSs, they were able to confirm that CFSs colocalize with SDR regions and regions of replication termination.

1.3 Description of our method

In order to precisely map sites of MiDAS, we developed a method (Fig. 1) that took advantage of the incorporation of EdU, a thymidine analog, to label and capture DNA that is replicated during mitosis (Ji et al., 2020; Macheret et al., 2020). A major challenge was to enrich the cell population

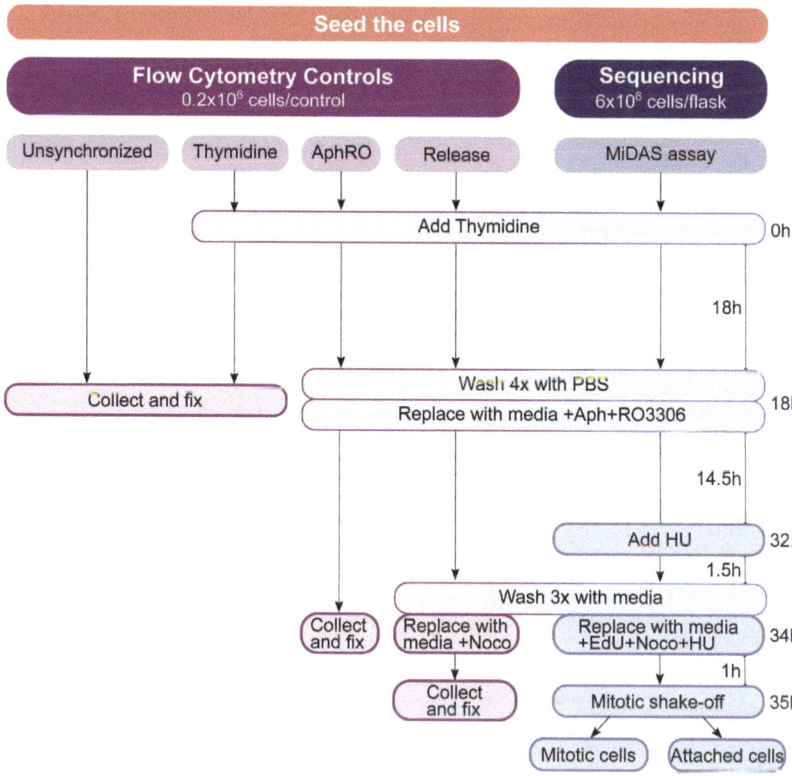

Fig. 1 Outline of the experimental protocol. Together with the cells subjected to high throughput sequencing, parallel cell cultures are used to monitor cell cycle synchronization (flow cytometry controls). The timing indicated on the right side of the graph has been optimized for U2OS and HeLa cells. Aph, aphidicolin; RO, RO-3306; HU, hydroxyurea; Noco, nocodazole.

performing MiDAS and to specifically collect these cells. To do so, cells were synchronized at the G1/S boundary, using a thymidine block; after release from the block, the cells were exposed to low doses of aphidicolin to slow down DNA replication and were also treated with RO-3306, a CDK1 inhibitor, to prevent entry into mitosis (Vassilev et al., 2006). When the cells reached late S phase/G2, aphidicolin and the CDK1 inhibitor were washed away to allow the cells to enter mitosis. At this point, EdU was added to the media to label the newly synthesized DNA and the mitotic cells were collected 1 h later by mitotic shake-off. The cells were fixed with 90% methanol and the EdU-labeled DNA was isolated and sequenced following the EdU-seq protocol. We found that the signal-to-noise ratio could be significantly improved by treating the late S phase cells with hydroxyurea (HU), a ribonucleotide reductase inhibitor. This treatment effectively depleted the nucleotide pools of the late S phase cells, so that any such cells contaminating the pool of mitotic cells would not make a significant contribution to the EdU-seq signal.

The cell synchronization and the efficient isolation of mitotic cells are crucial for the method to discriminate MiDAS from late S phase DNA replication. It is therefore important to find the optimal duration and concentration for each drug treatment to ensure that the cell population is enriched at the appropriate phase of the cell cycle. After the synchronization with thymidine, the majority of cells should be at the G1/S boundary, whereas, after release into S phase, progress through the cell cycle should be monitored to ensure that the cells reach late S/G2, before aphidicolin and the CDK1 inhibitor are removed from the media (Fig. 2A). The optimal concentrations and exposure times for the various inhibitors need to be optimized for each cell line.

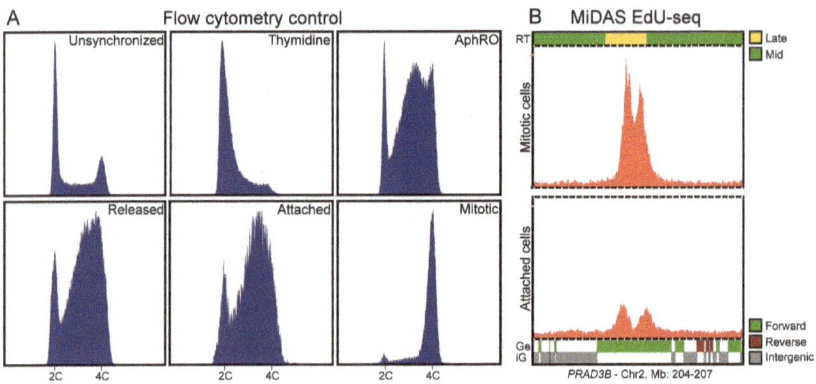

Fig. 2 Anticipated outcomes. (A) Flow cytometry controls. DNA content profiles of propidium iodide-stained unsynchronized U2OS cells and U2OS cells collected at the indicated steps of the MiDAS EdU-seq protocol, as described in Fig. 1. (B) Sequencing results showing nascent DNA synthesis in mitotic and attached U2OS cells subjected to the MiDAS EdU-seq protocol. Both plots are shown at the same scale. The genomic region corresponding to the *PRAD3B* gene (Chr. 2, Mb: 204–207; genome assembly: GRCh37/hg19) is shown. The peaks indicate EdU incorporation. Top: Replication timing (RT) showing the genomic regions that are replicated in mid S (green) and late S (yellow) phase. Bottom: gene annotation. Genes (Ge) are colored green or red, according to the direction of transcription (forward and reverse, respectively). Intergenic regions (iG) are colored gray.

2. List of materials

Reagent or resource	Source	Identifier
Chemicals		
EdU (5-ethynyl-2′-deoxyuridine)	Invitrogen	A10044
RO-3306	Sigma-Aldrich	SML0569
Thymidine	Sigma-Aldrich	T1895
Nocodazole	Tocris	1228
DMEM, high glucose, no glutamine	Thermo Fisher Scientific	11960044
Penicillin–streptomycin–glutamine	Thermo Fisher Scientific	10378016
FBS	Thermo Fisher Scientific	10500064
Dulbecco's PBS (DPBS), no calcium, no magnesium	Thermo Fisher Scientific	14190094
Trypsin-EDTA (0.05% (wt/vol))	Thermo Fisher Scientific	25300054
Triton X-100	PanReacAppliChem	A1388,2500
Tween 20	PanreacAppliChem	A4974,0100
Copper (II) sulfate	Acros Organics	197711000
(+) Sodium L-ascorbate	Sigma-Aldrich	A7631
Cleavable biotin–azide conjugate (azide-PEG(3 + 3)-S-S-biotin)	Jena Biosciences	CLK-A2112-10
SDS	Sigma-Aldrich	L3771
Tris base	Roche	000000010708976001
EDTA	Sigma Life Sciences	E5134
Sodium chloride	Sigma-Aldrich	S3014
Dimethyl sulfoxide	Sigma-Aldrich	D2650
Hydroxyurea (HU)	Sigma-Aldrich	H8627

Continued

—cont'd

Reagent or resource	Source	Identifier
Aphidicolin from Nigrospora sphaerica	Sigma-Aldrich	A0781
Methanol	Sigma-Aldrich	34860
Ethanol	Sigma-Aldrich	51976
Phenol/chloroform/isoamyl alcohol (25:24:1)	Sigma-Aldrich	P3803
Chloroform/isoamyl alcohol (24:1)	Sigma-Aldrich	C0549
Proteinase K (recombinant, PCR grade)	Roche	03115844001
Glycogen for molecular biology	Roche	10901393001
β-Mercaptoethanol	Sigma-Aldrich	M6250
RNAse, DNAse-free	Roche	11119915001
Propidium iodide	Sigma-Aldrich	81845
Dynabeads MyOne streptavidin C1	Invitrogen	65001
Critical commercial assays		
Qubit dsDNA BR Assay Kit	Thermo Fisher Scientific	Q32850
Qubit dsDNA HS Assay Kit	Thermo Fisher Scientific	Q32851
Experimental models: cell lines		
U2OS		
HeLa		

3. Materials and equipment

Equipment

- Incubator 20% O_2 5% CO_2
- Cell culture hood
- Hemocytometer
- Bioruptor Pico sonication device (Diagenode, cat. no. B01060003)
- Centrifuge

- Micro centrifuge
- Hemocytometer
- Tabletop centrifuge (centrifuge 5810 R; Eppendorf, cat. no. B1_022628181)
- Vortex
- UV lamp
- Electrophoresis system
- Qubit 4 fluorometer (Thermo Fisher Scientific, cat. no. Q33226)
- Rotating wheel
- Water bath
- DynaMag™-2 (Invitrogen, cat. no. 12321D)
- HiSeq sequencing system (Illumina)
- Flow cytometer (we typically use a Gallios 8-color/2-laser flow cytometer; Beckman Coulter)
- (optional) Multitron incubation shaker

Material

- 1.5-mL Microcentrifuge tubes
- 15- and 50-mL Conical tubes
- Tissue culture flasks and dishes
- V-bottom 96-well plates (Böttger, cat. no. 05-021-0100)
- 5PRIME Phase Lock Gel, Heavy, 1.5 mL (Quantabio, cat. no. 2302830)
- 1.5-mL MaxyMum Recovery (clear) microtubes (Axygen Scientific, cat. no. MCT 150-L-C)
- Barrier tips

Reagents

- Complete media: DMEM supplemented with 10% FBS and 1 × penicillin–streptomycin–glutamine. Store at 4 °C for some days
- Thymidine 200 mM diluted in water. Aliquots can be stored at −20 °C for several months
- Aphidicolin 4 mM dissolved in DMSO. Aliquots can be stored at −20 °C for several months
- RO-3306 9 mM dissolved in DMSO. Aliquots can be stored at −20 °C for several months
- HU 2 M dissolved in water. Aliquots can be stored at −20 °C for several months
- Nocodazole 1 mg/mL dissolved in DMSO. Aliquots can be stored at −20 °C for several months
- EdU 10 mM dissolved in DMSO. Aliquots can be stored at −20 °C for several months

- Permeabilization buffer: PBS supplemented with 0.2% Triton. Store at 4 °C for several months
- Tris-HCl 100 mM pH 8 and Tris–HCl 1 M pH 7.5 dissolved in water. Store at room temperature (RT) for several months
- Cleavable biotin-azide 10 mM dissolved in DMSO. Aliquots can be stored at −20 °C for several months
- Na ascorbate 1 M dilute in water. Aliquots can be stored at −20 °C for several months
- Copper $CuSO_4$ 100 mM in water. Aliquots can be stored at 4 °C for several months
- TE10: 10 mM Tris pH 8 with 10 mM EDTA
- Propidium iodide (PI) 0.5 mg/mL
- TE (Tris-EDTA) is 1 mM EDTA with 10 mM Tris–HCl pH 8 in water. Store at RT for months

Buffers to prepare just before use

- *Lysis buffer* (200 μL/sample):
  For 1 mL: Add 890 μL of TE10, 100 μL of SDS (5%) and 10 μL of Proteinase K (20 mg/mL)
- *2× Bead and wash* (B&W): (350 μL/sample: 100 μL of 2× B&W and 450 μL of 1 × B&W dilute in water):
  For 1 mL: Add 10 μL of 1 M Tris–HCl pH 7.5, 2 μL of EDTA 0.5 M, 400 μL of NaCl 5 M, 10 μL of Tween 20 and 578 μL of water
- *Click-it solution* (250 μL/sample):
  For 1 mL: Add 855 of 100 mM Tris pH 8, 40 μL CuSO4 100 mM, 100 μL sodium ascorbate 1 M and 5 μL of azide-biotin S-S 10 mM
- *Propidium iodide* (PI) solution (150 μL/sample):
  For 150 μL: combine 150 μL PBS with 5 μL of PI (0.5 mg/mL) and 1 μL of RNAse (DNAse-free)

4. Step-by-step method details

4.1 Before starting

To accurately perform MiDAS EdU-seq an adherent cell line in which you can do a mitotic shake off is needed. The protocol present below is optimized for HeLa and U2OS cell lines but length of treatment might be adapted for each cell lines. We suggest studying the length of the different phases of the cell cycle of the cell line of interest to find the optimal treatment time of thymidine and subsequently aphidicolin to increase the number of cells performing MiDAS. Thymidine treatment should allow the majority of the cell population to be blocked near the G1/S boundary and to enter S-phase directly after being washed. The aphidicolin and RO-3306 treatment was approximately 1.5 times of the normal S-phase. In any case we strongly recommend running a preliminary experiment to verify that most of the cells are found in late S-phase after the aphidicolin and RO-3306 treatment. This can be performed by testing different times and concentrations of drug treatment and check the DNA content of the cells by flow cytometry after a PI staining (see Fig. 2A and the flow cytometry controls in the section expected outcome).

Optimization of Mitotic shake-off by shaker can be performed using an unsynchronized cell population submit to an 8–10 h nocodazole treatment. Efficiency of mitotic shake-off can be assessed by flow cytometry and PI staining. A proper mitotic shake-off should have at least 80% of the cells having 4C DNA content.

4.2 MiDAS induction in cells

Duration: 3 days

1. Plate the cells on a flask to have them at a confluence of ~70% at the time of Thymidine addition.

   Controls: Seed 4 wells of a 6 well plate with 0.2×10^6 cells (one well per control: Unsynchronized, Thymidine, AphRO and Released)

   Note: We generally seed 6×10^6 U2OS cells and 4×10^6 HeLa cells per T-175 flask and use 4 flasks per condition. After 12 h, cells are at an appropriate confluence for Thymidine treatment.

Timing: 0 h, duration: 2 min

2. Add 2 mM Thymidine to the cells, mix well by rotating the flasks, and place them back to the incubator for 18 h.

   Controls: add 2 mM Thymidine in 3 out of the 4 well (Thymidine, AphRO and Released).

Timing: 18 h, duration: 10–15 min

3. Aspirate the media with Thymidine of the flasks and wash 4× with warm PBS (with ~3/4 of the volume of the media). Replace with media containing 0.4 μM of Aphidicolin and 9 μM of RO-3306 and place the flasks in the incubator for 14.5 h.

   Controls: collect and fix (see 14–18) the cells in wells Unsynchronized and Thymidine block, instead of 500 μL, resuspend with 100 μL of PBS and add 1 mL of 100% ice-cold ethanol.

Wash the AphRO and Released controls 4 × with warm PBS (with ~3/4 of the volume of the media) and replace with media containing 0.4 μM of Aphidicolin and 9 μM of RO-3306. Incubate for 14.5 h with the flasks.

Timing: 32.5–32.8 h, duration: 2 min

4. Add 2 mM of HU to the flasks and mix well by rotating the flasks. Return the cells to the incubator for 1.5 h.

Timing: 34 h, duration: 10–15 min

5. Aspirate the media containing the drugs (Aphidicolin, RO-3306 and HU) and wash 3 × with warm media (with ~3/4 of the volume of the culturing media). Replace with media containing 200 ng/mL Nocodazole, 2 mM HU and 10 μM of EdU and put it back in the incubator for 1 h.

   Controls: collect and fix the AphRO control (see 14–18), instead of 500 μL, resuspend with 100 μL of PBS and add 1 mL of 100% ice-cold ethanol.

   Wash the Released control 3 × with warm media (with ~3/4 of the volume of the culturing media). Replace with media containing 200 ng/mL Nocodazole, 2 mM HU and 10 μM of EdU and put it back in the incubator for 1 h.

Timing: 35 h, duration: 30–40 min

6. Carefully take the flask out of the incubator as mitotic cells may detach.

   Note: mitotic cells are visible under the microscope as they have a rounded shape.

   Controls: collect and fix the Release control (see 14–18), instead of 500 μL, resuspend with 100 μL of PBS and add 1 mL of 100% ice-cold ethanol.
7. Aspirate slowly the media and add PBS (~3/4 of the volume of the media).

   Critical: Remove carefully the media to avoid aspirating mitotic cells
8. Perform the mitotic shake-off:
   - **a.** For U2OS cells: place the flask in a shaker for 1.5 min at 170 rpm
   - **b.** For HeLa cells: place the flask in a shaker for 0.5 min at 240 rpm
   - **c.** Or hit multiple times the flask by hand to detach most of the mitotic cells
9. Check under the microscope that the majority of rounded-shape cells, representing the mitotic cells, are detached but not the others.
10. Collect the mitotic cells in falcon and spin at 350 g for 5 min.
11. Aspirate the PBS and resuspend in 500 μL of PBS.

12. Fix by adding 4.5 mL of 100% ice-cold methanol, dropwise, while vortexing (speed ~3–5) the tube.
13. Keep on ice at least 30 min and then store at −20 °C until the click-it, at least 4 h.
14. To collect the attached cells, put trypsin (~1/10 of the volume of media) on the cells and return the flask to 37 °C for 3–5 min.
15. Add media (4 × the volume of trypsin) to the trypsinized cells, collect them in a falcon and spin at 350 g for 5 min.
16. Aspirate the supernatant and wash once by resuspending the pellet with PBS (with a similar volume as what was aspirated). Spin again at 350 g for 5 min.
17. Aspirate the PBS and resuspend in 500 μL of PBS.
18. Fix by adding 4.5 mL of 100% ice-cold methanol, dropwise, while vortexing (speed ~3–5) the tube.

Pause Point: Samples can be stored for several weeks at −20 °C.

4.3 EdU click-it

Duration: 2.5–3 h

19. Spin down your cells and controls samples at 350 × *g*, 4 °C for 10 min.
20. Take out the supernatant and resuspend the cells with 500 μL of PBS. Transfer them in a 1.5 mL Eppendorf tube and take 50 μL as a control that will be process with the other controls

    Controls: take out the supernatant and resuspend the cells with 150 μL of PBS. Transfer them in wells of a 96 well plate.
21. Spin down at 350 × *g*, 4 °C for 10 min.
22. Carefully remove the PBS and add 250 μL of PBS 0.2% Triton, 150 μL for the controls, for permeabilization. Resuspend by flicking and incubate 30 min at RT in the dark.
23. Spin down at 350 × *g*, 4 °C for 10 min, remove supernatant and wash once with PBS.
24. Prepare Click-it reaction buffer (see *Buffers to prepare just before use*): add 250 μL per samples, flick to resuspend and keep 30 min at RT in the dark.

    Controls: Prepare PI buffer (see *Buffers to prepare just before use*): resuspend in 150 μL per samples and incubate for 4 h or overnight at RT. Process them by flow cytometry.

25. Spin down at 350 × g, 4 °C for 10 min and wash the pellet once with 500 μL PBS. Spin again at 350 × g, 4 °C for 10 min.
26. Remove PBS and in each samples add 200 μL of lysis buffer (see *Buffers to prepare just before use*). Incubate 4 h at 55 °C or overnight.

4.4 DNA extraction

Duration: 30 min

27. In a pre-spun phase lock tube add 200 μL of Phenol-chloroform. Add 1 μL of glycogen in your sample and transfer them in the phase lock. Flick the tube until the liquid appears white and let sit 5 min at RT.
28. Spin 5 min at 13,000 × *g* at RT. Transfer the aqueous phase in a new pre-spun phase lock tube and add 200 μL of chloroform. Mix by inverting and spin 5 min at 13,000 × *g*, RT.
29. Transfer the aqueous phase in a maximum recovery 1.5 mL tube and add 8 μL of 5 M NaCl. Add 416 μL of ice-cold Ethanol 100%. Store at −20 °C for 1 h at least.

    Critical: Be careful to not aspirate the gel phase as it could affect the efficiency of the precipitation.

Pause Point: Samples can be stored at −20 °C for at least several hours, overnight or maximum 1 month.

4.5 DNA precipitation

Duration: 1–1.5 h

30. Spin your sample 45 min at 13,000 × *g*, 4 °C. Carefully remove the supernatant.
31. Wash with 200 μL of ice-cold 70% ethanol and spin again for 20 min at 13 × 000 *g* at 4 °C.
32. Carefully remove ethanol and let the pellet dry approximatively 5 min.

    Critical: Make sure the pellet is not drying as it could result in a lower resuspension.
33. Add 55 μL of ultra-pure water and let the DNA dissolve for 1 h at 37–42 °C or overnight at RT.

Pause Point: DNA can be resuspended overnight at RT or 2–3 days at 4 °C.

4.6 EdU pull-down

Duration: 4 h

34. Measure the DNA concentration using the BR Qbit kit.

35. Take 2–10 μg of DNA in a novel Maximum recovery tube and complete with water to reach a final volume of 105 μL.
36. Sonicate your samples following the program: 15″ on 90″ off, 6 cycles.
    Note: run an agarose gel with 1 μL of your sample. DNA smear should appear between 100 and 500 bp, otherwise repeat for 1–2 cycles until the ideal range sized is reached.
37. Prepare 350 μL of 2 × Bead and Wash buffer per sample, 100 μL of 2 × and 450 μL of 1 × will be needed per sample.
38. For each sample, take 50 μL of beads (Dynabeads MyOne Streptavidin C1) in an Eppendorf.
    Note: The beads stick at the bottom of the stock vial, vortex slowly until beads are well resuspended.
39. Perform a quick spin to drop down the liquid, place the Eppendorf on a magnet and wait 2–3 min until the beads bind to the magnet.
40. Carefully remove the supernatant, keeping the Eppendorf on the magnet, to save only the beads.
41. Add 50 μL per sample of freshly made 1 × B&W buffer and resuspend by vortexing slowly.
42. Repeat steps 38–39-40 twice for a total of 3 washes, to equilibrate the beads.
43. Resuspend the beads in twice the volume of the beads (100 μL for one sample) of 2 × B&W buffer. Add 100 μL of beads to the sonicated sample and incubate 15 min on a rocking wheel at RT.
    Critical: longer time may increase non-specific binding between DNA and beads.
44. Spin down and place the samples on a magnet.
45. Remove slowly the buffer to just retain the beads and add 100 μL of 1 × B&W buffer. Quickly vortex and spin down. Wash 3 times with 100 μL of 1 × B&W buffer and one time with 100 μL of TE.
46. Remove the TE carefully and resuspend in 55 μL of Tris 10 mM supplemented with 2% β-mercaptoethanol.
47. Quickly vortex and spin down.
48. Let sit for 1 h at RT in the dark.
49. Place the samples on a magnet and collect the liquid phase.
50. Measure your final DNA concentration with Qbit HS DNA kit, using 3 μL of your sample.

Pause Point: Samples can be stored at −20 °C for several days.

51. Continue to standard library preparation.

5. Expected outcomes

The Unsynchronized control represents the steady state of the cell population. Therefore, in the flow cytometry profile, cells should be caught at the different phases of the cell cycle. In U2OS cells, G1 is the longest phase, thus most of the cells can be found in the G1 phase (2C DNA content); a clear fraction of cells is also visible at the G2-M phases (4C DNA content), whereas the cells progressing through S phase have a DNA content between 2C and 4C (Fig. 2A, Unsynchronized control). In the thymidine control, an enrichment of cells at the G1-S boundary should be observed and the peak corresponding to G2 should disappear (Fig. 2A, Thymidine control). In the AphRO and Released controls the majority of the cells should be in late S phase and G2-M, as they have progressed through S phase. Finally, the mitotic shake-off allows to specifically separate the mitotic cells from the rest of the cell population. Thus, an enrichment of 4C cells is expected in the mitotic fraction. At least 80% of the cells should be in mitosis to ensure a good selection and therefore an optimal sequencing of MiDAS (Fig. 2A, Mitotic control). In contrast, the DNA content profile of the remaining cells (Fig. 2A, Attached control) should resemble that of the AphRO and Release controls with a relative depletion of the 4C cells.

After sequencing of the EdU-labeled DNA and genomic alignment of the sequencing reads, signal should be seen at most of the CFSs, mainly as single, double or merge peaks. Depending on the cell type used, a supplementary set of peaks could be identified in addition to the one present at CFSs. Cancer cell lines have higher levels of endogenous DNA replication stress and are, therefore, more likely to enter mitosis with unreplicated DNA and to perform MiDAS. It is therefore expected to find a higher number of MiDAS sites in cancer cell lines than in normal cells.

As previously characterized, MiDAS sites are expected to be localized in late replication domains and in large, transcribed genomic regions. The sequencing profile of mitotic cells should show clear peaks over background (Fig. 2B, top). Depending on the size of the genomic region undergoing MiDAS, either single or double peaks will be evident (Fig. 2B, bottom), corresponding to the sites where replication re-started when the cells entered mitosis.

6. Advantages and limitations

The EdU-seq technique allows to precisely capture DNA synthesis events, independently of the state of the cell cycle and the nature of the event (normal DNA replication in S phase or MiDAS). To explore MiDAS, cells were synchronized and subjected to replication stress induced by aphidicolin. Following mitotic shake-off, we isolated the pool of mitotic cells that are performing MiDAS. To discriminate EdU incorporation due to MiDAS in mitotic cells from late S phase replication in cells that had not entered mitosis, the cells were exposed to HU before removing the Cdk1 inhibitor from the media. HU depletes the deoxyribonucleotide pools in the cells that are still in S phase and, thus, suppresses EdU incorporation in the non mitotic cells, whereas the mitotic cells, which are consuming much lower levels of deoxyribonucleotides, are much less affected by the presence of HU.

Moreover, thanks to the mitotic shake-off, the mitotic cells can be significantly enriched. This is a second step to ensure that the observed EdU incorporation corresponds to MiDAS. We note, of course, that mitotic shake-off can only be performed with adherent cells and not cells grown in solution.

Working with EdU allows the nascent DNA fragments to be biotinylated using a biotin-azide and a click-it reaction. Therefore, no intermediates, such as antibodies, are needed to capture the nascent DNA, which improves the sensitivity and selectivity of the method. A second advantage of EdU is that it can be administered as a pulse to allow the monitoring of DNA synthesis with high temporal precision.

One disadvantage of the method is that it requires a significant number of cells, which must progress through the cell cycle in a synchronous manner. Thus, the synchronization protocol needs to be optimized.

To summarize, the MiDAS EdU-seq technique allows the identification of nascent DNA events, triggered by BIR, in mitotic cells. The genomic regions being replicated in these cells correspond to CFSs. Therefore, both the biology of CFSs and the mechanisms by which cells maintain genome stability at these sites can be investigated.

7. Optimization and troubleshooting

Problem	Possible origin	Solution
Cells are not well synchronized in G1/S (control Thymidine)	Problem with the Thymidine bock	Adjust the time and the concentration to find the optimal condition for your cells
Cells are not well synchronized in G2 (control AphRO)	The S phase length may be different than the U2OS or the drugs not efficient enough	Adjust the time and the concentration to find the optimal condition for your cells
	The thymidine may not have been well washed	Make sure the liquid is completely aspirated between the washes and that the volume is sufficient. S phase progression can be verified by collecting cells released from Thymidine block at different time point and examined by flow cytometry
Too few cells in mitosis or cells are not well enriched in mitosis after mitotic shake off (control Release and Mitotic)	Cells were not properly released from RO-3306 and Aphidicolin	Make sure the liquid is completely aspirated between the washes and that the volume is sufficient. Check if rounded shape cells corresponding to mitotic cells are visible.
	Cells were not well enriched in G2 in the previous step	The aphidicolin and RO-3306 treatment should be elongated. Check if cells from the AphRO control are properly enriched in 4C (flow cytometry)
	Mitotic cells are aspirated with the media just before mitotic shake-off	Check under the microscope whether rounded-shape cells (mitotic cells) are visible before and after aspirating the media. Handle the flask carefully to make sure the mitotic cells are not detaching prematurely
	The mitotic shake-off is too strong or too weak	Check under the microscope before and after the mitotic shake-off to see the mitotic and non-mitotic cells. Optimize the method as proposed in the section Before starting

Cells are dying	Drugs are toxic	Reduce the time of drug treatment or, eventually, of the concentration but make sure the block is still efficient enough
Cells do not synchronize well	Drugs are not effective	Increase the time exposure of the drugs or the concentration. Try to find an alternative drug, ex colcemid instead of nocodazole
Too few DNA	Too few cells	Optimize the synchronization (incubation time and concentration) steps to ensure ending with as much cells as possible entering mitosis Start the experiment with more cells
	Loosing cells in the intermediate steps	Increase the centrifugation time and/or the speed Be careful not to aspirate cells while removing the supernatant
Too much background after sequencing	Too few DNA	Too few cells
	Poor synchronization	Check the flow cytometry control and optimize the time and/or concentration of drugs at the time points
	Poor mitotic shake-off	Check the enrichment of mitotic cells by flow cytometry and see above to optimize

8. Safety considerations and standards

Some of the reagents are toxic. It is therefore necessary to work with caution. We recommend wearing gloves during the entire protocol for safety but also to preserve the samples from any contamination. Moreover, it is preferable to avoid inhaling and swallowing any compound. Working with the following reagents are particularly dangerous and we strongly preconize to work under a hood: β-Mercaptoethanol, Phenol/chloroform/isoamyl alcohol and Chloroform/isoamyl alcohol. In addition, it is important to point out that some compounds are flammable: Ethanol, β-Mercaptoethanol, and Methanol. It is therefore necessary to work in an adequate environment.

9. Summary

Mitotic DNA Synthesis (MiDAS) is a repair mechanism not yet fully understood and characterized. It takes place during mitosis at unreplicated genomic regions, encompassing Common Fragile Sites (CFSs). In this chapter, we describe a method that allows the identification of genomic regions where MiDAS occurs, at high resolution. This chapter reports the specific conditions that are required to ensure optimal sequencing data for the precise mapping of MiDAS sites. The most challenging aspect of this protocol is to obtain large and synchronous cell populations to accommodate for the low number of reads that are inherently associated with MiDAS.

References

Bhowmick, R., Minocherhomji, S., & Hickson, I. D. (2016). RAD52 facilitates mitotic DNA synthesis following replication stress. *Molecular Cell*, *64*, 1117–1126.

Blin, M., Le Tallec, B., Nähse, V., Schmidt, M., Brossas, C., Millot, G. A., et al. (2019). Transcription-dependent regulation of replication dynamics modulates genome stability. *Nature Structural & Molecular Biology*, *26*, 58–66.

Brison, O., Gnan, S., Azar, D., Schmidt, M., Koundrioukoff, S., El-Hilali, S., et al. (2020). Unscheduled origin building in S-phase upon tight CDK1 inhibition suppresses CFS instability. *BioRxiv*. 2020.11.19.390054.

Dereli-Öz, A., Versini, G., & Halazonetis, T. D. (2011). Studies of genomic copy number changes in human cancers reveal signatures of DNA replication stress. *Molecular Oncology*, *5*, 308–314.

Di Marco, S., Hasanova, Z., Kanagaraj, R., Chappidi, N., Altmannova, V., Menon, S., et al. (2017). RECQ5 helicase cooperates with MUS81 endonuclease in processing stalled replication forks at common fragile sites during mitosis. *Molecular Cell*, *66*. 658-671.e8.

Glover, T. W., Berger, C., Coyle, J., & Echo, B. (1984). DNA polymerase? Inhibition by aphidicolin induces gaps and breaks at common fragile sites in human chromosomes. *Human Genetics*, *67*, 136–142.

Helmrich, A., Ballarino, M., & Tora, L. (2011). Collisions between replication and transcription complexes cause common fragile site instability at the longest human genes. *Molecular Cell*, *44*, 966–977.

Howlett, N. G., Taniguchi, T., Durkin, S. G., D'Andrea, A. D., & Glover, T. W. (2005). The Fanconi anemia pathway is required for the DNA replication stress response and for the regulation of common fragile site stability. *Human Molecular Genetics*, *14*, 693–701.

Ji, F., Liao, H., Pan, S., Ouyang, L., Jia, F., Fu, Z., et al. (2020). Genome-wide high-resolution mapping of mitotic DNA synthesis sites and common fragile sites by direct sequencing. *Cell Research*, *30*, 1009–1023.

Le Tallec, B., Dutrillaux, B., Lachages, A.-M., Millot, G. A., Brison, O., & Debatisse, M. (2011). Molecular profiling of common fragile sites in human fibroblasts. *Nature Structural & Molecular Biology*, *18*, 1421–1423.

Letessier, A., Millot, G. A., Koundrioukoff, S., Lachagès, A.-M., Vogt, N., Hansen, R. S., et al. (2011). Cell-type-specific replication initiation programs set fragility of the FRA3B fragile site. *Nature*, *470*, 120–123.

Macheret, M., Bhowmick, R., Sobkowiak, K., Padayachy, L., Mailler, J., Hickson, I. D., et al. (2020). High-resolution mapping of mitotic DNA synthesis regions and common fragile sites in the human genome through direct sequencing. *Cell Research*, *30*, 997–1008.

Macheret, M., & Halazonetis, T. D. (2018). Intragenic origins due to short G1 phases underlie oncogene-induced DNA replication stress. *Nature*, *555*, 112–116.

Minocherhomji, S., Ying, S., Bjerregaard, V. A., Bursomanno, S., Aleliunaite, A., Wu, W., et al. (2015). Replication stress activates DNA repair synthesis in mitosis. *Nature*, *528*, 286–290.

Mrasek, K., Wilhelm, K., Quintana, L. G., Theuss, L., Liehr, T., Leskovac, A., et al. (2015). BAC-probes applied for characterization of fragile sites (FS). In K. Narayanan (Ed.), *Bacterial artificial chromosomes* (pp. 289–298). New York, NY: Springer.

Okamoto, Y., Iwasaki, W. M., Kugou, K., Takahashi, K. K., Oda, A., Sato, K., et al. (2018). Replication stress induces accumulation of FANCD2 at central region of large fragile genes. *Nucleic Acids Research*, *46*, 2932–2944.

Paradee, W., Mullins, C., He, Z., Glover, T., Wilke, C., Opalka, B., et al. (1995). Precise localization of aphidicolin-induced breakpoints on the short arm of human chromosome 3. *Genomics*, *27*, 358–361.

Park, S. H., Bennett-Baker, P., Ahmed, S., Arlt, M. F., Ljungman, M., Glover, T. W., et al. (2021). Locus-specific transcription silencing at the FHIT gene suppresses replication stress-induced copy number variant formation and associated replication delay. *Nucleic Acids Research*. https://doi.org/10.1093/nar/gkab559.

Pentzold, C., Shah, S. A., Hansen, N. R., Le Tallec, B., Seguin-Orlando, A., Debatisse, M., et al. (2018). FANCD2 binding identifies conserved fragile sites at large transcribed genes in avian cells. *Nucleic Acids Research*, *46*, 1280–1294.

Sarni, D., Sasaki, T., Miron, K., Tur-Sinai, M. I., Rivera Mulia, J. C., Magnuson, B., et al. (2019). Replication timing and transcription identifies a novel fragility signature under replication stress (Preprint). *Genetics*.

Sotiriou, S. K., Kamileri, I., Lugli, N., Evangelou, K., Da-Ré, C., Huber, F., et al. (2016). Mammalian RAD52 functions in break-induced replication repair of collapsed DNA replication forks. *Molecular Cell*, *64*, 1127–1134.

Vassilev, L. T., Tovar, C., Chen, S., Knezevic, D., Zhao, X., Sun, H., et al. (2006). Selective small-molecule inhibitor reveals critical mitotic functions of human CDK1. *Proceedings of the National Academy of Sciences*, *103*, 10660–10665.

Weise, A. (2010). Global screening and extended nomenclature for 230 aphidicolin-inducible fragile sites, including 61 yet unreported ones. *International Journal of Oncology*, *36*. https://doi.org/10.3892/ijo_00000572.

Wilke, C. M., Guo, S.-W., Hall, B. K., Boldog, F., Gemmill, R. M., Chandrasekharappa, S. C., et al. (1994). Multicolor FISH mapping of YAC clones in 3p14 and identification of a YAC spanning both FRA3B and the t(3;8) associated with hereditary renal cell carcinoma. *Genomics*, *22*, 319–326.

Wilson, T. E., Arlt, M. F., Park, S. H., Rajendran, S., Paulsen, M., Ljungman, M., et al. (2015). Large transcription units unify copy number variants and common fragile sites arising under replication stress. *Genome Research*, *25*, 189–200.

Zhao, P. A., Sasaki, T., & Gilbert, D. M. (2020). High-resolution Repli-Seq defines the temporal choreography of initiation, elongation and termination of replication in mammalian cells. *Genome Biology*, *21*, 76.

CHAPTER FOURTEEN

Loop-seq: A high-throughput technique to measure the mesoscale mechanical properties of DNA

Aakash Basu*
Johns Hopkins University School of Medicine, Baltimore, Maryland, United States
*Corresponding author: e-mail address: abasu10@jhmi.edu

Contents

Abstract

The local mechanical properties of the DNA polymer influence molecular processes in biology that require mechanical deformations of DNA. Lack of suitable high-throughput experimental techniques had precluded measuring how these properties might vary with sequence along the vast lengths of genomes. Here, we present a detailed protocol for a recently developed experimental technique called loop-seq, which measures at least one local mechanical property of DNA—its propensity to cyclize—in genome-scale throughput. Loop-seq has been used to obtain experimentally derived genome-wide maps of a physical property of DNA. Such measurements have revealed that diverse DNA-deforming processes involved in chromatin organization at various genomic loci are regulated by the genetically encoded, sequence-dependent variations in the mechanical properties of DNA.

Methods in Enzymology, Volume 661
ISSN 0076-6879
https://doi.org/10.1016/bs.mie.2021.08.011

1. Introduction

Local DNA deformations such as bending, twisting, stretching, occur as part of diverse DNA–protein interactions (Afek et al., 2020; Battistini et al., 2019; Garcia et al., 2007; Luger, Mäder, Richmond, Sargent, & Richmond, 1997) associated with the packaging, replication, transcription, and repair of genetic information (Basu, Bobrovnikov, & Ha, 2021). The local mechanical properties of the DNA polymer, such as persistence length (Bouchiat et al., 1999) or torsional rigidity (Lebel, Basu, Oberstrass, Tretter, & Bryant, 2014), can impact such DNA deforming processes by setting the energetic penalty needed to deform DNA. Low-throughput dynamic measurements (Basu et al., 2021; Geggier & Vologodskii, 2010; Rief, Clausen-Schaumann, & Gaub, 1999; Rosanio, Widom, & Uhlenbeck, 2015), molecular dynamics simulations (Beveridge, Cheatham, & Mezei, 2012; Reymer, Zakrzewska, & Lavery, 2018), or structural studies (Balasubramanian, Xu, & Olson, 2009), have suggested that local sequence modulates the local mechanical properties of DNA, and thus the energetics of diverse DNA–protein interactions genome-wide. However, because of low throughput, these methods have typically been limited to studying the mechanical properties of DNA at specific loci, or at specific protein binding sites.

DNA cyclization assays have been combined with single-molecule Fluorescence Resonance Energy Transfer (smFRET) detection to measure the mechanical bendability of short DNA segments (Vafabakhsh & Ha, 2012). Fluorophores acting as FRET pairs are placed at the ends of DNA constructs involving a central 50–100 bp duplex DNA region flanked by complementary 10 nucleotide (nt) single-stranded (ss) overhangs. These constructs can cyclize via annealing of the complementary single-stranded ends, leading to an increase in FRET signal. Thus, FRET measurements reveal the rate at which such DNA constructs cyclize, which in turn serves as a measure of the mechanical bendability of the sequence. Such experiments have revealed that short DNA duplexes can be surprisingly flexible. This may be owing to the structural complexity of looped molecules, which can involve kinking (Wiggins, Phillips, & Nelson, 2005), bubble formation (Yan & Marko, 2004), and mismatches (Jeong & Kim, 2019). It may also be owing to the properties of the long single-stranded overhangs (Vologodskii, Du, & Frank-Kamenetskii, 2013).

DNA cyclization assays have also directly revealed that the mesoscale cyclizability of DNA can depend on sequence (Basu et al., 2021; Geggier & Vologodskii, 2010; Vafabakhsh & Ha, 2012)—more than an order of magnitude differences in looping rates have been attributed to differences in sequence alone. This has therefore called for the development of high-throughput DNA cyclization assays that can explore the sequence-dependence of DNA cyclizability. Systematic Evolution of Ligands by Exponential Enrichment (SELEX) has been used to design experiments where a large library of unknown sequences is enriched for the highly loopable ones by several cycles of enzymatic and gel-electrophoresis based removal of molecules that do not loop rapidly, followed by PCR amplification of the surviving molecules (Rosanio et al., 2015). High-throughput sequencing of the enriched molecules identify certain short sequence features that are enriched in highly cyclizable molecules. However, beyond identifying short sequence features that contribute to high cyclizability, the method has not been used to quantify the cyclizability of any given sequence. The method also suffers from an artifact that favors the enrichment of certain sequence features at specific locations along the DNA strand, based on the fixed adapter sequences at the ends of molecules, off which PCR is primed.

Here, we present the detailed protocol of a recently developed high-throughput experimental procedure called loop-seq that measures the cyclizabilities of a large number (~100,000) of pre-specified 50-bp reference DNA sequences. These reference sequences can be designed to, for instance, tile a large region of the genome. DNA molecules in a chemically synthesized library containing multiple copies of each of these reference sequences are permitted to rapidly cyclize. Those molecules that fail to cyclize in a short amount of time are enzymatically digested. Via deep-sequencing, the probability of each sequence rapidly cyclizing (and therefore surviving digestion) is calculated. It serves as a measure of the mechanical bendability of the sequence.

In a recent study (Basu et al., 2021), loop-seq was applied to understand how sequence-encoded mechanical properties of DNA along the yeast genome aid in the global organization of nucleosomes and the activities of nucleosome remodeling enzymes (Basu et al., 2021). Nucleosomes are formed throughout eukaryotic genomes by the binding of 150 bp of DNA in ~2 turns on the surface of an octamer of histone proteins (Luger et al., 1997). Nucleosome formation involves extensive DNA bending

(Luger et al., 1997), and is thus likely to be facilitated by flexible or bendable DNA sequence. Just upstream of Transcription Start Sites (TSSs) lies a Nucleosome Depleted Region (NDR) (Lai & Pugh, 2017). Low nucleosome occupancy here likely permits proper assembly of the transcription apparatus. By contrast, there exists a well-ordered array of nucleosomes downstream of TSSs, which is also essential for proper transcription (Lai & Pugh, 2017). Via loop-seq, DNA at the NDRs of a yeast genes was shown to be unusually rigid (Fig. 1A), which likely favors nucleosome depletion. Further, loop-seq and other biochemical experiments were performed to show that this region of unusually rigid DNA is precisely positioned to serve as a mechanical landmark guiding chromatin remodeling enzymes like INO80 in accurately positioning the first nucleosome downstream of the TSS (Basu et al., 2021). This is mechanistically possible because the sliding activity of INO80 was shown to be sensitive to the local rigidity of DNA.

Loop-seq also reveals that nucleosomes downstream of the TSS may be positioned, in part, because of the favorable mechanical properties of DNA as encoded in sequence—there are local peaks in intrinsic cyclizability at the centers of downstream nucleosomes (Fig. 1A). Surprisingly, these local peaks in DNA cyclizability at the centers of nucleosomes was shown to be more pronounced for nucleosomes deeper in the gene body (Basu et al., 2021). This likely suggests an increased reliance on the sequence-encoded mechanical properties of DNA, as opposed to the action of nucleosome remodeling enzymes (which operate near TSSs), in the positioning of deeper gene-body nucleosomes. It was also shown that the specific choice of codons along yeast genes (from among the various synonymous codons possible that would have encoded for the same polypeptide) has been optimized by evolution to establish local peaks in DNA cyclizability at the known canonical locations of gene-body nucleosomes (Basu et al., 2021). This suggests the broader possibility that pressure to preserve sequence-encoded modulations in the mechanical properties of DNA has impacted the evolution of genomes.

2. Performing loop-seq to measure DNA cyclizability

2.1 Overview of the loop-seq workflow

The purpose of the loop-seq assay is to measure the cyclizabilities of ~100,000 short 50-bp reference DNA sequences simultaneously. The reference sequences are first designed in silico, with two 25 base adaptor

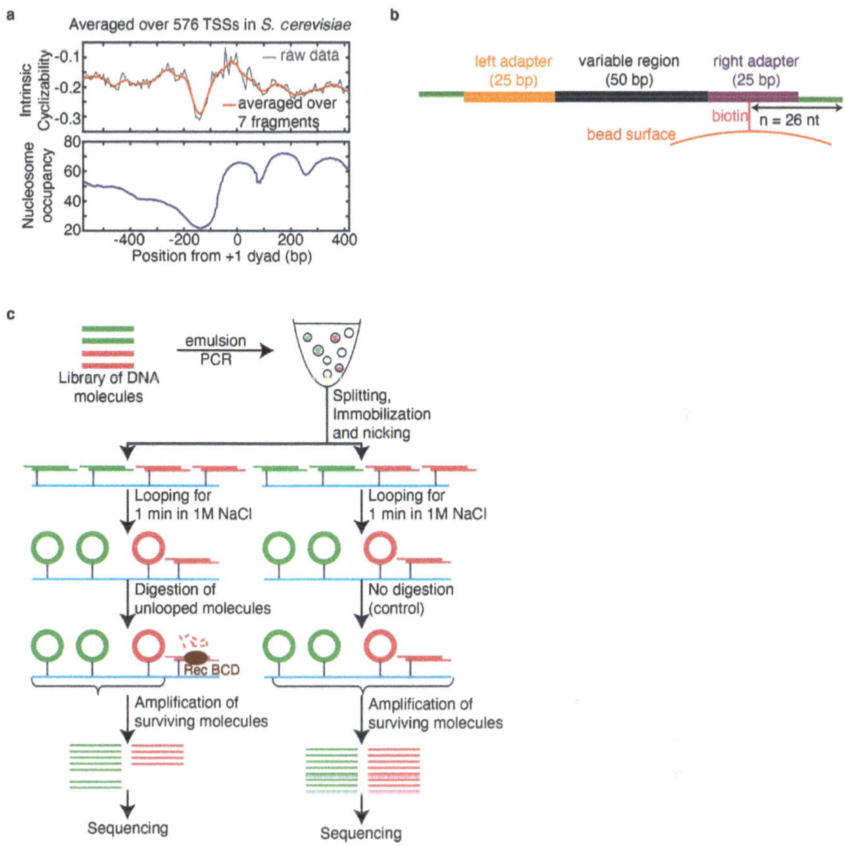

Fig. 1 (A) Top panel: Measured intrinsic cyclizability of DNA in the region around the dyad (center) of the first nucleosome downstream of the TSS of yeast genes (Basu et al., 2021). Data is average across 576 randomly chosen TSSs. There is a sharply defined dip in intrinsic cyclizability upstream of the dyad of the first nucleosome that lies downstream of the TSS (position 0). Bottom panel: measured nucleosome occupancy in the same region, averaged across the same set of genes (Chereji, Ramachandran, Bryson, & Henikoff, 2018). The three peaks correspond to the canonical locations of the dyads of the first three nucleosomes downstream of the TSS. The upstream dip corresponds to the nucleosome depleted region, and is co-centric with the rigid DNA region seen in the top panel. (B) Schematic of a single DNA construct in the library prior to looping (Basu et al., 2021). Each DNA construct has a central 50-bp duplex region which is variable. This region is flanked by two 25-bp duplex adapters that are constant across all constructs in the library. They are in-turn flanked by 10-nt complementary single-stranded overhangs, which are also constant across all constructs in the library. Each construct is further immobilized on the surface of streptavidin-coated beads via a biotin moiety attached to a modified thymine base. The value of n (in this example, $n = 26$ nt) reflects the distance of the biotin tether from the closer end of the DNA construct. (C) Schematic

(Continued)

sequences attached to either end of every sequence to permit subsequent PCR amplification. This set of ~100,000 100-nucleotide (nt) sequences can be chemically synthesized commercially as an oligo pool of single-stranded (ss) DNA. The 100-nt ssDNA sequences in the pool are converted to 120-bp double stranded (ds) DNA with a biotin molecule attached to a modified thymine base via PCR amplification using primers that add the overhangs and introduce the biotin. The amplified dsDNA library is immobilized on the surface of streptavidin-coated beads. Subsequently, the immobilized library is chemically modified, resulting in each construct having a central 100-bp duplex region (of which the central 50 bp are variable, and the flanking 25-bp regions are identical among every molecule), flanked by two complementary 10-nt ssDNA overhangs (Fig. 1B). As shown in Fig. 1C, the bead-immobilized pool of dsDNA with flanking ssDNA overhangs is then exposed to 1 M NaCl for 1 min. This high salt concentration permits looping of individual constructs by allowing for stable annealing of the 10-nt ssDNA overhangs. Within 1 min, most copies of the very flexible sequences in the pool would have looped, while most copies of the most rigid sequences would have remained unlooped. After a minute, RecBCD is introduced, which digests the as-yet unlooped molecules. This results in the library being enriched for the more cyclizable sequences. A side-by-side control is performed involving all steps except RecBCD digestion. The two resulting libraries (RecBCD digested and control) are subject to deep-sequencing to count the relative population of every sequence in each of the libraries. Finally, cyclizability of each sequence is defined as the log of the ratio of the relative population of that sequence in the RecBCD digested library, to than in the control library. It is thus a measure of the probability

Fig. 1—Cont'd of the loop-seq protocol (Basu et al., 2021). In this example, there are only two reference sequences—green and pink—in the original oligo pool. After emulsion PCR amplification, immobilization on beads, generation of 10-nt overhangs, the beads are split into two identical fractions called "sample" (left column) and "control" (right column). DNA constructs in each fraction are permitted to loop for 1 min. This is followed by RecBCD digestion of unlooped molecules in the sample fraction, while the digestion step is skipped in the control fraction. Finally, surviving molecules in either fraction are PCR amplified and sequenced. As per the definition of cyclizability, the cyclizability of green is log $(8/8) = 0$, while that of pink is log $(4/8) = -0.69$. Thus green is more cyclizable than pink. *All three figure panels have been adapted from reference Basu, A. et al. Measuring DNA mechanics on the genome scale. Nature 589, 462–467 (2021). https://doi.org/10.1038/s41586-020-03052-3.*

that a DNA molecule bearing that sequence would have looped in under a minute, and thus survived the subsequent digestion step.

2.2 Consideration for biotin position

DNA constructs in the library are immobilized on the surface of streptavidin coated beads via a biotin molecule attached near the ends of the DNA constructs. This biotin is introduced via the primer used during PCR amplification. Surface immobilization is necessary to only permit intramolecular looping and prevent formation of multimers. We have noticed that the location of the biotin tether influences the measured value of cyclizability (Basu et al., 2021). It is possible that based on the inherent poloidal angle preference of the DNA construct, different biotin attachment points along the DNA might result in the biotin facing the inside or outside of the closed circle in the looped configuration (Supplementary note 7 in this reference (Basu et al., 2021)). Steric considerations might favor an "outside" orientation over an "inside" one, thereby leading to a dependence of cyclizability on the location of the base along the DNA to which the biotin moiety is attached.

To account for this effect (Basu et al., 2021), we prepare the original immobilized library for three different biotin locations, defined by the number of nucleotides n from the end of the construct, $n=26, 29, 31$. We perform the loop-seq protocol for each, and obtain a value of cyclizability for each reference sequence, for all three biotin positions. We use the data to solve the following set of equations for each i:

$$C^i(n) = C_0^i + A^i(n)\sin\left(kn + \varnothing^i\right), n = 26, 29, 31 \tag{1}$$

Here, $C^i(n)$ is cyclizability of the ith reference sequence in the oligo pool, measured when the biotin is located at position n, C_0^i is a mean term which we define to be the intrinsic cyclizability of the ith sequence, $A^i(n)$ and $\varnothing^i$ are amplitude and phase terms associated with the ith sequence, respectively, and $k = \frac{2\pi}{10.4}$ reflects the helical periodicity of DNA. Further, from empirical considerations as reported earlier (Basu et al., 2021), we assume that:

$$A^i(29) = A^i(26)/0.82 \tag{2}$$

$$A^i(31) = A^i(26)/0.7 \tag{3}$$

Thus, for each reference sequence i, there are three unknowns: C_0^i, $A^i(26)$, and $\varnothing^i$ (because $A^i(29)$ and $A^i(31)$ are related to $A^i(26)$ by Eqs. (2) and (3)). There are also three sets of equations in (1) for the three values of n for which we have measured $C^i(n)$. We solve the equations to obtain the value of C_0^i for each reference sequence *i* in the oligo pool. C_0^i is the quantity we use as a proxy for the mechanical bendability of the sequence.

2.3 Equipment

1. Nano drop 2000 (Thermo Scientific)
2. Centrifuge 5424 (Eppendorf)
3. Rare earth magnetic stand (NEB S1506S)
4. Trio Thermocycler (Biometra)
5. *E*-gel iBASE version 1.4.0 and E-gel EX Agarose Gels, 2% (Invitrogen, catalog number G402022)
6. Nextseq 500 Illumina sequencer (Illumina) or MiSeq Illumina sequencer (Illumina)

2.4 DNA oligos, reagents, and buffers

1. An oligo pool containing up to ~100,000 different 100 nt ss reference DNA sequences. The central 50 nt of each sequence is variable. The two flanking 25-nt adapters are constant across all sequences in the library. Each sequence should be present at a concentration of ~10 pM. The library can be designed in silico, and chemically synthesized commercially (Genscript). The 25-nt left and right adapter sequences are:

   5' – TTTCTTCACTTATCTCCCACCGTCC – 3'
   5' – GGCAGAAGACAAGGGAACGAAATAG – 3'

   This is an example of a 100-nt reference sequence in the oligo pool:

   5' –
   TTTCTTCACTTATCTCCCACCGTCCGGAATGGTTACCGAAAACGGGCTTCTTCTCAT
   CCA
   TATCCATAATTTCATGGCAGAAGACAAGGGAACGAAATAG – 3'

   here, the adapters are in red (meaning they are same for other sequences in the pool) and the 50-nt variable region is in black (meaning it is unique to this oligo pool member only).

2. Amplification primers:

   P1: 5' – CAGAATCCGTCGAAGAGCCTTATCTCCCACCGTCC – 3'
   P2: 5' – ACGGATTCTGCGAAGAGCTTCCCTTGTCTTCTGCC – 3'
   P2-Biotin-26: 5' – ACGGATTCTGCGAAGAGCTTCCCTTG/iBiodT/CTTCTGCC – 3'
   P2-Biotin-29: 5' – ACGGATTCTGCGAAGAGCTTCCCTTGTCT/iBiodT/CTGCC – 3'
   P2-Biotin-31: 5' – ACGGATTCTGCGAAGAGCTTCCCTTGTCTTC/iBiodT/GCC – 3'

   PCR amplification of sequences in the initial oligo pool is carried out using primers P1 and one of the three P2-Biotin primers. Amplification with these primers adds overhangs to the sequences in the oligo pool, resulting in 120-bp dsDNA sequences with a biotin moiety attached at a distance of 26, 29, or 31 nt from one end of the molecule (depending on which P2-Biotin primer is used) (Fig. 2). Blue represents regions

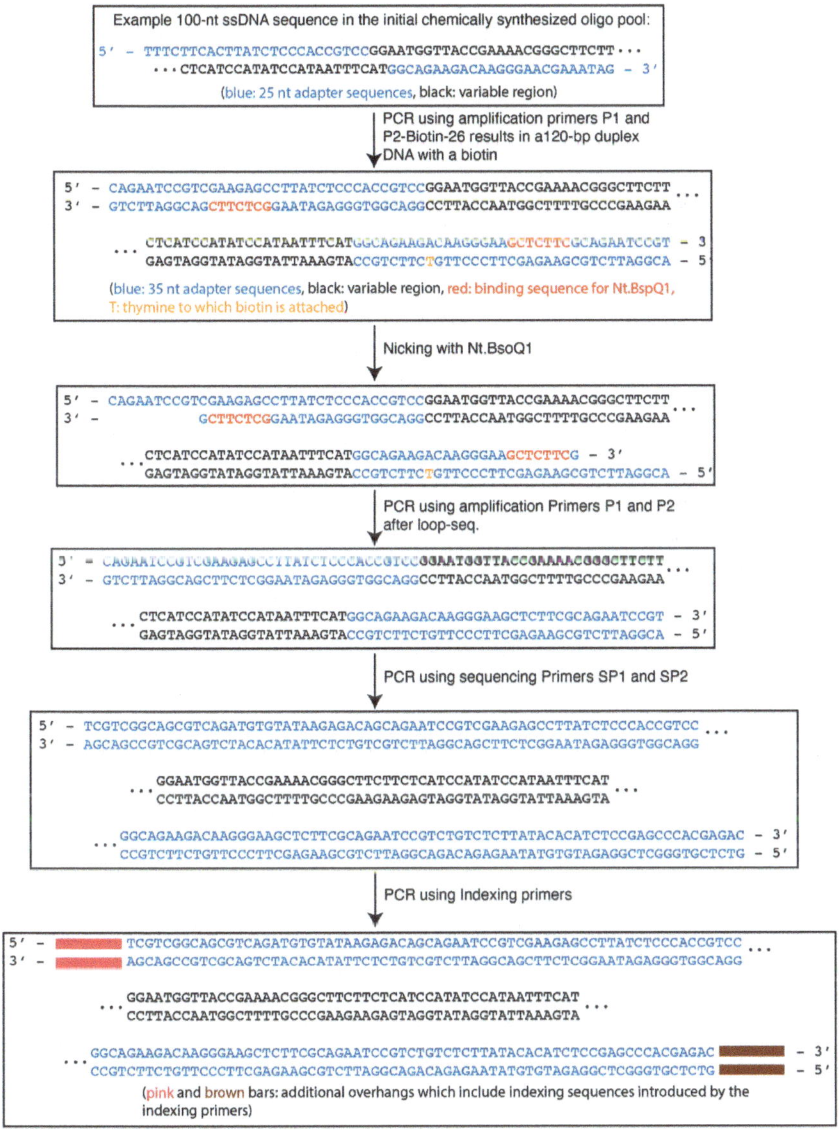

Fig. 2 Schematic of the results of the four rounds of PCR amplification which are involved in the overall protocol.

that are complementary to the ends of the reference sequences in the chemically synthesized oligo pool. Black represents overhangs that get incorporated after PCR. /iBiodT/ represents a biotin attached to a modified thymine base.

3. Sequencing Primers:

SP1: 5' – TCGTCGGCAGCGTCAGATGTGTATAAGAGACAGCAGAATCCGTCGAAGA GCCTTATCTCCCACCGTCC – 3'

SP2: 5' – GTCTCGTGGGCTCGGAGATGTGTATAAGAGACAGACGGATTCTGCGAAG AGCTTCCCTTGTCTTCTGCC – 3'

4. Nextera XT Index Kit (24 Indices 96 Samples) (Illumina, catalog number FC-131-1001).
5. Emulsion PCR (ePCR) kit: Micellular DNA Emulsion & Purification (ePCR) Kit (CHIMERx, Catalog number 3600)
6. PCR purification kit: QIAquick PCR purification kit (Qiagen, catalog number 28106)
7. PCR purification beads: Ampure XP (Beckman Coulter)
8. Streptavidin coated magnetic beads: Dynabeads MyOne Streptavidin T1 (Thermo Fisher Scientific, catalog number 65601)
9. Polymerasees for PCR:
   (1) KAPA Hifi Hot Start ReadyMix (Roche)
   (2) Phusion High-Fidelity PCR Master Mix with HF Buffer (New England Biolabs, catalog number M0431S)
10. Nicking enzyme: Nt.BspQ1 and NEBuffer 3.1 (New England Biolabs, catalog number R0644S)
11. Digestion enzyme: RecBCD and NEBuffer 4 (New England Biolabs, catalog number: M0345S)
12. Buffers:

    T50BSA: 47.5 mM NaCl, 9.5 mM tris HCl pH 8, 1 mg/mL BSA

    T10BSA: 9.5 mM NaCl, 9.5 mM tris HCl pH 8, 1 mg/mL BSA

    T2.5BSA: 2.375 mM NaCl, 9.5 mM tris HCl pH 8, 1 mg/mL BSA

    Looping buffer: 1 M NaCl, 20 mM tris HCl pH 8, 1 mg/mL BSA

    Digestion buffer: 1 × NEBuffer 4 (New England Biolabs), 1 mM ATP, 0.333 units/mL RecBCD (New England Biolabs, catalog number M0345S)

    Dummy digestion buffer: 1 × NEBuffer 4 (New England Biolabs), 1 mM ATP
13. Software: MATLAB (Matworks), Bowtie, SAMtools

3. Protocol

The goal of this protocol is to obtain the intrinsic cyclizabilities of each of the reference sequences present in the chemically synthesized oligo pool.

3.1 Oligo pool amplification

This PCR amplifications amplifies the initial oligo pool, converting 100-nt ssDNA sequences into 120-bp dsDNA sequences with a biotin molecule 26 nt from one end. The additional overhangs also introduce the recognition sequence of the site-specific nicking enzyme Nt.BspQ1, which will be used later to generate complementary single-stranded overhangs. See Fig. 2.

1.1 Prepare a PCR mix as follows:

2 × KAPA polymerase: 25 μL

Amplification primer P1 (100 μM): 1 μLAmplification primer P2-Biotin-26 (100 μM): 1 μL

Template: 0.4 μL of chemically synthesized oligo pool at ~0.3 ng/μL

Water: 22.6 μL

1.2 Prepare 300 μL of oil mixture as per the emulsion ePCR kit manufacturer's instructions. Perform ePCR and purification as per the instructions, using the following thermocycler settings:

Step 1: 95 °C: 3 min

Step 2: 95 °C: 20 s

Step 3: 65 °C: 15 s

Step 4: 72 °C: 15 s

(loop over Steps 2–4 for 20 cycles)

Step 5:72 °C: 2 min

Step 6: hold at 4 °C

After ePCR, purify the PCR product as per the kit manufacturer's instructions. Elute in 50 μL of the manufacturer's elution buffer.

1.3 Repeat the above twice more, substituting Amplification Primer P2-Biotin-26 with Amplification Primer P2-Biotin-29 and Amplification Primer P2-Biotin-31

1.4 Call the three amplification products A1–26, A1–29, A1–31. In this nomenclature, "An" refers to the nth amplification round (here $n = 1$ because this step represents the first of 4 amplification rounds).

The number that follows An represents the distance of the biotin moiety from the end of the DNA construct, as introduced by the P2 amplification primers

3.2 Oligo pool quantification

This is an empirical means of roughly quantifying the concentration of DNA obtained after the oligo pool amplification step.

2.1 Take 11 μL Streptavidin coated magnetic beads in a 1.5 mL tube and pull down the beads over a magnet. Remove the supernatant. Wash the beads twice with 200 μL T50BSA and resuspend in 11 μL T50BSA

2.2 Add 1 μL of a 1:50 dilution of A1–26 to 4 μL of water and mix well. Add 10 μL of streptavidin coated magnetic beads resuspended in T50BSA from the above step to the mixture and mix immediately. Let it incubate for 5 min for DNA to bind to the beads. Pull down the beads over a magnet, remove the supernatant, wash the beads twice with 200 μL T50BSA

2.3 Prepare the following PCR mixture:
2 × KAPA polymerase: 25 μL
Amplification primer P1 (100 μM): 1 μL
Amplification primer P2 (100 μM): 1 μL
Water: 23 μL

2.4 Pull down the magnetic beads and resuspend them in the above PCR mixture. PCR amplify the mixture using the same thermocycle protocol as in the library amplification section, except looping for 16 cycles

2.5 Pull down the beads and remove the supernatant

Purify the supernatant using the Qiagen PCR Purification kit and following the manufacturer's protocol. Elute in 30 mL of elution buffer.

2.6 Measure the concentration of eluded DNA using a nanodrop. Say the concentration is c26 ng/μL. Calculate the value of 50/c26

2.7 Repeat the above using A1–29 and A1–31. Accordingly calculate the values of 50/c29 and 50/c31

3.3 Performing loop-seq

Here, the amplified oligo pool is immobilized on the surface of streptavidin-coated beads and nicked using the site-specific nicking enzyme Nt.BspQ1 to generate 10-nt complementary overhangs at either end (Fig. 2). After that, the sample is split into two fractions. Both fractions are subject to looping for

1 min in the presence of 1 M NaCl. After that, unlooped DNA in one of the two fractions is digested, while all DNA is retained in the other fraction.

3.1 Resuspend the streptavidin coated magnetic beads well. Extract 22 μL of the beads in a 1.5 mL tube, pull down the beads over magnets and remove the supernatant, wash the beads twice with 200 μL T50BSA and resuspend the beads in 22 μL T50BSA

3.2 Add 2 μL of 1:c26 dilution of A1–26 to 5 μL water and mix. To it, add 20 μL of the resuspended magnetic beads and mix immediately. Incubate for 5 min for DNA to bind

3.3 Wash the beads twice with 200 μL T50BSA, followed by once with T10BSA

3.4 Prepare Nicking Mixture in a PCR tube as follows
84 μL water
10 μL 10 × NEB buffer 3.1 (New England Biolabs)
6 μL Nt.BspQ1 (10 units/μL) (New England Biolabs)

3.5 Incubate the PCR tube containing the nicking mixture at 50 °C for 5 min in a thermal cycler

3.6 Pull down the magnetic beads, remove the supernatant, resuspend in the heated nicking mixture, transfer to a PCR tube, and immediately incubate at 37 °C for 25 min

3.7 After the 25-min incubation, pull down the magnetic beads and wash twice with T10BSA pre-heated to 50 °C, followed by once with 200 μL T10BSA at room temperature, and finally resuspend in 200 μL in T2.5BSA. The wash with heated T50BSA removes the complementary 10-nt fragments after nicking, revealing 10-nt complementary overhangs at either end of the molecule (Fig. 2).

3.8 Incubate the beads on a rotator at room temperature for 1.5 h. This step is necessary to unloop any molecules that may have already looped because of the presence of Mg^{2+} ions in the nicking mixture

3.9 Remove from the rotator and split the bead solution into two 95 μL fractions. Label one "sample" and the other "control."

3.10 Pull down the beads in "sample" by placing the tube on the magnetic stand and remove the supernatant. Remove the tube from the magnetic stand. Add 200 μL looping buffer to the beads, start a timer, and immediately mix well

3.11 When the timer reads 40 s, place the tube on the magnetic stand. When the timer reads 1 min, carefully remove bubbles (if any) from the top surface of the liquid in the tube while keeping the tube on the magnets

3.12 When the timer reads 1 min 15 s, remove all remaining liquid in the tube (while it is still on the magnetic stand), then remove the tube from the magnetic stand, immediately add 200 μL digestion buffer, and mix immediately. Incubate for 20 min

3.13 After the 20 min of incubation in digestion buffer, place the tube on the magnetic stand for 1 min, remove the supernatant, remove the tube from the stand, add 200 μL of looping buffer and mix immediately. Re-introduction of looping buffer prevents digestion beyond 20 min as the high salt inhibits RecBCD

3.14 Repeat steps 3.9–3.13 above for the "control" fraction, but using the dummy digestion buffer instead of the digestion buffer

3.4 Library preparation

Here, we will perform three additional rounds of PCR amplification on both the "sample" and the "control" fractions. The first round simply amplifies the DNA from the beads after the loop-seq protocol. The second round adds sequencing adapters to the ends of the DNA molecules. The third round adds indexing adapters, so that all fractions can be pooled and sequenced together. All PCR steps are described Fig. 2.

4.1 Prepare the following PCR mixture:

2 × KAPA polymerase: 52.5 μL
Amplification primer P1 (100 μM): 2.1 μL
Amplification primer P2 (100 μM): 2.1 μL
Water: 48.3 μL

4.2 Place the "sample" and "control" tubes on the magnetic stand and remove the supernatants. Resuspend the beads in each tube in 50 μL of the above PCR mixture. Transfer the two bead solutions to two PCR tubes and PCR amplify according to the specifications described in the Oligo Pool Amplification section, except looping for 16 cycles.

4.3 After amplification, purify the DNA using the Qiagen PCR purification kit. Elute the "sample" and "control" DNA in 30 μL of elution buffer each. Label the "sample" and "control" A2–26-S and A2–26-C (here "S" and "C" refer to sample and control).

4.4 Measure the concentrations of A2–26-S and A2–26-C using a nanodrop. Say the concentration of A2–26-S is c2. Calculate the number 20/c2.

4.5 Run A2–26-S and A2–26-C on a 2% agarose *E*-gel to confirm proper PCR amplification of a 120-bp product. For equal loading volume, calculate the factor by which the intensity of the A2–26-C band is brighter than the A2–26-S band. Call the factor f, which should be in the range 3–10.

At this point, we will perform two additional PCR amplifications of sample and control DNA in order to sequentially add sequencing and indexing adapters for sequencing on an Illumina platform.

4.6 Prepare the two following PCR mixtures for the sample and control in order to add sequencing adapters:

PCR mix for the sample:

2× KAPA polymerase: 25 μL
Sequencing primer 1 (200 μM): 0.5 μL
Sequencing primer 2 (200 μM): 0.5 μL
Template: 20/c2 μL of A2–26-S
Water: up to a total volume of 50 μL

PCR mix for the control:

2× KAPA polymerase: 25 μL
Primer SP1 (200 μM): 0.5 μL
Primer SP2 (200 μM): 0.5 μL
Template: 20/c2 μl of a 1:f dilution of A2–26-C (this roughly ensured the same amount of template as used in the case of A2 26-S).
Water: up to a total volume of 50 μL

4.7 Perform PCR amplification using the specifications described in the Oligo Pool Amplification section, except loop for 8 cycles.

4.8 Purify each PCR mix separately using Ampure XP beads (Beckman Coulter) as follows:

1. Vortex the beads well
2. Take 56 mL of the Ampure beads in a 1.5 mL tube
3. Add 50 μL PCR mixture to it and mix immediately
4. Incubate for 5 min
5. Place the tube for 2 min on the magnetic stand
6. Remove the supernatant and add 200 mL of 80% ethanol. Continue of keep the tube on the magnets
7. Incubate for 30 s
8. Repeat steps 6–7 again
9. Remove as much of liquid as possible while the tube stays on the magnetic stand. Let it dry on the magnetic stand for 10 min
10. Add 27.5 μL 10 mM tris pH 8.5

11. Remove the tube from the magnetic stand and mix well. Let it incubate for 2 min

12. Place the tube on magnets again and remove 25 μl of solution. Discard the beads and store the solution. Label it as A3–26-S or A3–26-C, depending of whether the "sample" or "control" was being purified

4.9 Run A3–26-S and A3–26-C on a 2% agarose gel. Also measure their concentrations on a nanodrop. The concentrations of both should be roughly in the range of 60–80 ng/μL. If outside this range, repeat the PCR above adjusting the amount of template DNA.

The next PCR adds indexing adapters. Use indexing primers from the Nextera XT Index Kit (24 Indices 96 Samples) Nextera XT Index Kit (24 Indices 96 Samples).

4.10 Prepare the two following PCR mixtures for the sample and control in order to add indexing adapters.

DMSO: 1.5 μL
2 × Phusion High-Fidelity PCR Master Mix with HF Buffer: 25 μL
Nextera XT Index 1 primer (N7XX): 5 μL
Nextera XT Index 2 primer (S5XX): 5 μL
Template: 2 μl of 1:17 dilution of A3–26-S or A3–26-C.
Water: 11.5 μL

4.11 Perform each PCR using the following specifications:

Step 1: 98 °C for 30 s.
Step 2: 98 °C for 10 s
Step 3: 58 °C for 20 s
Step 4: 72 °C for 12 s
Loop over Steps 2–4 8 times
Step 5: 72 °C for 1 min
Step 6: hold at 4 °C.

4.12 Purify the PCR mixture using Ampure beads using the protocol described for the purification of the previous PCR in step 4.8. Label the products A4–26-S and A4–26-C.

4.13 Repeat the Loop-Seq and Library preparation protocols starting with the other two amplified oligo pools as well (i.e., with A1–29 and A1–31). In each case, add appropriate indexing primers such that all 6 PCR products can be pooled and sequenced simultaneously. This will result in six final products: A4–26-S, A4–26-C, A4–29-S, A4–29-C, A4–31-S, A4–31-C.

4.14 Pool the six fractions from the previous step and sequence using an Illumina MiSeq platform with 25% phiX spiked in if the initial oligo pool had fewer than 20,000 sequences. Otherwise, use a HiSeq platform (with 25% phiX spiked in) or a NextSeq 500 platform (with 50% phiX spiked in). Perform single-ended sequencing for 61 cycles. Say the sequencing results are stored in 6 FASTQ files called 1.fastq, 2.fastq, …, 6.fastq.

3.5 Analysis of sequencing data

The goal of sequencing data analysis is to obtain, for each reference sequence, how many times it occurs in the reads reported in the six fastq files in the previous step. To achieve this, we will align the reads in the six fastq files to the reference sequences in the original oligo pool. The following section describes a procedure involving using Bowtie, SAMtools, and MATLAB, though the same can be accomplished via any other computational means.

5.1 Create a text file called "Sequences.fa" as follows:

1. The first line contains the characters ">1" (without the quotations).
2. Construct a string involving the letters "CAGAATCCGTCGAA GAGCCTTATCTCCCAC CGTCC" followed by the first 26 nucleotides of the central 50 bp variable region of the first reference sequence in the original oligo pool (the nucleotides pasted before the first 26 nucleotides of the variable region were introduced during the PCR step using primers SP1 and SP2).
3. The string constructed in step 2 above forms the second line of the text file (i.e., the line right below ">1")
4. The third line reads ">2"
5. Repeat step 2, but this time for the second reference sequence. The string so constructed is now the fourth line of the text file, i.e., below ">2"
6. Continue like this for all reference sequences in the oligo pool

5.2 Build a bowtie index file by running the MATLAB command "bowtiebuild('Sequences.fa', SEQ)" in the same directory where the "Sequences.fa" file is located. This builds the bowtie index file. The index file can also be built using any other means.

5.3 Run the following set of four commands in the Terminal in a MacOSX environment. If using a different environment, modifications may be necessary

```
bowtie -S -q -v 3 SEQ 1.fastq PC1.sam
```

```
samtools sort -O sam -T PC1_sort -o PC1_sort.sam PC1.sam
grep -v '^@' PC1_sort.sam > PC1_sort_noheader.sam
cut -f3 PC1_sort_noheader.sam > PC1_sort_cut.sam
```

5.4 After running these commands, the file "PC1_sort_cut.sam" will contain a list of the reference sequence numbers to which the reads in the file 1.fastq align

5.5 Run the set of four commands in step 5.3 above replacing "1" with "2," "3," ..., "6."

5.6 Run the following set of MATLAB commands to count how many times each reference sequence is listed in the files "PC1_sort_cut.sam," "PC2_sort_cut.sam," ..., "PC6_sort_cut.sam":

```
% N in an integer variable storing the number of reference
% sequences
for i=1:6
    filename=['PC',num2str(i),'_sort_cut.sam']
    fid = fopen(filename,'r');
    A=fscanf(fid,'%u');
    d=diff(A);
    r=find(d>0);
    r=[0;r];
    u=unique(A);
    w=diff(r);
    w=[w;numel(find(A==u(end)))];
    eval(['NumIn',num2str(i),'=zeros([1 N]);']);
    eval(['NumIn',num2str(i),'(u)=w;']);
    fclose(fid);
end
```

After running the above code, the arrays NumIn1, NumIn2, ..., NumIn6 will be generated. The number of elements in each of these arrays is equal to the number of reference sequences. The elements in NumIn1, in order, are the number of times each reference sequence occurrs in the reads in the file 1.fastq. Likewise for NumIn2, ..., NumIn6.

5.7 For each reference sequence, intrinsic cyclizability measured at biotin position 26 ($n=26$ in Eq. 1) is defined as the log of the ratio of the relative counts of that reference sequence in the file 1.fastq to that in the file 2.fastq. The following MATLAB commands calculate intrinsic cyclizability:

```
% The variables NumIn1, ..., NumIn6 are obtained from step 5.6
% above
```

```
n1=NumIn1+1
nn1=n1./sum(n1)
n2=NumIn2+1
nn2=n2./sum(n1)
C26=log(nn1./nn2)
n3=NumIn3+1
nn3=n3./sum(n3)
n4=NumIn4+1
nn4=n4./sum(n4)
C29=log(nn3./nn4)
n5=NumIn5+1
nn5=n5./sum(n5)
n6=NumIn6+1
nn6=n6./sum(n6)
C31=log(nn5./nn6)
```

After running the above lines of MATLAB code, three arrays called C26, C29, and C31 are generated. The elements of the array C26 store, in order, the intrinsic cyclizabilities of all reference sequence measured at biotin position 26. Likewise for C29 and C31.

5.8 We now need to obtain intrinsic cyclizability. For each reference sequence i, we solve the three equations

$$C^i(26) = C_0^i + A^i(26)\ \sin\left(26k + \varnothing^i\right),$$
$$C^i(29) = C_0^i + A^i(29)\ \sin\left(29k + \varnothing^i\right),$$
$$C^i(31) = C_0^i + A^i(31)\ \sin\left(31k + \varnothing^i\right).$$

for C_0^i, $A^i(26)$, and $\varnothing^i$, assuming $A^i(29) = A^i(26)/0.82$ and $A^i(31) = A^i(26)/0.7$. These assumptions stem from empirical estimates as reported earlier (Basu et al., 2021).

Create a MATLAB file called "fun.m" with the following lines of code in it:

```
function F=fun(x,C26,C29,C31,i)
a = C26(i);
b = C29(i);
c = C31(i);
k=2.*pi./10.4;
F(1) = x(1) + x(2).*sin(x(3)) - a;
F(2) = x(1) + (x(2)./.82).*sin(k.*3 + x(3)) - b;
F(3) = x(1) + (x(2)./.7).*sin(k.*5 + x(3)) - c;
```

Now run the following MATLAB commands (while noting the comments):

```
% create an integer variable N and set it equal to the number of
% reference sequences
% The variables C26, C29, C31 are obtained from step 5.7 above.
XX=[];
for i=1:N
    evalc('f=@(x) fun(x,C26,C29,C31,i);');
    evalc('X=fsolve(f,[0,0,0]);');
    XX=[XX;X];
end
```

The MATLAB variable XX stores a matrix with three rows and N columns, where N is the number of references sequences in the original oligo pool. The values of C_o^i, $A^i(26)$, and $\varnothing^i$ are the three numbers in the three columns of the ith row of XX.

C_o^i is the intrinsic cyclizability of the ith sequence.

4. Caveats and future directions

Loop-seq provides an experimental platform to assay a mechanical property of DNA—its intrinsic cyclizability—in high throughput. The interpretation of intrinsic cyclizability, however, is not fully transparent. As measured, it reflects on how quickly ends of the short DNA fragments can approach, and then be stabilized by subsequent annealing. However, loop-seq does not report on the shape of DNA in the looped fraction. Looped molecules could be smoothly bent, but could also include other geometries such as kinks, base flipping, or melting. These considerations have been acknowledged and discussed earlier (Basu et al., 2021). Further, intrinsic cyclizability may have contributions from both dynamic flexibility or equilibrium bends present in the molecule, and loop-seq does not provide an immediate way to tease apart these two contributions. Finally, there is likely useful structural or mechanical information in the other two parameters that loop-seq measures—the amplitude (A) and phase ($\varnothing$) of each sequence. Physical interpretations of these parameters are as yet lacking.

Applications of loop-seq have demonstrated that DNA can store information with broad regulatory impact in the complex nature of its mesoscale mechanical properties. Future experiments that alter the mechanical properties of DNA in genomic regions where such properties are functionally important might shed better light on how DNA mechanics impacts molecular interactions with DNA. Further, comparisons of the mechanical properties of DNA across organisms might reveal the extent to which the evolution of genomes has been impacted by the requirement to preserve functionally important DNA mechanical properties. Finally, bending or cyclization is just one mode of DNA deformation. Stretching, twisting, or supercoiling of DNA occur ubiquitously and are part of several critical processes. Development of similar high-throughput techniques that report on the sequence-dependence of various categories of the mechanical properties of DNA would be necessary to fully decipher the mechanical code of the genome and understand its biological impact.

Recent applications of loop-seq have suggested that epigenetic modifications of DNA such as CpG methylation can alter intrinsic cyclizability (Basu, Bobrovnikov, Cieza, Qureshi, & Ha, 2020), with likely consequences for nucleosome organization and transcription. Advancing the loop-seq protocol to make it possible to measure the mechanical properties of natively extracted DNA fragments bearing all epigenetic and chemical modifications might shed light on how such modifications, which are part of developmental processes or diseases, might achieve a part of their biological effects by altering the physical properties of the DNA polymer.

Acknowledgment

I acknowledge Taekjip Ha for providing a conducive laboratory setting and Bailey Liu for a critical reading of the manuscript. This work was supported by the National Science Foundation grants PHY-1430124 and EFMA 1933303 to Taekjip Ha.

References

Afek, A., et al. (2020). DNA mismatches reveal conformational penalties in protein–DNA recognition. *Nature*, *587*, 291–296. https://doi.org/10.1038/s41586-020-2843-2.

Balasubramanian, S., Xu, F., & Olson, W. K. (2009). DNA sequence-directed Organization of Chromatin: Structure-based computational analysis of nucleosome-binding sequences. *Biophysical Journal*, *96*, 2245–2260.

Basu, A., Bobrovnikov, D. G., Cieza, B., Qureshi, Z., & Ha, T. (2020). Deciphering the mechanical code of genome and epigenome. *bioRxiv*. https://doi.org/10.1101/2020.08.22.262352.

Basu, A., Bobrovnikov, D. G., & Ha, T. (2021). DNA mechanics and its biological impact. *Journal of Molecular Biology*, *433*, 166861.

Basu, A., et al. (2021). Measuring DNA mechanics on the genome scale. *Nature*, *589*, 462–467. https://doi.org/10.1038/s41586-020-03052-3.

Battistini, F., et al. (2019). How B-DNA dynamics decipher sequence-selective protein recognition. *Journal of Molecular Biology*, *431*, 3845–3859.

Beveridge, D. L., Cheatham, T. E., & Mezei, M. (2012). The ABCs of molecular dynamics simulations on B-DNA, circa 2012. *Journal of Biosciences*, *37*, 379–397.

Bouchiat, C., et al. (1999). Estimating the persistence length of a worm-like chain molecule from force-extension measurements. *Biophysical Journal*, *76*, 409–413.

Chereji, R. V., Ramachandran, S., Bryson, T. D., & Henikoff, S. (2018). Precise genome-wide mapping of single nucleosomes and linkers in vivo. *Genome Biology, 19*, 19.

Garcia, H. G., et al. (2007). Biological consequences of tightly bent DNA: The other life of a macromolecular celebrity. *Biopolymers, 85*, 115–130.

Geggier, S., & Vologodskii, A. (2010). Sequence dependence of DNA bending rigidity. *Proceedings of the National Academy of Sciences, 107*, 15421–15426.

Jeong, J., & Kim, H. D. (2019). Base-pair mismatch can destabilize small DNA loops through cooperative kinking. *Physical Review Letters, 122*, 218101.

Lai, W. K. M., & Pugh, B. F. (2017). Understanding nucleosome dynamics and their links to gene expression and DNA replication. *Nature Reviews. Molecular Cell Biology, 18*, 548–562.

Lebel, P., Basu, A., Oberstrass, F. C., Tretter, E. M., & Bryant, Z. (2014). Gold rotor bead tracking (AuRBT) for high-speed measurements of DNA twist, torque, and extension. *Nature Methods, 11*, 456–462.

Luger, K., Mäder, A. W., Richmond, R. K., Sargent, D. F., & Richmond, T. J. (1997). Crystal structure of the nucleosome core particle at 2.8 a resolution. *Nature, 389*, 251–260.

Reymer, A., Zakrzewska, K., & Lavery, R. (2018). Sequence-dependent response of DNA to torsional stress: A potential biological regulation mechanism. *Nucleic Acids Research, 46*, 1684–1694.

Rief, M., Clausen-Schaumann, H., & Gaub, H. E. (1999). Sequence-dependent mechanics of single DNA molecules. *Nature Structural Biology, 6*, 346–349.

Rosanio, G., Widom, J., & Uhlenbeck, O. C. (2015). In vitro selection of DNAs with an increased propensity to form small circles. *Biopolymers, 103*, 303–320.

Vafabakhsh, R., & Ha, T. (2012). Extreme bendability of DNA less than 100 base pairs long revealed by single-molecule cyclization. *Science, 337*, 1097–1101.

Vologodskii, A., Du, Q., & Frank-Kamenetskii, M. D. (2013). Bending of short DNA helices. *Artif. DNA PNA XNA, 4*, 1–3.

Wiggins, P. A., Phillips, R., & Nelson, P. C. (2005). Exact theory of kinkable elastic polymers. *Physical Review E, 71*, 021909.

Yan, J., & Marko, J. F. (2004). Localized single-stranded bubble mechanism for cyclization of short double helix DNA. *Physical Review Letters, 93*, 108108.

CHAPTER FIFTEEN

Characterization of the telomerase modulating activities of yeast DNA helicases

David G. Nickens and Matthew L. Bochman*
Molecular & Cellular Biochemistry Department, Indiana University, Bloomington, IN, United States
*Corresponding author: e-mail address: bochman@iu.edu

Contents

Abstract

Eukaryotes with linear chromosomes circumvent the end replication problem via the action of a specialized ribonucleoprotein reverse transcriptase known as telomerase. Cells lacking telomerase activity will senesce when their chromosome ends shorten to a critical length. In contrast, cancer cells can become immortalized by upregulating telomerase to lengthen telomeres during each cycle of DNA replication. Thus, the regulation of telomerase is critical for normal telomere homeostasis. Of the various known ways that telomerase activity is modulated in vivo, recent studies have demonstrated that DNA helicases are involved. In *Saccharomyces cerevisiae*, the Hrq1 and Pif1 helicases act in a pathway that regulates telomerase extension at telomeres and at DNA double-strand DNA breaks. In vitro analysis demonstrates that when these helicases are combined in reactions, they synergistically inhibit or stimulate telomerase activity depending on which helicase is catalytically active. Here, we describe the methods for the overproduction and purification of Hrq1 and Pif1. We also report the preparation of partially purified cell extracts with telomerase activity and how the effects of these helicase on telomerase activity can be assessed in vitro.

Methods in Enzymology, Volume 661
ISSN 0076-6879
https://doi.org/10.1016/bs.mie.2021.08.005

1. Introduction

Due to the semi-discontinuous nature of DNA replication, the duplication of linear chromosomes must overcome the end replication problem, i.e., the progressive shortening of chromosome ends as the RNA primers for terminal Okazaki fragment synthesis are removed (Bonnell, Pasquier, & Wellinger, 2021). Prokaryotes with linear chromosomes have developed several ways to deal with this challenge (Casjens, 1999; Kobryn & Chaconas, 2001; Volff & Altenbuchner, 2000; Yang et al., 2002). However, a single conserved mechanism is utilized by most eukaryotic genomes. Organisms from yeast to humans have evolved specialized chromosome ends called telomeres and a dedicated ribonucleoprotein complex called telomerase that together protect chromosome ends from degradation and solve the end replication problem (Bonnell et al., 2021).

In the majority of studied eukaryotes, telomeric DNA is comprised of short repeats of TG-rich sequences, such as TG_{1-3} in *Saccharomyces cerevisiae* and TTAGGG in *Homo sapiens* (Bonnell et al., 2021). The composition of such sequences can lead to the formation of inter- and intramolecular G-quadruplex structures in the context of single-stranded DNA (Bochman, Paeschke, & Zakian, 2012), which is likely a protective mechanism to inhibit nuclease activity from degrading naked telomeric DNA (Smith et al., 2011). Under most circumstances though, telomeric DNA is highly proteinaceous, being chromatinized by a number of transcriptional silencing proteins and sequence-specific DNA binders (Bonnell et al., 2021). This protects telomeres from degradation and being recognized as DNA double-strand breaks.

The telomerase holoenzyme is comprised of several subunits, including an RNA that acts as the template for a reverse transcriptase, with other proteins providing regulatory or structural roles. These factors in *S. cerevisiae* are the *TLC1* RNA, Est2 reverse transcriptase, and the Est1 and Est3 proteins, respectively (Wellinger & Zakian, 2012). However, many other proteins bind to the large *TLC1* molecule, and the holoenzyme interacts with many other proteins, forming a mega-complex in vivo (Lin et al., 2015). This array of protein–protein and protein–nucleic acid interactions, as well as a complex set of post-translational modifications of the protein players involved, regulate telomerase to avoid problems associated with too much or too little telomerase activity (Bonnell et al., 2021; Wellinger & Zakian, 2012).

It is known that many DNA helicases localize to telomeres and perform roles in DNA replication, recombination, and repair that are necessary for telomere maintenance (Bochman, 2014). One well-studied example is the *S. cerevisiae* Pif1 helicase, which is a catalytic inhibitor of telomerase (reviewed in Bochman, Sabouri, & Zakian, 2010). Evidence suggests that Pif1 uses its helicase activity to unwind the telomeric DNA-*TLC1* RNA hybrid formed by telomerase and thereby evict telomerase from its sites of action (Boule, Vega, & Zakian, 2005). In contrast, the second PIF1 family helicase in budding yeast, Rrm3, is not known to directly affect telomerase but instead is necessary for efficient DNA replication through telomeric DNA (Ivessa, Zhou, Schulz, Monson, & Zakian, 2002). Similarly, RecQ family helicases, such as Sgs1 in *S. cerevisiae*, can participate in telomerase-independent telomere maintenance via DNA recombination (Azam et al., 2006; Cohen & Sinclair, 2001; Huang et al., 2001; Johnson et al., 2001).

Recently, it has been shown that the second RecQ helicase in *S. cerevisiae*, Hrq1 (a homolog of human RECQL4), is also involved in multiple facets of telomere maintenance (Bochman, Paeschke, Chan, & Zakian, 2014). For instance, telomere addition to double-strand breaks increases in the absence of Hrq1, Hrq1 is necessary for Type I survivor formation in cells lacking telomerase, and telomeres hyper-lengthen in *hrq1Δ pif1* double mutants. These effects also appear to be direct, as Hrq1 binding is enriched at telomeres in vivo. Using biochemistry, we have further characterized the synergistic in vitro activity of recombinant Hrq1 and Pif1 to regulate telomerase extension of a model telomeric DNA substrate (Nickens, Rogers, & Bochman, 2018). Hrq1 alone stimulates telomerase activity at high helicase concentrations, Pif1 alone stimulates telomerase at low helicase concentrations but inhibits its activity at high concentrations, and the combination of helicases significantly inhibits telomerase at all concentrations tested (Fig. 1A) (Nickens et al., 2018). By using either catalytically inactive Hrq1-K318A or Pif1-K264A recombinant helicase in the dual-helicase reactions, the combination either acts as a strong inhibitor of telomerase (Fig. 1B) or a strong stimulator of telomerase (Fig. 1C). We hypothesize that the cell regulates Hrq1 and Pif1 activity—and thus their regulation of telomerase—to establish a homeostatic telomere length.

The methods used to perform the in vitro telomerase activity assays with and without added helicase(s) are described in detail below. However, the basic assay is amenable to many modifications. One could assess the effects of different DNA sequences and structures on telomerase,

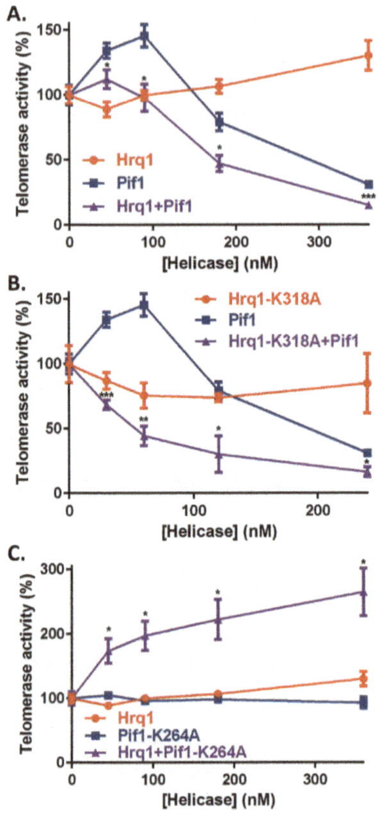

Fig. 1 The effects of Hrq1 and/or Pif1 on in vitro telomerase activity. (A) The wild-type helicases alone display differing effects on telomerase, but in combination, they inhibit telomerase primer extension significantly better than Pif1 alone. (B) The inactive Hrq1-K318A mutant has no effect on telomerase activity alone but, together with wild-type Pif1, is a potent telomerase inhibitor. (C) The inactive Pif1-K264A mutant has no effect on telomerase activity alone but, in the presence of wild-type Hrq1, greatly stimulates telomerase activity. $^{*}P < 0.01$, $^{**}P < 0.001$, and $^{***}P < 0.0001$. *This figure was adapted from Nickens, D. G., Rogers, C. M., & Bochman, M. L. (2018). The Saccharomyces cerevisiae Hrq1 and Pif1 DNA helicases synergistically modulate telomerase activity in vitro.* The Journal of Biological Chemistry, *293(37), 14481–14496. doi:10.1074/jbc.RA118.004092.*

screen small molecules for impacts on the reaction, test different or additional helicases, and/or determine the effects of other recombinant proteins known or suspected to be involved in telomere maintenance.

2. Before you begin

Like many biochemical methods, measuring telomerase activity in vitro is frontloaded. It can take quite a while to generate and assemble all of the necessary reagents, but once that is accomplished, data can quickly be generated. In our experience, it is best not to rush the preparation of any reagent and to, for instance, perform DNA unwinding assays with the recombinant helicases to verify their activity before moving forward with telomerase assays. We have also found that careful buffer preparation and titration of each telomerase preparation is essential for repeatable experiments. Thus, we recommend reserving 10% of any reagent (protein, buffer, etc.) before completely depleting the stocks in the event that subsequent preparations of reagents produce variable results. Having samples of the original materials on hand will enable troubleshooting to determine the root of the problem. The following are additional tips and tricks that should be considered before performing the methods described below:

1. After cells cultures are harvested, chill flasks and centrifuge bottles on ice prior to the centrifugation steps. Similarly, when thawing harvested cells for purification, have all equipment cleaned and ready to use, keeping all tubes and centrifuge bottles at 4 °C.
2. Before grinding yeast "BBs," add and grind dry ice to the grinder mill to chill it before processing cells.
3. Telomerase preparations can take up to 11 h to complete, but it is best to finish the process as quickly as possible in 1 day to preserve activity. Two people sharing part of the processing time can be helpful.
4. Although the previously published versions of the in vitro telomerase primer extension assay (Boule et al., 2005; Cohn & Blackburn, 1995) work to assess the effects of the Pif1 helicase on telomerase activity, we found that they were insufficient for examining the effects of Hrq1 alone or in combination with Pif1 (Nickens et al., 2018; Nickens, Sausen, & Bochman, 2019). This is because the Hrq1 helicase has a larger single-stranded DNA binding site size than Pif1, necessitating a longer oligonucleotide substrate in the reactions. Thus, substrate design and selection should reflect the alterations that one may wish to make with this assay, such as including a different helicase.

3. Materials and equipment

3.1 General equipment

- Vertical polyacrylamide gel electrophoresis equipment (CBS Scientific Vertical Mini-gel System)
- Large vertical gel electrophoresis equipment (Bethesda Research Laboratories, Model S2)
- Heated water baths: 30 and 42 °C

- Standing incubators: 27, 30, and 37 °C
- Shaking incubators capable or heating (≥37 °C) and cooling (≤18 °C)
- Centrifuges (micro, table-top, super-speed, and ultra) capable of cooling to 4 °C
- Sonicator or alternative method for bacterial cell lysis
- Medium-pressure liquid chromatography instrument (AKTA Pure)
- Liquid nitrogen Dewar
- Cell grinder (electric coffee/spice grinder mill)
- Phosphorimager plate
- Phosphorimager
- pH meter

3.1.1 Strains

- *S. cerevisiae* strain JBBY26, a derivative of BCY123 (*MATa can1 ade2 trp1 ura3-52 his3 leu2-3,112 pep4::HIS3 prb1::LEU2 bar1::HISG lys2::pGAL1/10-GAL4*) (Zhao et al., 1994) harboring a dual over-expression plasmid for *TLC1* and Est2 (Boule et al., 2005)
- *Escherichia coli* strain Rosetta 2(DE3)pLysS (F^- *ompT* $hsdS_B(r_B^- m_B^-)$ *gal dcm* (DE3) pLysSRARE2 (Cam^R))
- *Spodoptera frugiperda* cell line Sf9

3.1.2 Reagents and chemicals

- Protease inhibitor mix: 600 nM leupeptin, 2 μM pepstatin A, 2 mM benzamidine, and 1 mM phenylmethanesulfonyl fluoride (PMSF). Commercially available protease inhibitor mixes are also appropriate
- Urea
- NaCl
- Glycerol
- Lysozyme
- β-mercaptoethanol
- HEPES
- EDTA
- Tris
- HCl
- NaOH
- NaOAc
- $MgCl_2$
- Triton-X100
- NP-40 or NP-40 substitute

- $Mg(OAc)_2$
- $Zn(OAc)_2$
- NH_4OAc
- Dithiothreitol (DTT)
- Tween-20
- Spermidine
- dGTP
- dTTP
- RNase inhibitor, murine (New England BioLabs)
- SDS
- Proteinase K
- [α-^{32}P]TTP (10 μCi/μL; Perkin-Elmer)
- [γ-^{32}P]ATP (Perkin-Elmer)

3.1.3 Homemade buffers

- Pif1 Lysis Buffer: 50 mM sodium phosphate buffer (pH 7.5), 300 mM NaCl, 5 mM β-mercaptoethanol, 10% (w/v) glycerol, protease inhibitor mixture, and 0.5 mg/mL lysozyme
- Pif1 IEX Buffer: 50 mM sodium phosphate buffer (pH 6.8), 10% (w/v) glycerol, 0.5 mm EDTA (pH 8.0), and 4 mM β-mercaptoethanol
- Pif1 Storage Buffer: 25 mM HEPES (pH 7.5), 150 mM NaCl, 2 mM β-mercaptoethanol, 0.1 mM EDTA (pH 8.0), and 30% (w/v) glycerol
- L Buffer: 40 mM Tris–HCl (pH 8.0), 500 mM NaOAc, 1.1 mM $MgCl_2$, 0.1 mM EDTA, 0.1% (w/v) Triton X-100, 0.2% (w/v) NP-40, 10% (w/v) glycerol, and 0.1 mM PMSF
- TMG30 Buffer: 10 mM Tris–HCl (pH 8.0), 30 mM NaOAc, 2.2 mM $MgCl_2$, 0.1 mM EDTA, 0.1% (w/v) Triton X-100, 0.2% (w/v) NP-40, 10% (w/v) glycerol, and 0.1 mM PMSF
- TMG500 Buffer: 10 mM Tris–HCl (pH 8.0), 500 mM NaOAc, 2.2 mM $MgCl_2$, 0.1 mM EDTA, 0.1% (w/v) Triton X-100, 0.2% (w/v) NP-40, 10% (w/v) glycerol, and 0.1 mM PMSF
- TMG900 Buffer: 10 mM Tris–HCl (pH 8.0), 900 mM NaOAc, 2.2 mM $MgCl_2$, 0.1 mM EDTA, 0.1% (w/v) Triton X-100, 0.2% (w/v) NP-40, 10% (w/v) glycerol, and 0.1 mM PMSF
- Hrq1 Lysis Buffer: 50 mM Na-HEPES (pH 7.6), 0.4 M NaCl, 20 mM imidazole, and 10% (w/v) glycerol
- Hrq1 Storage Buffer: 25 mM Na-HEPES (pH 8.0), 30% (w/v) glycerol, 300 mM NaOAc (pH 7.6), 25 mM NaCl, 5 mM $Mg(OAc)_2$, 1 mM DTT and 0.1% (*v*/v) Tween-20

- Telomerase Reaction Buffer: 20 mM Tris–HCl (pH 8.0), 20 mM NaCl, 5 mM $MgCl_2$, 1 mM DTT, 1 mM spermidine, 1 μL of 10 μM telomeric primer, 50 μM dGTP, 5 mM TTP, 1 μL of [α-^{32}P]TTP (10 μCi/μL), RNase inhibitor (1 U/μL) and 1–3 μL of telomerase-enriched extract
- Stop Buffer: 20 mM Tris–HCl (pH 8.0), 1 mM EDTA, 0.5% (w/v) SDS, 250 μg/mL proteinase K, and ~1000 cpm of [γ-^{32}P]ATP-labeled loading control oligonucleotide
- Formamide load buffer: 10 mM NaOH, 95% formamide, 0.05% bromophenol blue, and 0.05% xylene cyanol

3.1.4 Oligonucleotides

- Tel15: 5′-TGTGGTGTGTGTGGG-3′. This is the standard substrate used to standardize telomerase activity levels and to determine the effects of Pif1 on telomerase activity (Boule et al., 2005).
- Tel30: 5′-CGCCATGCTGATCCGTGTGGTGTGTGTGGG-3′. This substrate can be used with helicases such as Hrq1 that have binding footprints >15 nt and/or in telomerase displacement assays (Nickens et al., 2018).
- Tel50: 5′-GTGTGGGTGTGGTGTGGGTGTGGTGTGGGTGTG TGGTGTGGTGTGTGTGG-3′. This substrate is used to determine the effects of Hrq1 or the combination of Hrq1 and Pif1 on telomerase.
- Loading control: We used ^{32}P-labeled Tel15 as the loading control for telomerase assays with Tel30 or Tel50 as the extended primers. For assays in which Tel15 was the extended primer, we used ^{32}P-labeled Tel50 as the loading control.

4. Step-by-step method details

4.1 Over-expression and purification of the nuclear isoform of Pif1

Despite being a large eukaryotic protein with an N-terminal domain that is predicted to be natively disordered (Andis, Sausen, Alladin, & Bochman, 2018), *S. cerevisiae* Pif1 is amenable to over-expression and purification from *E. coli*. Many groups have successfully generated recombinant Pif1 in this manner, and we favor the expression constructs and basic protocols of the Raney lab, which produce excellent yields of the nuclear isoform of Pif1 with no affinity tags (Ramanagoudr-Bhojappa, Blair, Tackett, & Raney, 2013). The catalytically inactive Pif1-K264A mutant can be purified in the same manner, as well as various other substitution and domain truncation mutants (Nickens et al., 2018, 2019; Ononye, Sausen, Balakrishnan, & Bochman, 2020).

1. Transform the expression strain of *E. coli* with plasmid pSUMO-Pif1 (Ramanagoudr-Bhojappa et al., 2013). We have achieved excellent over-expression with both Rosetta 2(DE3) pLysS cells (Novagen) and NiCo21(DE3) cells (New England Biolabs) that also carry the pLysS plasmid. Select for transformants on LB medium supplemented with the appropriate antibiotics, e.g., 50 μg/mL kanamycin and 34 μg/mL chloramphenicol for the above-suggested strains, at 37 °C.
2. Inoculate 5-mL of LB medium supplemented with the same antibiotics with a fresh transformant colony, and grow overnight at 30 °C with aeration. Only fresh transformants should be used as we have observed little-to-no expression from transformants stored on plates for extended periods of time or from glycerol stocks of expression strains carrying the pSUMO-Pif1 plasmid.
3. Back-dilute the overnight culture into 2x YT medium supplemented with 10 mM $MgSO_4$, 1 mM $Zn(OAc)_2$, and the same antibiotics listed above, and grow the cells to mid-log phase at 24 °C with aeration. A 1-L culture is generally sufficient to produce 0.5 mg of purified protein.
4. Induce SUMO-Pif1 over-expression by adding IPTG to a final concentration of 0.2 mM, and incubate the cells for 14–18 h at 18 °C with aeration. All subsequent steps should be performed at 4 °C or on ice using instruments and equipment that is likewise chilled.
5. Harvest the cells by pelleting in a centrifuge at 7000 × *g* for 10 min, decant the growth medium, and resuspend the cells in Lysis Buffer.
6. Lyse cells, and subject the lysate to ultracentrifugation at 150,000 × *g* for 1 h. We have used a Fisher Scientific FB505 sonicator, a Microfluidics microfluidizer, an Avestin EmulsiFlex-C3 emulsifier, and a high-pressure cell cracker to lyse cells with comparable results, so any lysis method that does not introduce excessive amounts of heat is likely to be acceptable.
7. Apply the ultracentrifugation supernatant to a 5-mL Talon HiTrap column, wash with 10 column volumes (CVs) of Lysis Buffer, 20 CVs of Lysis Buffer supplemented with 15 mM imidazole, and elute with Lysis Buffer containing 200 mM imidazole using an AKTA FPLC or similar medium-pressure liquid chromatography instrument.
8. Identify the peak SUMO-Pif1-containing fractions by SDS-PAGE and Coomassie staining analysis of 10-μL aliquots of each elution fraction,

pool the peak fractions, and dialyze them against 20-fold excess lysis buffer to remove imidazole.

9. Treat the SUMO-Pif1 with Ulp1 SUMO protease to remove the N-terminal SUMO tag, which itself is 6 × His tagged, and reapply to the Talon HiTrap column. The His-tagged Ulp1 protease, uncleaved SUMO-Pif1, and contaminants from the elution in Step 7 will be retained on the Talon column, and untagged Pif1 will elute in the column flow-through.
10. To further purify the untagged nuclear isoform of Pif1 diluting it 1:1 with IEX Buffer and applying it to a Resource S. Elute with a NaCl gradient from 150 mM to 1 M. Once again, identify Pif1-containing fractions by SDS-PAGE and Coomassie staining, and pool the peak fractions.
11. Concentrate the Pif1 with Amicon Ultra-4 centrifugal filters with a 30,000 cutoff, and store at −80 °C in Storage Buffer.

4.2 Over-expression and purification of Hrq1

S. cerevisiae Hrq1 is a larger enzyme (~125 kDa) than Pif1, and although it can be over-expressed and purified from *E. coli* in a similar manner as that described for Pif1 above (Bochman et al., 2014), the yields are low. Even using the high-density autoinduction method (Studier, 2005), only ~100 μg of purified recombinant Hrq1 can be expected from 100 g of cell paste (Fig. 2A). Thus, we typically generate Hrq1 using baculovirus-infected insect cell culture, which has equivalent in vitro activity to Hrq1 over-expressed in *E. coli* but is produced at much higher yields (Rogers et al., 2017; Rogers & Bochman, 2017) (Fig. 2B). The catalytically inactive Hrq1-K318A mutant can be purified in the same manner to that described below for wild-type Hrq1.

1. Infect a 500-mL culture of Sf9 insect cells with Hrq1 baculovirus, and incubate at 27 °C for 5 days with gentle aeration.
2. Harvest cells by centrifugation, decant the medium, and snap-freeze the pellet in liquid nitrogen.
3. Thaw the frozen cells (~10 g) on ice, and lyse at 4 °C by stirring the cell pellet in 100 mL of Hrq1 Lysis Buffer supplemented with fresh 5 mM β-mercaptoethanol, protease inhibitor mix, and 20 μg/mL DNase I.
4. Clarify the cell lysate was by centrifugation at 4 °C for 30 min at 14,000 rpm, and transfer the supernatant into a 50-mL conical tube containing 2.5 mL HIS-Select Nickel Affinity Gel (Sigma) pre-equilibrated with Hrq1 Lysis Buffer.

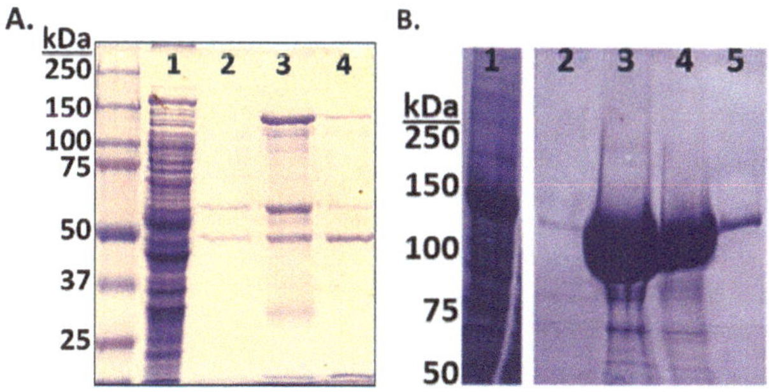

Fig. 2 Generation of recombinant Hrq1. Hrq1 (~125 kDa) was over-expressed in *E. coli* (A) or insect cells (B), and cell lysates were subjected to one step of purification over a Streptactin Sepharose column or in batch mode in the presence of Ni^{+2} affinity resin, respectively. In both images, Lane 1 contains a sample of clarified cell lysate, and the following lanes contain 10-μL samples of elution fractions. Despite coming from 10-fold less cell mass, the insect cells yielded far more Hrq1 protein than *E. coli*.

5. Gently mix the contents of the tube for 30 min at 4 °C. This can be accomplished with a nutator, rocking platform, tube roller/rotisserie, or an orbital shaker used at low speed.
6. Centrifuge the lysate and resin at 2000 × *g* for 5 min at 4 °C, and decant most of the supernatant. Then, transfer the remaining supernatant and resin slurry to a 30-mL gravity column, and allow the resin to settle as the supernatant flows through.
7. Wash the resin three times (~30 mL each) with Hrq1 Lysis Buffer, and elute 10 × His-Hrq1 with 5 CVs of Hrq1 Lysis Buffer containing 500 mM imidazole.
8. Pool the purest fractions, as judged by SDS-PAGE and Coomassie staining, and dialyze against Hrq1 Storage Buffer.

4.3 Preparation of telomerase extract

To date, no one has purified biochemically useful amounts of recombinant *S. cerevisiae* telomerase holoenzyme (i.e., the Est1-3 proteins and the *TLC1* RNA), but many groups have successfully produced yeast extracts enriched for telomerase activity (Boule et al., 2005; Cohn & Blackburn, 1995; Nickens et al., 2018, 2019). Though largely undefined, these extracts are useful to determine the effects of various reagents (e.g., recombinant proteins or small molecules) on telomerase extension of model DNA substrates using the telomerase primer extension assay described in the next section.

1. Over-expression of *S. cerevisiae* Est2/*TLC1* *via* galactose induction. Grow cultures at 30 °C to an OD_{600} of ~1, and add dry galactose to a final concentration of 2%. Incubate cultures at 30 °C for an additional 16h. Yeast telomerase has also been over-produced approximately 10-fold by co-galactose induction of GFP-Est1, GFP-Est2, Est3, and *TLC1* in the W303 genetic background (Lin et al., 2015), making this an excellent strain to use for this protocol.
2. Harvest the cells by centrifugation at 3000 × *g* for 10 min at 4 °C.
3. Resuspend the cells in 1 mL of L Buffer per mL of cell paste in a 50-mL conical tube, and flash-freeze the cells as "BBs" by puncturing the tube with an 18-gauge needle and dripping the slurry into one or more 50-mL conical tubes containing liquid nitrogen.
4. Lyse the cells by grinding the frozen BBs with dry ice. This step can be accomplished with a mortar and pestle or more rapidly with an inexpensive electric coffee grinder. Only use enough dry ice to keep the cell BBs and lysate powder frozen during grinding because you need to wait for it to sublimate away (e.g., overnight at −80 °C) before proceeding to the next step.
5. Slowly thaw the lysate at 4 °C or in an ice water bath, and clarify it by centrifugation at 10,000 × *g*.
6. Pre-equilibrate DEAE-Sepharose Fast Flow resin (GE Healthcare) with TMG500 Buffer, and incubate it with the clarified cell lysate for 30 min at 4 °C.
7. Gently centrifuge the resin for 1 min at 800 × *g*, decant the buffer, and wash it three times with TMG500 Buffer.
8. Elute telomerase with TMG900 buffer, and concentrate the eluate to ~4 mL using an Amicon Ultra-4 centrifugal filter with a 30,000 Da cutoff.
9. Desalt the telomerase preparation using a 4-mL Zeba spin column (ThermoFisher Scientific), and further concentrate the sample to ~100 μL/L of starting culture.
10. Dilute the samples 1:1 with 100% glycerol, flash-freeze with liquid nitrogen, and store at −80 °C.
11. Each telomerase preparation should be titrated to standardize activity levels before use in experiments (Fig. 3). Control experiments should also be performed to ensure that extension activity is RNase A sensitive.

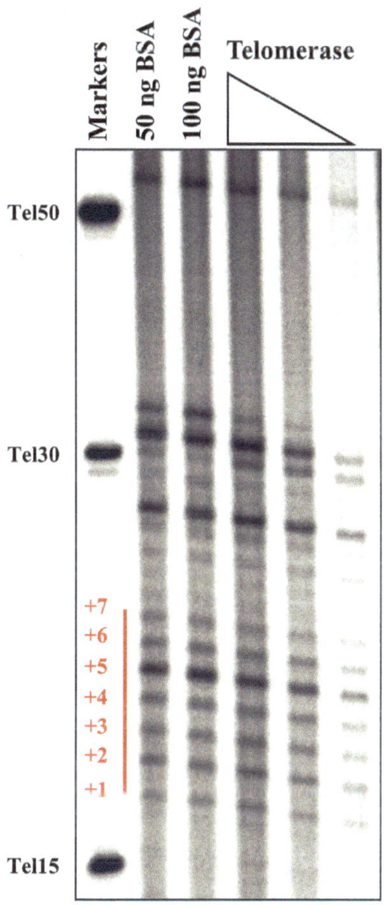

Fig. 3 Titrating a fresh telomerase preparation. The ability of increasing amounts of a telomerase extract to perform primer extension of the Tel15 substrate is tested for each batch of extract prepared to standardize activity levels with previous experiments. The +1 to +7 nucleotide additions are noted in red, and we often see a prominent +5 extension. The markers are ^{32}P-labeled Tel15, Tel30, and Tel50 oligonucleotides. The addition of two amounts of BSA to the primer extension assay was also tested in this experiment to control for the effects of added protein to the reaction.

4.4 Telomerase primer extension assay

This assay utilizes one or more single-stranded DNA substrates comprised of *S. cerevisiae* telomeric repeat sequence (TG_{1-3}) and radiolabeled TTP to measure the reverse transcriptase activity of the yeast telomerase extract produced above.

1. Perform telomerase reactions in 10 μL of 1× Telomerase Reaction Buffer, and incubate at 30 °C for 45 min. In reactions with two primers, each is used at 0.5 μM. For combined helicase and telomerase assays, the desired helicase is diluted into 1 × Hrq1 Storage Buffer so that a 10-fold dilution into reactions achieves the desired helicase concentration. Control reactions lacking added helicase receive 1 μL of 1× Hrq1 Storage Buffer without added protein. We have found that a range of 0–360 nM Hrq1 and/or Pif1 is informative.
2. Stop reactions by adding 25 volumes of Stop Buffer and incubating at 30 °C for 45 min.
3. Extract samples twice with phenol/chloroform (pH 7.0), and precipitate DNA by the addition of 1 volume of 4 M NH_4OAc and 2.5 volumes of ice-cold 100% ethanol.
4. Chill tubes at −80 °C for ≥30 min, and centrifuged them at 21,130 × *g* to pellet reaction products.
5. Resuspend pellets in 6 μL of 95% (v/v) formamide load buffer, heat to 95 °C for 5 min, and separate reaction products on 16% 19:1 acrylamide/bisacrylamide gels containing 6 M urea run at 2500 V for 120 min.
6. Dry the gels using a vacuum gel drier, expose the dried gel to a phosphorimager plate for ~24 h, and image using a Typhoon 9500 scanner or similar phosphorimager. The gels can also be frozen at −20 °C for phosphorimager or film exposure. In either case, to avoid gel cracking during drying or freezing, the gels should be poured with a minimal amount of TEMED (e.g., 10 μL/100 mL of degassed acrylamide-TBE solution) and allowed to slowly polymerize.
7. Quantify total activity by measuring the densitometry for each band on a gel using ImageQuant, ImageJ, or similar software. The sum of the measured values for each band in a lane is reported as the total activity. Bands should be corrected for the number of dT residues (i.e., the amount of [α-^{32}P]TTP incorporation) and normalized to the loading control to generate corrected pixel values. The distribution of label is determined by adding the corrected pixels for each individual band to obtain total cumulative activity. Each individual band is then divided by the total activity to determine the percent of label in each band, referred to as label distribution. Telomerase extensions T51–T58 from experiments with the Tel50 substrate are quantified as a group due to low resolution of the bands. Processivity is calculated by submitting the corrected pixel data to the following calculation: $(T_I + 1 \ldots T_n)/(T_I + \ldots T_n)$, where T_I is the corrected pixel value of the lowest molecular weight telomerase extension band, and T_n is the highest molecular weight band in a series.

References

Andis, N. M., Sausen, C. W., Alladin, A., & Bochman, M. L. (2018). The WYL domain of the PIF1 helicase from the thermophilic bacterium Thermotoga elfii is an accessory single-stranded DNA binding module. *Biochemistry, 57*(7), 1108–1118. https://doi.org/10.1021/acs.biochem.7b01233.

Azam, M., Lee, J. Y., Abraham, V., Chanoux, R., Schoenly, K. A., & Johnson, F. B. (2006). Evidence that the S. cerevisiae Sgs1 protein facilitates recombinational repair of telomeres during senescence. *Nucleic Acids Research, 34*, 506–516. 34/2/506 [pii] https://doi.org/10.1093/nar/gkj452.

Bochman, M. L. (2014). Roles of DNA helicases in the maintenance of genome integrity. *Molecular & Cellular Oncology, 1*(3). https://doi.org/10.4161/23723548.2014.963429, e963429.

Bochman, M. L., Paeschke, K., Chan, A., & Zakian, V. A. (2014). Hrq1, a homolog of the human RecQ4 helicase, acts catalytically and structurally to promote genome integrity. *Cell Reports, 6*(2), 346–356. https://doi.org/10.1016/j.celrep.2013.12.037.

Bochman, M. L., Paeschke, K., & Zakian, V. A. (2012). DNA secondary structures: Stability and function of G-quadruplex structures. *Nature Reviews. Genetics, 13*(11), 770–780. https://doi.org/10.1038/nrg3296.

Bochman, M. L., Sabouri, N., & Zakian, V. A. (2010). Unwinding the functions of the Pif1 family helicases. *DNA Repair (Amst), 9*, 237–249. doi: S1568-7864(10)00011-X [pii] https://doi.org/10.1016/j.dnarep.2010.01.008.

Bonnell, E., Pasquier, E., & Wellinger, R. J. (2021). Telomere replication: Solving multiple end replication problems. *Frontiers in Cell and Development Biology, 9*, 668171. https://doi.org/10.3389/fcell.2021.668171.

Boule, J., Vega, L., & Zakian, V. (2005). The Yeast Pif1p helicase removes telomerase from DNA. *Nature, 438*, 57–61. Retrieved from http://www.nature.com/nature/journal/v438/n7064/abs/nature04091.html.

Casjens, S. (1999). Evolution of the linear DNA replicons of the Borrelia spirochetes. *Current Opinion in Microbiology, 2*(5), 529–534. https://doi.org/10.1016/s1369-5274(99)00012-0.

Cohen, H., & Sinclair, D. (2001). Recombination-mediated lengthening of terminal telomeric repeats requires the Sgs1 DNA helicase. *Proceedings of the National Academy of Sciences of the United States of America, 98*, 3174–3179.

Cohn, M., & Blackburn, E. H. (1995). Telomerase in yeast. *Science, 269*, 396–400.

Huang, P., Pryde, F., Lester, D., Maddison, R., Borts, R., Hickson, I., et al. (2001). SGS1 is required for telomere elongation in the absence of telomerase. *Current Biology, 11*, 125–129.

Ivessa, A. S., Zhou, J.-Q., Schulz, V. P., Monson, E. M., & Zakian, V. A. (2002). *Saccharomyces* Rrm3p, a 5′ to 3′ DNA helicase that promotes replication fork progression through telomeric and sub-telomeric DNA. *Genes & Development, 16*, 1383–1396.

Johnson, F., Marciniak, R., McVey, M., Stewart, S., Hahn, W., & Guarente, L. (2001). The *Saccharomyces cerevisiae* WRN homolog Sgs1p participates in telomere maintenance in cells lacking telomerase. *The EMBO Journal, 20*, 905–913.

Kobryn, K., & Chaconas, G. (2001). The circle is broken: Telomere resolution in linear replicons. *Current Opinion in Microbiology, 4*(5), 558–564. https://doi.org/10.1016/s1369-5274(00)00251-4.

Lin, K. W., McDonald, K. R., Guise, A. J., Chan, A., Cristea, I. M., & Zakian, V. A. (2015). Proteomics of yeast telomerase identified Cdc48-Npl4-Ufd1 and Ufd4 as regulators of Est1 and telomere length. *Nature Communications, 6*, 8290. https://doi.org/10.1038/ncomms9290.

Nickens, D. G., Rogers, C. M., & Bochman, M. L. (2018). The Saccharomyces cerevisiae Hrq1 and Pif1 DNA helicases synergistically modulate telomerase activity in vitro. *The Journal of Biological Chemistry, 293*(37), 14481–14496. https://doi.org/10.1074/jbc.RA118.004092.

Nickens, D. G., Sausen, C. W., & Bochman, M. L. (2019). The biochemical activities of the Saccharomyces cerevisiae Pif1 helicase are regulated by its N-terminal domain. *Genes (Basel)*, *10*(6). https://doi.org/10.3390/genes10060411.

Ononye, O. E., Sausen, C. W., Balakrishnan, L., & Bochman, M. L. (2020). Lysine acetylation regulates the activity of nuclear Pif1. *The Journal of Biological Chemistry*. https://doi.org/10.1074/jbc.RA120.015164.

Ramanagoudr-Bhojappa, R., Blair, L. P., Tackett, A. J., & Raney, K. D. (2013). Physical and functional interaction between yeast Pif1 helicase and Rim1 single-stranded DNA binding protein. *Nucleic Acids Research*, *41*(2), 1029–1046. https://doi.org/10.1093/nar/gks1088.

Rogers, C. M., & Bochman, M. L. (2017). Saccharomyces cerevisiae Hrq1 helicase activity is affected by the sequence but not the length of single-stranded DNA. *Biochemical and Biophysical Research Communications*, *486*(4), 1116–1121. https://doi.org/10.1016/j.bbrc.2017.04.003.

Rogers, C. M., Wang, J. C., Noguchi, H., Imasaki, T., Takagi, Y., & Bochman, M. L. (2017). Yeast Hrq1 shares structural and functional homology with the disease-linked human RecQ4 helicase. *Nucleic Acids Research*, *45*(9), 5217–5230. https://doi.org/10.1093/nar/gkx151.

Smith, J. S., Chen, Q., Yatsunyk, L. A., Nicoludis, J. M., Garcia, M. S., Kranaster, R., et al. (2011). Rudimentary G-quadruplex-based telomere capping in Saccharomyces cerevisiae. *Nature Structural & Molecular Biology*, *18*(4), 478–485. https://doi.org/10.1038/nsmb.2033.

Studier, F. W. (2005). Protein production by auto-induction in high density shaking cultures. *Protein Expression and Purification*, *41*(1), 207–234. Retrieved from http://www.ncbi.nlm.nih.gov/pubmed/15915565.

Volff, J. N., & Altenbuchner, J. (2000). A new beginning with new ends: Linearisation of circular chromosomes during bacterial evolution. *FEMS Microbiology Letters*, *186*(2), 143–150. https://doi.org/10.1111/j.1574-6968.2000.tb09095.x.

Wellinger, R. J., & Zakian, V. A. (2012). Everything you ever wanted to know about Saccharomyces cerevisiae telomeres: Beginning to end. *Genetics*, *191*(4), 1073–1105. https://doi.org/10.1534/genetics.111.137851.

Yang, C. C., Huang, C. H., Li, C. Y., Tsay, Y. G., Lee, S. C., & Chen, C. W. (2002). The terminal proteins of linear Streptomyces chromosomes and plasmids: A novel class of replication priming proteins. *Molecular Microbiology*, *43*(2), 297–305. Retrieved from https://www.ncbi.nlm.nih.gov/pubmed/11985710.

Zhao, S., Douglas, N. W., Heine, M. J., Williams, G. M., Winther-Larsen, H. C., & Meaden, P. G. (1994). The STL1 gene of Saccharomyces cerevisiae is predicted to encode a sugar transporter-like protein. *Gene*, *146*(2), 215–219. https://doi.org/10.1016/0378-1119(94)90295-x.

CHAPTER SIXTEEN

Single-molecule studies of yeast Rad51 paralogs

Upasana Roy[a], Youngho Kwon[b], Patrick Sung[b], and Eric C. Greene[a,*]
[a]Department of Biochemistry & Molecular Biophysics, Columbia University, New York, NY, United States
[b]Department of Biochemistry and Structural Biology, University of Texas Health Science Center at San Antonio, TX, United States
*Corresponding author: e-mail address: ecg2108@cumc.columbia.edu

Contents

Abstract

Homologous recombination (HR) is a conserved mechanism essential for the accurate repair of DNA double stranded breaks and the exchange of genetic information during meiosis. The key steps in HR are carried out by the RecA/Rad51 class of recombinases, which form a helical filament on single-stranded DNA (ssDNA) and catalyze homology search and strand exchange with a complementary duplex DNA target. In eukaryotes, assembly of the Rad51–ssDNA filament requires regulatory factors called mediators, including Rad51 paralogs. A mechanistic understanding of the role of Rad51 paralogs

Methods in Enzymology, Volume 661
ISSN 0076-6879
https://doi.org/10.1016/bs.mie.2021.08.006

in HR has been hampered by the transient and diverse nature of intermediates formed with the Rad51–ssDNA filament, which cannot be resolved by traditional ensemble methods. The biochemical characterization of Rad51 paralogs, including the *S. cerevisiae* complex Rad55–Rad57 has also been limited by their propensity to aggregate. Here we describe the preparation of monodisperse GFP-tagged Rad55–Rad57 complex and the methodology for its analysis in our single-molecule DNA curtain assay.

1. Introduction

1.1 Homologous recombination

Homologous recombination (HR) is a conserved process essential for the error-free repair of DNA double stranded breaks (DSBs). Defects in HR lead to genomic instability and chromosomal rearrangements, as observed in multiple cancers and cancer-prone syndromes (Kass, Moynahan, & Jasin, 2016; Prakash, Zhang, Feng, & Jasin, 2015; San Filippo, Sung, & Klein, 2008). Repair by HR relies on the use of an undamaged homologous DNA template, and occurs during S/G2 phases of the cell cycle where the sister chromatid is available as a template (San Filippo et al., 2008). Here we focus on studies directed at mechanistically delineating the HR pathway in *S. cerevisiae*, which has served as the model for understanding conserved HR mechanisms in higher eukaryotes as well.

At the onset of HR, DNA around the DSB site is resected by nucleases to expose long 3' single-stranded DNA (ssDNA) tails that are bound by replication protein A (RPA) (Fig. 1A–D). Subsequently, the Rad51 recombinase displaces RPA from the ssDNA with the help of mediator proteins Rad52 and Rad55–Rad57 (Fig. 1E). Rad51 assembles into a helical filament on ssDNA, termed the presynaptic complex, which searches the genome for homologous sequences and invades the target duplex to form a displacement loop (D-loop) structure (Fig. 1F). In this D-loop intermediate, the invading strand primes DNA synthesis to replace the sequence lost during DNA end resection. This extended D-loop intermediate can then be processed by one of several possible pathways, to produce either crossover or non-crossover products (Fig. 1G).

1.2 Single-molecule studies of HR

HR proceeds via multiple short-lived, heterogenous intermediates, which are difficult to capture by ensemble biochemical analyses and have therefore, remained largely elusive. Single-molecule studies offer a solution to this problem, as they allow direct examination of such transient species to furnish

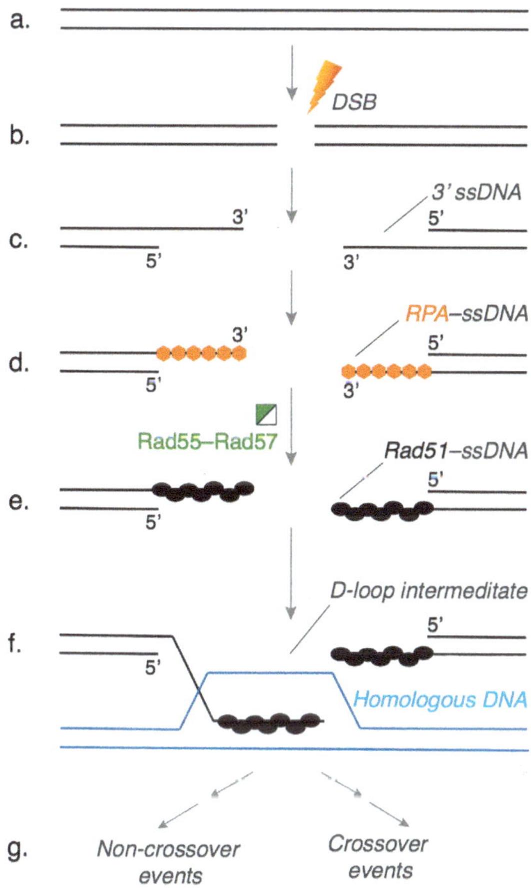

Fig. 1 Early steps of homologous recombination (HR). A simplified schematic of the early steps of the HR pathway is depicted. Genomic DNA (A) can be damaged by a variety of agents to generate a double stranded break (DSB), (B) nucleases carry out DNA end resection exposing 3′ single-stranded DNA (ssDNA) overhangs, (C) these overhangs are bound by the ssDNA binding protein RPA, (D) Rad51 recombinase displaces RPA to form the presynaptic complex, (E) the presynaptic complex searches the genome for a homologous DNA template, (F) DNA strand invasion catalyzed by the presynaptic complex yields a D-loop intermediate, and (G) repair DNA synthesis ensues, with the resulting extended D-loop intermediate being resolved via one of several mechanistically distinct pathways (not shown) to produce crossover or non-crossover events.

insights into the dynamics and regulation of each reaction intermediate. Several recent single-molecule studies have greatly advanced our understanding of the complexity of HR, including regulation of the DNA resection step (Myler et al., 2017; Xue et al., 2019), regulation of dynamics of the Rad51–ssDNA filament (Belan et al., 2021; Gibb et al., 2014; Ma, Gibb,

Kwon, Sung, & Greene, 2017; Taylor et al., 2015) and defining the action of anti-HR helicases (Crickard et al., 2019; Kaniecki et al., 2017; Qiu et al., 2013; Xue et al., 2021).

We have developed a single-molecule imaging method called "DNA curtain," which allows us to visualize hundreds of DNA-protein events in real-time (De Tullio, Kaniecki, & Greene, 2018; Gibb, Silverstein, Finkelstein, & Greene, 2012; Greene, Wind, Fazio, Gorman, & Visnapuu, 2010; Ma, Steinfeld, & Greene, 2017). Specifically, total internal reflection fluorescence microscopy (TIRFM) is used to monitor the behavior of fluorescently labeled HR factors as they interact with individual DNA molecules within a microfluidic chamber. The microfluidics setup makes the injection or removal of proteins and buffers possible, enabling precise control over the generation and disassembly of reaction intermediates. DNA curtains can be assembled either with ssDNA or dsDNA, depending on the nature of the HR stage under investigation (Gibb et al., 2012; Greene et al., 2010; Ma, Steinfeld, & Greene, 2017; Meir, Kong, Xue, & Greene, 2020; Soniat, Myler, & Finkelstein, 2019). Together, these properties render DNA curtain a versatile method for detailed single-molecule analyses of HR regulation.

1.3 Rad55–Rad57: A Rad51 paralog complex

Rad51 paralogs share 20–30% amino acid sequence identity with the ATPase core domain of Rad51, but otherwise possess divergent N and C terminal domains (Sullivan & Bernstein, 2018). These Rad51 paralogs are indispensable for the timely execution of HR, as cells depleted of any of the Rad51 paralogs become sensitized to DNA damaging agents. Importantly, deletion of Rad51 paralogs in mammalian species often leads to embryonic lethality, and mutations in the human RAD51 paralogs are associated with cancer (Bonilla, Hengel, Grundy, & Bernstein, 2020; Sullivan & Bernstein, 2018). In *S. cerevisiae*, Rad55, Rad57, Shu1, Csm2 and Psy3 have been identified as Rad51 paralogs, and they function in two distinct complexes: as the Rad55–Rad57 heterodimer, and the SHU complex comprising of Shu1–Shu2–Csm2–Psy3 (Bonilla et al., 2020). The Rad55–Rad57 complex acts at an early stage of HR—as a mediator during Rad51 filament assembly, by relieving the inhibitory effect of RPA in this step and facilitating assembly of the Rad51–ssDNA filament (Liu et al., 2011; Sung, 1997).

Purification of Rad51 paralog complexes has been a technical challenge due to their low expression levels and propensity to aggregate, limiting biochemical analyses of their activity. We have developed procedures for the purification of mono-dispersed GFP-tagged Rad55–Rad57 complex, and have carried out ssDNA curtain analyses to furnish evidence that it acts as a molecular chaperone in facilitating Rad51 presynaptic complex assembly on RPA-coated ssDNA (Roy et al., 2021). Here, we describe the purification of the GFP–Rad55–Rad57 complex, and the workflow procedure of our ssDNA curtain assay involving the use of RPA coated ssDNA as substrate.

2. Purification of GFP–Rad55–Rad57

2.1 General comments

We generated a N-terminal GFP-tagged version of Rad55 for TIRFM-based imaging of the Rad55–Rad57 complex (Roy et al., 2021). The GFP–Rad55–Rad57 complex is overexpressed in the *S. cerevisiae* yRP654 strain transformed with the pESC–URA vector containing $(His)_6$–GFP–RAD55/FLAG–RAD57 genes under the control of the GAL1/GAL10 promoters respectively. Protein complex expression is induced by adding 2% galactose to the culture medium. Two affinity purification steps followed by size exclusion chromatography are employed for protein purification. Importantly, ATP–Mg^{2+} is included in all the buffers during purification to reduce protein aggregation (Gaines et al., 2015; Roy et al., 2021).

2.2 Equipment

1. Incubator shaker (large capacity that can accommodate 5L flasks, temperature controlled)
2. Bead beater (Biospec) or equivalent homogenizer capable of disrupting yeast cells
3. 0.5 mm diameter glass beads (Biospec), prewashed according to manufacturer's protocol
4. High speed centrifuge with rotors (capable of spinning 1 L bottles, 15 mL and 250 mL tubes)
5. Ultracentrifuge and rotor (i.e., Beckman Optima L-90K, Ti45)
6. Rotary incubator
7. Empty polypropylene column (~10 mL size) with a stopcock valve
8. Liquid chromatography system (i.e., AKTA pure)
9. Superdex 200 increase 10/300 GL (GE Healthcare)
10. Protein gel electrophoresis (Biorad mini-PROTEAN) setup
11. Coomassie staining system

2.3 Buffers and reagents

S. cerevisiae yRP654 strain (*MATα ura3–52 trp1Δ leu2Δ1 his3Δ200 pep4:: HIS3 prb1Δ1.6R can1 GAL*) (Johnson, Prakash, & Prakash, 2006) or equivalent

1. pESC–Ura GFP–Rad55–Rad57 plasmid (Roy et al., 2021)
2. SC–Ura medium (2% glucose, 7.2 mg/mL Synthetic Complete (SC) medium with Uracil dropout formulation) (Sherman, 1991)
3. SC–Ura agar plate (2% glucose, 7.2 mg/mL SC–Ura, 1.7% agar)
4. 2% galactose medium (2% galactose, 3% lactic acid, 3% glycerol, 7.2 mg/mL SC–Ura)
5. Cell breakage buffer (CBB): 50 mM Tris–HCl (pH 7.5), 600 mM KCl, 10% sucrose, 2 mM EDTA, 0.01% IGEPAL CA–630 (Sigma, Cat No. I8896), 1 mM DTT, 1 mM phenylmethylsulfonyl fluoride, and 1.25 μg/mL each of protease inhibitors aprotinin, chymostatin, leupeptin and pepstatin), 2 mM ATP and 5 mM $MgCl_2$
6. K-300 (20 mM K_2HPO_4, pH 7.5, 10% glycerol, 0.5 mM EDTA, 1 mM DTT, 0.01% IGEPAL, 300 mM KCl, 2 mM ATP, 5 mM $MgCl_2$)
7. K-1000 (20 mM K_2HPO_4, pH 7.5, 10% glycerol, 0.5 mM EDTA, 1 mM DTT, 0.01% IGEPAL, 1 M KCl, 2 mM ATP, 5 mM $MgCl_2$)
8. 4 Protease inhibitor mixture (4PI, 1000x) 0.5 mg/mL each of protease inhibitors aprotinin, chymostatin, leupeptin and pepstatin
9. FLAG peptide (synthetic peptide, DYKDDDDK, 5 mg/mL stock)
10. 1 M imidazole (pH 7.5)
11. Ni-NTA beads (Qiagen)
12. anti-FLAG M2 affinity gel (Sigma)
13. Protein concentrator (Amicon Ultra-4 Centrifugal Filters (Millipore Sigma) or equivalent, 30 KDa cutoff)
14. SDS polyacrylamide gel (10% polyacrylamide or equivalent)
15. SDS running buffer (10x: 250 mM Tris base, 1.92 M Glycine, 1% SDS, pH 8.3)
16. Coomassie staining solution (1 mg/mL Coomassie Brilliant Blue R-250, 50% methanol, 10% acetic acid)
17. Coomassie destaining solution (50% methanol, 10% acetic acid)

2.4 GFP–Rad55–Rad57 overexpression

1. Transform the *S. cerevisiae* yRP654 strain with pESC $(His)_6$–GFP–RAD55/FLAG–RAD57 using the standard PEG-Li Acetate transformation method (Gietz & Woods, 2002).

2. Plate the transformed cells on SC–Ura plates until the transformants produce distinct colonies (usually in 2–3 days).
3. Pick a single colony and streak the cells on two SC–Ura plates. Once colonies are formed, streak them again on the same plate, and incubate until the transformants form a lawn on the plates.
4. Transfer the yeast cells into 2 L of SC–Ura medium and incubate at 30 °C for 24 h in a shaker.
5. Dilute the culture into 20 L (4 L of five flasks) of 2% galactose medium, and incubate at 30 °C for 24 h.
6. Harvest the cells with highspeed centrifugation at 5000 rpm for 10 min per run, and store at −80 °C until purification.

2.5 Protein purification

All the purification steps are carried out between 0 and 4°C.

2.5.1 Cell lysis and ammonium sulfate precipitation

1. Resuspend the cell pellets (~80 g) in three pellet volumes of CBB buffer.
2. Transfer the cell suspension into a bead-beater chamber with ½ full pre-chilled 0.5 mm diameter glass beads and set up the bead beater according to the manufacturer's instruction.
3. Cooling the chamber with ice, run the bead beater for 30 s, ~8 times with 1-minute interval each.
4. Transfer the cell lysate into ultracentrifugation tubes and centrifuge at 100,000 x *g* RCF for 1 h.
5. Transfer the clarified lysate into a beaker and measure the volume.
6. Slowly add and dissolve ammonium sulfate using a magnetic stirrer with gentle mixing over 1 h to make the lysate to 40% ammonium sulfate saturated solution (0.242 g/mL).
7. Centrifuge the turbid mixture at 18,000 x *g* for 15 min and remove the supernatant.
8. Process the pellets to the next step or store at −80 °C.

2.5.2 Protein purification

1. Resuspend the ammonium sulfate precipitation pellets in 30 mL of K-300 buffer supplemented with 20 mM imidazole.
2. Add 2 mL Ni-NTA beads prewashed with K-300, and incubate for 1 h using a rotary incubator.

3. Pour the mixture into a 15 mL empty column with a stopcock valve and capture the resin by draining the column while collecting the flow-through fraction. Close the stopcock valve before the resin is exposed to air.
4. Add 50 mL of K-1000 buffer to the column containing the resin and wash the resin by stirring with a stick. Open the stopcock valve and collect the wash fraction.
5. Repeat washing with 30 mL of K-300 buffer.
6. Add 4 mL buffer K-300 supplemented with 200 mM imidazole containing 1x 4PIs to the column, mix, and incubate for 5 min.
7. Open the stopcock valve and collect the elution fraction.
8. Repeat the elution five times.
9. Identify protein elute factions using SDS-PAGE followed by Coomassie staining (Fig. 2) or a Bradford assay.
10. Combine the fractions containing GFP–Rad55–Rad57 and add 1.5 mL of anti-FLAG affinity resin prewashed with K-300 buffer.
11. Mix and incubate for 1.5 h using a rotary incubator.
12. Pour the mixture into a 10 mL empty column and capture the resin.
13. Wash the resin with 50 mL of K-1000 buffer, followed by washing with 30 mL of K-300 buffer as above.

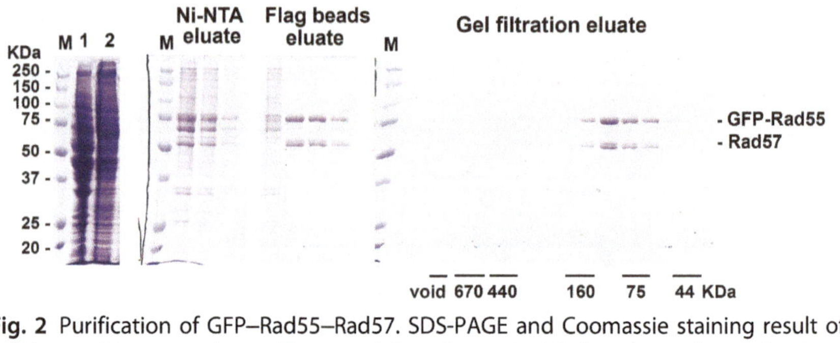

Fig. 2 Purification of GFP–Rad55–Rad57. SDS-PAGE and Coomassie staining result of cell lysate (1), ammonium sulfate precipitated protein (2), first three elution fractions from Ni-NTA beads and FLAG affinity resin, and fractions (7.4 to 15.8 mL) from the Superdex 200 gel filtration column. The elution profile of protein size markers (Thyroglobulin (670 KDa), Apoferritin (440 KDa), Aldolase (160 KDa), Conalbumin (75 KDa), and Ovalbumin (44 KDa) is indicated under the gel. *Panels adapted with permission from Roy, U., Kwon, Y., Marie, L., Symington, L., Sung, P., Lisby, M., & Greene, E.C. (2021). The Rad51 paralog complex Rad55–Rad57 acts as a molecular chaperone during homologous recombination.* Molecular Cell, 81, *1043–1057e1048.*

14. Add 1.5 mL of K-300 buffer containing 200 μg/mL FLAG peptide to the column and incubate for 5 min and collect the elution fraction.
15. Repeat the elution five times.
16. Identify protein elute factions using SDS PAGE analysis (Fig. 2).
17. Combine fractions containing GFP–Rad55–Rad57 and concentrate to ~400 μL with a protein concentrator.
18. Equilibrate Superdex 200 10/300 column with K-300 buffer.
19. Load the GFP–Rad55–Rad57 into the column and perform size exclusion chromatography using a flow rate of ~0.4 mL/min and collect every 600 μL.
20. Analyze fractions containing proteins by SDS–PAGE followed by Coomassie staining (Fig. 2).
21. Pool the peak fractions of GFP–Rad55–Rad57, concentrate with a protein concentrator, aliquot to 2~5μL per tube, and snap-freeze in liquid nitrogen, and store at –80 °C (*Note 5.1*).

3. Single molecule analysis of GFP–Rad55–Rad57

3.1 Overview of single-stranded DNA curtain assays

A schematic of the ssDNA curtain setup is shown in (Fig. 3). Briefly, chrome nanopatterns of barriers and pedestals are deposited on quartz slides using electron-beam (E-beam) lithography, and the patterned slide is used to assemble a flowcell (Fazio, Visnapuu, Wind, & Greene, 2008; Gorman, Fazio, Wang, Wind, & Greene, 2010; Ma, Steinfeld, & Greene, 2017) (Fig. 3A and B). The barriers and pedestals are located in the center of the microfluidic reaction chamber, which is flanked by two nanoports that allow us to flow in, and flush out buffers and proteins of choice. Biotinylated lipids are first flushed in to passivate the quartz surface within the reaction chamber (Fig. 3C). This prevents sticking of DNA and proteins to the surface, and provides a fluid medium for support of DNA molecules. The biotinylated groups serve as anchor points in the lipid bilayer, to which we attach biotinylated ssDNA via streptavidin. Using buffer flow, the DNA molecules are pushed along the lipid layer until they encounter the chrome barriers, which causes the ssDNA molecules to align at the barriers. Continued application of buffer flow with RPA, along with urea denaturation extends the ssDNA molecules, such that the free end gets attached to the downstream chrome pedestal via non-specific interactions with RPA

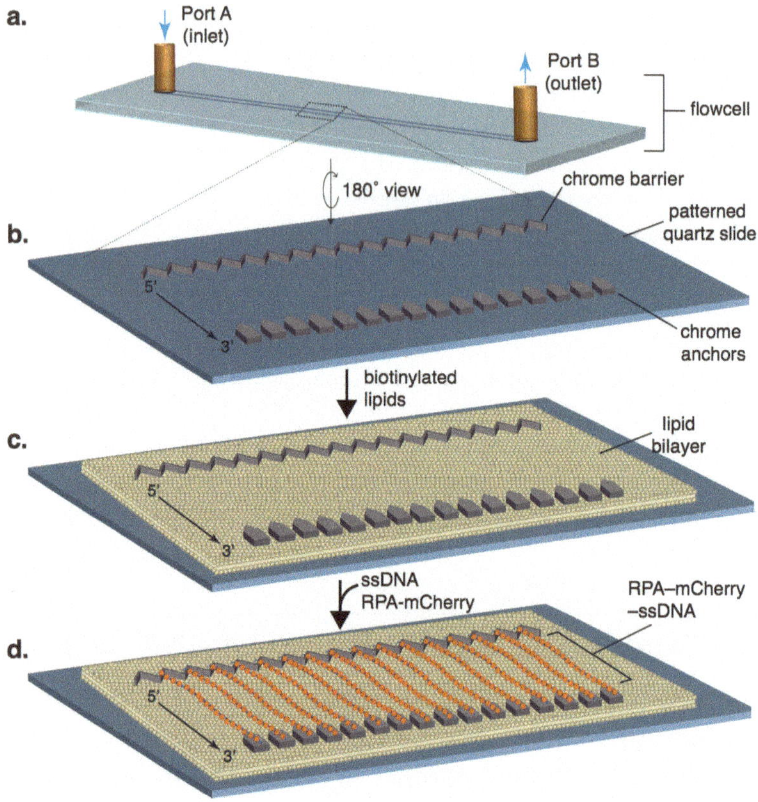

Fig. 3 ssDNA curtain schematic. (A) Flowcell with nanoports and central sample chamber. (B) The patterned quartz slide forms the base of the central sample chamber, and contains nano-scale chrome barriers and anchors deposited by electron-beam lithography. (C) Biotinylated lipids form a bilayer that passivates the quartz slide surface. (D) Biotinylated ssDNA is attached to the lipid layer via streptavidin and stretched to the downstream anchor points by buffer flow and RPA binding. ssDNA can be visualized using labeled RPA by TIRFM.

(Fig. 3D). This process allows us to align hundreds of individual ssDNA molecules in a "curtain-like" array and observe the protein dynamics on DNA molecules in real-time.

3.2 Equipment, buffers and reagents

Prepare all solutions in ultrapure MilliQ water and thoroughly clean tubing and connectors with ultrapure water before use.

1. Lipid buffer: 10 mM Tris–HCl [pH 8], 100 mM NaCl
2. BSA buffer: 40 mM Tris–HCl [pH 8], 2 mM $MgCl_2$, 1 mM DTT, 0.2 mg/mL BSA
3. HR buffer: 30 mM Tris-Ac [pH 7.5], 50 mM KCl, 20 mM $MgAc_2$, 1 mM DTT, 0.2 mg/mL BSA
4. 1 mg/mL Streptavidin solution
5. 2% Hellmanex solution
6. 1N NaOH
7. 1 mL syringes with slip-tip
8. 3 mL syringes with Luer-lock tip
9. 10 mL syringes with Luer-lock tip
10. 30 mL syringes with Luer-lock tip
11. Teflon PFA resin tubing 1/16" OD x 0.02" ID
12. PEEK Headless short 10–32 coned connector for 1/16" OD
13. Two piece fingertight fitting, 10–32 coned, for 1/16" OD
14. PEEK Female to female Luer adapter, 10–32
15. Kd Scientific Dual Syringe Pump (KDS-201)
16. High-pressure switch valve
17. 1 μM yeast RPA–mCherry stock in 30 mM $NaHPO_4$ [pH 7.5], 250 mM KCl, 10% glycerol, 0.02% Tween-20, 1 mM TCEP, 0.5 mM EDTA, 0.25% myoinositol (De Tullio et al., 2018)
18. Yeast Rad51 stock in 25 mM Tris Cl [pH 7.5], 10% Glycerol, 200 mM NaCl, 1 mM DTT (Gibb et al., 2014)
19. Yeast GFP–Srs2 stock in 40mM NaHPO4 [pH 7.5], 300 mM KCl, 10% glycerol, 0.01% Tween-20, 1 mM TCEP, 0.5 mM EDTA, 0.125% myoinositol (De Tullio et al., 2018)
20. ssDNA stock solution in 50 mM Tris–HCl [pH 7.5], 4 mM DTT, 10 mM ammonium sulfate, 10 mM $MgCl_2$ (De Tullio et al., 2018)
21. Biotinylated lipid stock solution: 10 mM Tris–HCl [pH 8], 100 mM NaCl, 10 mg/mL DOPC, 1 mg/mL PEG-2000 DOPE, 0.05 mg/mL biotinylated DOPE, (prepared as in De Tullio et al., 2018)

3.3 Preparation of ssDNA

Long ssDNA substrates (~30–40 kb) for ssDNA curtain experiments are generated by rolling circle replication of the M13 circular ssDNA template using a 5' biotinylated primer.

1. The 5' biotinylated primer (5'-BIO-TEG-TTT TTT TTT TTT TTT TTT TTT TTT TTT TTT GTA AAA CGA CGG CCA GT) is

annealed to the circular ssDNA M13mp18 template (New England Biolabs, Cat. No. N4040S) in a 100 μL annealing reaction: 10 mM Tris–HCl (pH 7.5), 50 mM NaCl, 10 mM $MgCl_2$, 10 μg M13mp18 and 40 nM primer. Heat the annealing reaction to 95 °C and gradually cool it down to 4 °C in a thermocycler over at least 30 mins. The annealing reaction can be stored at 4 °C until the rolling circle reaction.
2. Each rolling circle reaction (500 μL) is assembled by mixing 50 μL 10X rolling circle buffer (500 mM Tris–HCl (pH 7.5), 100 mM $MgCl_2$, 100 mM ammonium acetate, 40 mM DTT), 5 μL annealed reaction, 5 μL purified phi29 polymerase (10 μM stock, see Gibb et al., 2012; De Tullio et al., 2018), 10 μL dNTP solution (10 mM stock, NEB N0447S), 5 μL BSA (20 mg/mL stock) and 425 μL water. Mix the components by gentle pipetting (Note 5.2) and incubate the reaction at 30 °C for 25 mins.
3. Stop the reaction by adding EDTA to a final concentration of 25 mM and mix gently by inverting the tube 2–3 times.
4. Aliquot the ssDNA stock carefully into 30 μL aliquots using a wide bore pipette tip, and flash freeze with liquid nitrogen. The ssDNA aliquots can be stored at –80 °C until use for DNA curtain experiments.

3.4 Assembly of ssDNA curtains

Detailed procedures on preparation of flowcells and liposome stock solutions has been described previously (De Tullio et al., 2018; Ma, Steinfeld, & Greene, 2017). Here we describe the workflow for assembly of the ssDNA curtains.

1. The flowcell is first flushed with water to remove any air bubbles. For this, use two 10-cm pieces of tubing (IDEX, Cat. No. 1907L). Attach a connector (IDEX, Cat. No. F-333NX) on one end of each tubing (for connecting to nanoports on the flowcell). On the other end of each tubing, attach a connector (IDEX, Cat. No. F-300X) and then a female-to-female Luer peek adaptor (IDEX, Cat. No. P-659). This end is used to connect the syringes.
2. Connect the prepared pieces of tubing to the flowcell through nanoports A and B (Idex, Cat. No. N-333). Fill a 10 mL syringe (BD, Cat. no. 302995) with MilliQ water and attach to tubing on port A. Push 3–4 mL of water, flushing out all air bubbles in the flowcell and tubing (Fig. 3A).
3. Fill a second 10 mL syringe with MilliQ water and attach to the tubing on port B of the flowcell using drop-to-drop connections to avoid introduction of air bubbles.

4. Gently push water back and forth between the two syringes, making sure to remove any air bubbles trapped in the flowcell and tubing.
5. Fill a 3 mL syringe (BD, Cat. No. 309657) with lipid buffer (10 mM Tris–HCl (pH 8), 100 mM NaCl). Replace the syringe connected on port A with the 3 mL syringe containing lipid buffer, being careful to make drop-to-drop connections. Importantly, all syringe attachments henceforth should be done carefully using drop-to-drop connections to exclude air bubbles (Note 5.3).
6. Disconnect the 10 mL syringe on port B, and push 2.5 mL lipid buffer, flushing out water in the flowcell and tubing, and replacing it with lipid buffer.
7. Make a lipid dilution by mixing 40 μL of liposome stock solution (10 mg/mL DOPC, 1 mg/mL PEG-2000 DOPE, 0.05 mg/mL biotinylated DOPE, 10 mM Tris–HCl [pH 8], 100 mM NaCl, prepared as detailed in De Tullio et al., 2018) with 960 μL lipid buffer (10 mM Tris–HCl [pH 8], 100 mM NaCl). Fill a 1 mL syringe (BD, Cat. No. 309659) with this lipid dilution, being careful to exclude air bubbles, and attach it to port B. Disconnect the syringe attached to port A.
8. Gently push ~300 μL of the lipid solution into the flowcell, and incubate 5–8 mins. Repeat two more times until most of the lipid solution is used up.
9. Fill a 3 mL syringe with lipid buffer and attach to port A, then disconnect syringe on port B. Gently flush ~2.5 mL lipid buffer through the flowcell (Note 5.4). Incubate for 20 mins to allow the lipids to passivate the surface.
10. Fill a 3 mL syringe with BSA buffer (10 mM Tris–HCl (pH 8), 2 mM $MgCl_2$, 1 mM DTT, 0.2 mg/mL BSA) and attach to port B. Disconnect syringe on port A and gently flush the flowcell with BSA buffer.
11. Prepare streptavidin dilution by mixing 10 μL streptavidin stock (1 mg/mL) with 990 μL BSA buffer. Fill a 1 mL syringe with the streptavidin dilution and attach to port A. Disconnect syringe on port B. Gently push ~400 μL streptavidin dilution through the flowcell. Incubate for 5 mins. Repeat with another 400 μL of the streptavidin dilution.
12. Fill a 3 mL syringe with BSA buffer and attach to port B, then disconnect syringe on port A. Gently push ~2.5 mL buffer through the flowcell to wash off unbound streptavidin.

13. Dilute 30 μL of the ssDNA stock by adding 970 μL BSA buffer. Fill a 1 mL syringe carefully with the ssDNA solution (Note 5.5) and attach to port A. Gently push ~200 μL ssDNA solution into the flowcell (do not remove the syringe attached to port B, as it will be required when making connections to the microfluidic setup on the microscope.) Incubate 5–8 mins, and repeat twice more until most of the ssDNA has been injected.
14. Once all the ssDNA has been injected, mount the flowcell on the microscope, and connect to the microfluidics system using drop-to-drop connections. The flowcell is maintained at 30 °C for experiments with yeast recombinant proteins.

3.5 Rad51 filament assembly with GFP-Rad55-Rad57

The DNA is easily visualized by injecting mCherry labeled RPA (Gibb et al., 2012; Ma, Steinfeld, & Greene, 2017), which has a high affinity for ssDNA and pre-occupies ssDNA before Rad51 assembly *in vivo*. Secondary structures in the ssDNA are disrupted by injecting RPA and 7 M urea with a high buffer flow rate of 0.8 mL/min. Together these factors allow the full extension and coating of the ssDNA with mCherry labeled RPA for imaging. The 3' free end of RPA–ssDNA attaches to the downstream pedestal (Fig. 3D).

1. Fill two 30 mL syringes (BD Cat. no. 302832) with HR buffer (30 mM Tris-Ac [pH 7.5], 50 mM KCl, 20 mM $MgAc_2$, 1 mM DTT, 0.2 mg/mL BSA), supplementing one with 0.1 nM RPA–mCherry (RPA solution), and the other with 2 mM ATP (ATP solution). Attach both syringes to the syringe pump [Kd Scientific Dual Syringe Pump (KDS-201)], ensuring that the RPA solution is connected to the flowcell and the ATP solution goes to discard (Note 5.6).
2. Inject 500 μL of 7 M urea into the sample chamber while flowing the RPA solution at a flow rate of 0.8 mL/min. Continue flowing the RPA solution for 8–10 mins until the ssDNA is fully coated with RPA–mCherry and is extended until the downstream pedestals. The DNA can now be visualized using a low laser power (~20 mW) and TIRFM (Note 5.7). Select a field of view with well resolved ssDNA molecules (Note 5.8).
3. Switch buffer flow to flush the sample chamber with the ATP solution at a flow rate of 0.8 mL/min for at least 2 mins.

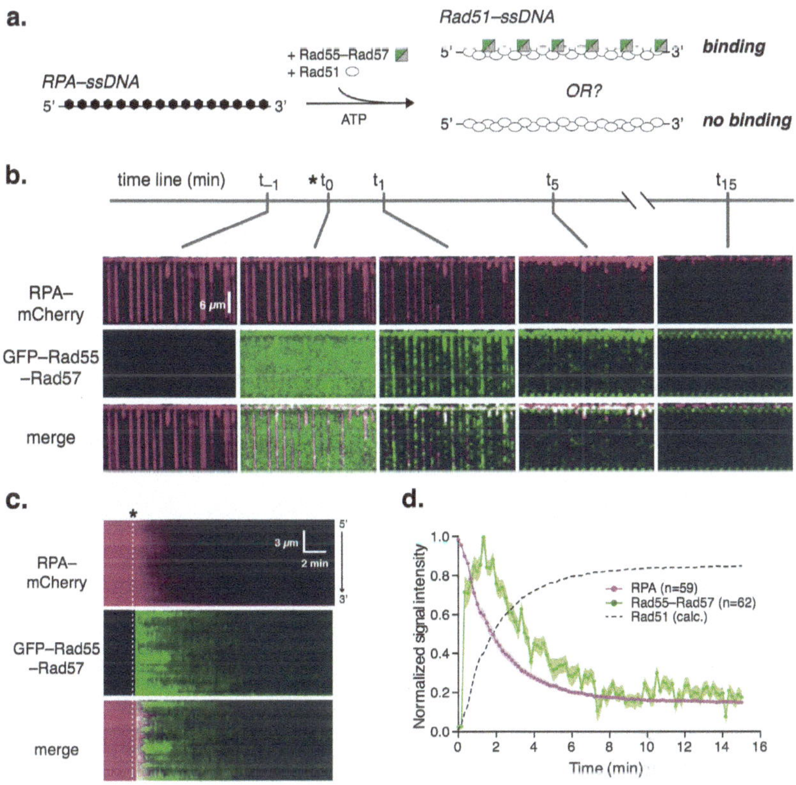

Fig. 4 Rad55–Rad57 binds in a transient burst during Rad51 filament assembly. (A) Schematic for the experimental setup. (B) Wide-field view of RPA–mCherry (magenta) bound ssDNA at indicated time points. Rad51 (unlabeled) and GFP-tagged Rad55–Rad57 (green) injection (t0) is highlighted with an asterisk (*). (C) Kymograph showing loss of RPA–mCherry after Rad51 and GFP–Rad55–Rad57 injection. (D) Mean normalized intensities for RPA–mCherry and GFP–Rad55–Rad57; shaded area represents 95% CI. Calculated Rad51 assembly kinetics is plotted as [1–RPA] signal for comparison. *Panels reproduced with permission from Roy, U., Kwon, Y., Marie, L., Symington, L., Sung, P., Lisby, M., & Greene, E.C. (2021). The Rad51 paralog complex Rad55–Rad57 acts as a molecular chaperone during homologous recombination.* Molecular Cell, 81, *1043–1057e1048.*

4. Begin recording images 1–2 mins before injecting Rad51. Data in this chapter are acquired using 50 mW laser power, 100 ms integration time, and images captured every 10 s.
5. Reduce buffer flow rate to 0.5 mL/min and inject 2 μM Rad51, with 60 nM GFP–Rad55–Rad57, in 150 μL ATP solution. Once the sample enters the flowcell, stop buffer flow and incubate for 15–20 mins, capturing images every 10 s (Fig. 4A and B).

4. Data analysis

The loss of RPA–mCherry signal can be used to calculate the rate of Rad51 filament assembly (Ma, Gibb, et al., 2017; Roy et al., 2021). Images are acquired through Nikon NIS Elements software, and each frame of TIRFM data is exported as a raw TIFF image. These image files are analyzed using Fiji software (ImageJ 1.48b, Wayne Rasband, National Institutes of Health, USA). The image stack is used to generate kymographs of individual ssDNA molecules for data analysis. A kymograph is the 2D image of a selected ssDNA molecule, where the Y-axis represents position along the ssDNA, and the X-axis represents time.

1. Combine TIFF files into a single image stack for each channel (GFP and mCherry) using Fiji software.
2. Correct for drift in each channel using the "StackReg" function. Then merge the two channels (GFP and mCherry) into a single hyperstack using the "Color > Merge channels" function.
3. To generate a kymograph, draw a 1-pixel wide ROI (region of interest) over a single ssDNA molecule, and use the "Reslice" function in Fiji, selecting the "rotate by 90 degrees" option. This will produce a 2D image where the Y-axis is positions along the ssDNA molecule, and X-axis is time (Fig. 4C).
4. Draw a rectangular ROI selecting the entire kymograph (or selecting time window of choice). Use the "Plot Profile" function in Fiji to generate a graph and list of intensity values for the kymograph. Each intensity value in the list is an average over the entire length of the ssDNA molecule for that time point.
5. Normalize the intensity values for a given kymograph by dividing each value in the list by the maximum value in that list. This will produce a list of "normalized intensity values" for the kymograph. Repeat the analysis for all kymographs.
6. Calculate the mean and error of the distribution across all kymographs/molecules, and plot mean normalized intensities (y-axis) against time (x-axis) using Prism software (Graphpad). (Fig. 4D).
7. To determine the rate of Rad51 assembly, fit the RPA signal loss to a single exponential decay curve, and calculate the rate constant. The same analysis can be applied to calculate GFP–Rad55–Rad57 dissociation rate, by fitting the loss of GFP signal to a single exponential curve (Fig. 4D).

5. Notes

5.1. To avoid protein aggregation and loss of activity due to freeze thaw cycles, we recommend preparing and freezing single-use aliquots.

5.2. The long ssDNA (~40 kb) products of the rolling circle reaction are prone to breakage with harsh pipetting. It is recommended that pipetting be minimized, and done using wide bore pipette tips.

5.3. All connections to the flowcell must be made using drop-to-drop connections to avoid the introduction of air bubbles, especially after the addition of lipids. Air bubbles will destroy the lipid bilayer and lead to DNA getting stuck on the quartz surface.

5.4. Once the lipid layer has been formed, all subsequent washing steps should be done gently, to prevent stripping of the lipid layer.

5.5. The ssDNA dilution should be handled carefully, as harsh pipetting can lead to breakage of the DNA.

5.6. Flow 4 mL of RPA solution and ATP solution at a rate of 0.5 mL/min, while gently tapping the microfluidic lines to remove air bubbles and fill the tubing with buffers before connecting to the flowcell.

5.7. Care should be taken to minimize laser exposure at this stage, to avoid photobleaching of RPA–mCherry.

5.8. It is important to choose a set of barriers where the ssDNA is not too crowded, to avoid analysis of overlapping ssDNA molecules.

6. Conclusions and future directions

Rad51 paralogs are crucial for the successful execution of HR and DSB repair in cells, and mutations in genes that encode the human RAD51 paralogs are associated with multiple cancer types (Bonilla et al., 2020; Sullivan & Bernstein, 2018). We have overcome technical challenges to prepare mono-dispersed GFP–Rad55–Rad57 complex from *S. cerevisiae*, and examined its behavior relevant for its role in the assembly of the Rad51–ssDNA filament. Single-molecule assays, such as DNA curtains, are a powerful complement to genetic and bulk biochemical studies, and provide a detailed picture of how HR intermediates are generated and resolved.

In cells, multiple factors regulate the formation and stability of the Rad51–ssDNA filament (Kowalczykowski, 2015; Sung & Klein, 2006).

Importantly, the ssDNA curtain assay can be used to determine the dynamics and interplay of multiple regulators, such as Rad52 and the SHU complex, in the formation and regulation of the presynaptic complex of Rad51 bound to ssDNA. This approach can also be easily adapted for analysis of fluorescently labeled Rad51 paralog complexes from other organisms, such as the human RAD51C-XRCC3 and RAD51B-RAD51C-RAD51D-XRCC2 complexes.

Acknowledgments

This research was funded by NIH grants R35GM118026, R01CA221858, R01CA236606 (to E.C.G.), R35CA241801, RO1ES007061 (to P.S.), and PO1CA092584 (to P.S. and E.C.G.); a Gray Foundation Team Science Grant under the Basser Initiative (to P.S.), and a Wellcome Trust Collaborative Award in Science (grant 206292/D/17/Z to E.C.G.). P.S. is the recipient of a CPRIT REI award (RR180029) and holder of the Robert A. Welch Distinguished Chair in Biochemistry (AQ-0012).

References

Belan, O., Barroso, C., Kaczmarczyk, A., Anand, R., Federico, S., O'Reilly, N., et al. (2021). Single-molecule analysis reveals cooperative stimulation of Rad51 filament nucleation and growth by mediator proteins. *Molecular Cell, 81*(5), 1058–1073.

Bonilla, B., Hengel, S. R., Grundy, M. K., & Bernstein, K. A. (2020). RAD51 gene family structure and function. *Annual Review of Genetics, 54*, 25–46.

Crickard, J. B., Xue, C., Wang, W., Kwon, Y., Sung, P., & Greene, E. C. (2019). The RecQ helicase Sgs1 drives ATP-dependent disruption of Rad51 filaments. *Nucleic Acids Research, 47*(9), 4694–4706.

De Tullio, L., Kaniecki, K., & Greene, E. C. (2018). Single-stranded DNA curtains for studying the Srs2 helicase using total internal reflection fluorescence microscopy. *Methods Enzymology, 600*, 407–437.

Fazio, T., Visnapuu, M. L., Wind, S., & Greene, E. C. (2008). DNA curtains and nanoscale curtain rods: High-throughput tools for single molecule imaging. *Langmuir, 24*, 10524–10531.

Gaines, W. A., Godin, S. K., Kabbinavar, F. F., Rao, T., VanDemark, A. P., Sung, P., et al. (2015). Promotion of presynaptic filament assembly by the ensemble of S. cerevisiae Rad51 paralogues with Rad52. *Nature Communications, 6*, 7834.

Gibb, B., Silverstein, T. D., Finkelstein, I. J., & Greene, E. C. (2012). Single-stranded DNA curtains for real-time single-molecule visualization of protein-nucleic acid interactions. *Analytical Chemistry, 84*, 7607–7612.

Gibb, B., Ye, L. F., Kwon, Y., Niu, H., Sung, P., & Greene, E. C. (2014). Protein dynamics during presynaptic-complex assembly on individual single-stranded DNA molecules. *Nature Structural & Molecular Biology, 21*, 893–900.

Gietz, R. D., & Woods, R. A. (2002). Transformation of yeast by lithium acetate/single-stranded carrier DNA/polyethylene glycol method. *Methods Enzymology, 350*, 87–96.

Gorman, J., Fazio, T., Wang, F., Wind, S., & Greene, E. C. (2010). Nanofabricated racks of aligned and anchored DNA substrates for single-molecule imaging. *Langmuir, 26*, 1372–1379.

Greene, E. C., Wind, S., Fazio, T., Gorman, J., & Visnapuu, M. L. (2010). DNA curtains for high throughput single-molecule optical imaging. *Methods Enzymology*, *472*, 293–315.

Johnson, R. E., Prakash, L., & Prakash, S. (2006). Yeast and human translesion DNA synthesis polymerases: Expression, purification, and biochemical characterization. *Methods Enzymology*, *408*, 390–407.

Kaniecki, K., De Tullio, L., Gibb, B., Kwon, Y., Sung, P., & Greene, E. C. (2017). Dissociation of Rad51 presynaptic complexes and heteroduplex DNA joints by tandem assemblies of Srs2. *Cell Reports*, *21*, 3166–3177.

Kass, E. M., Moynahan, M. E., & Jasin, M. (2016). When genome maintenance goes Badly Awry. *Molecular Cell*, *62*, 777–787.

Kowalczykowski, S. C. (2015). An overview of the molecular mechanisms of recombinational DNA repair. *Cold Spring Harbor Perspectives in Biology*, 7, a016410.

Liu, J., Renault, L., Veaute, X., Fabre, F., Stahlberg, H., & Heyer, W. D. (2011). Rad51 paralogues Rad55-Rad57 balance the antirecombinase Srs2 in Rad51 filament formation. *Nature*, *479*, 245–248.

Ma, C. J., Gibb, B., Kwon, Y., Sung, P., & Greene, E. C. (2017). Protein dynamics of human RPA and RAD51 on ssDNA during assembly and disassembly of the RAD51 filament. *Nucleic Acids Research*, *45*, 749–761.

Ma, C. J., Steinfeld, J. B., & Greene, E. C. (2017). Single-stranded DNA curtains for studying homologous recombination. *Methods Enzymology*, *582*, 193–219.

Meir, A., Kong, M., Xue, C., & Greene, E. C. (2020). DNA curtains shed light on complex molecular systems during homologous recombination. *Journal of Visualized Experiments*, (160), e61320.

Myler, L. R., Gallardo, I. F., Soniat, M. M., Deshpande, R. A., Gonzalez, X. B., Kim, Y., et al. (2017). Single-molecule imaging reveals how Mre11-Rad50-Nbs1 initiates DNA break repair. *Molecular Cell*, *67*(5), 891–898.

Prakash, R., Zhang, Y., Feng, W., & Jasin, M. (2015). Homologous recombination and human health: the roles of BRCA1, BRCA2, and associated proteins. *Cold Spring Harbor Perspectives in Biology*, 7, a016600.

Qiu, Y., Antony, E., Doganay, S., Koh, H. R., Lohman, T. M., & Myong, S. (2013). Srs2 prevents Rad51 filament formation by repetitive motion on DNA. *Nature Communications*, *4*, 2281.

Roy, U., Kwon, Y., Marie, L., Symington, L., Sung, P., Lisby, M., et al. (2021). The Rad51 paralog complex Rad55-Rad57 acts as a molecular chaperone during homologous recombination. *Molecular Cell*, *81*(1043-1057), e1048.

San Filippo, J., Sung, P., & Klein, H. (2008). Mechanism of eukaryotic homologous recombination. *Annual Review of Biochemistry*, *77*, 229–257.

Sherman, F. (1991). Getting started with yeast. *Methods in Enzymology*, *194*, 3–21.

Soniat, M. M., Myler, L. R., & Finkelstein, I. J. (2019). Assembling the human resectosome on DNA curtains. In *DNA Repair* (pp. 225–244). New York, NY: Humana.

Sullivan, M. R., & Bernstein, K. A. (2018). RAD-ical new insights into RAD51 regulation. *Genes (Basel)*, *9*.

Sung, P. (1997). Yeast Rad55 and Rad57 proteins form a heterodimer that functions with replication protein A to promote DNA strand exchange by Rad51 recombinase. *Genes & Development*, *11*, 1111–1121.

Sung, P., & Klein, H. (2006). Mechanism of homologous recombination: mediators and helicases take on regulatory functions. *Nature Reviews Molecular Cell Biology*, 7(10), 739–750.

Taylor, M. R. G., Spirek, M., Chaurasiya, K. R., Ward, J. D., Carzaniga, R., Yu, X., et al. (2015). Rad51 paralogs remodel pre-synaptic Rad51 filaments to stimulate homologous recombination. *Cell*, *162*, 271–286.

Xue, C., Molnarova, L., Steinfeld, J. B., Zhao, W., Ma, C., Spirek, M., et al. (2021). Single-molecule visualization of human RECQ5 interactions with single-stranded DNA recombination intermediates. *Nucleic Acids Research*, *49*(1), 285–305.

Xue, C., Wang, W., Crickard, J. B., Moevus, C. J., Kwon, Y., Sung, P., et al. (2019). Regulatory control of Sgs1 and Dna2 during eukaryotic DNA end resection. *Proceedings of the National Academy of Sciences of the United States of America*, *116*(13), 6091–6100.

CHAPTER SEVENTEEN

Synthesis and polymerase bypass studies of DNA-peptide and DNA-protein conjugates

Suresh S. Pujari* and Natalia Tretyakova*
Department of Medicinal Chemistry and Masonic Cancer Center, University of Minnesota, Minneapolis, MN, United States
*Corresponding authors: e-mail address: spujari@umn.edu; trety001@umn.edu

Contents

Methods in Enzymology, Volume 661
ISSN 0076-6879
https://doi.org/10.1016/bs.mie.2021.09.005

Abstract

DNA-peptide (DpCs) and DNA-protein cross-links (DPCs) are DNA lesions formed when polypeptides and nuclear proteins become covalently trapped on DNA strands. DNA-protein cross-links are of enormous size and hence pose challenges to cell survival by blocking DNA replication, transcription, and repair. However, DPCs can undergo proteolytic degradation *via* various pathways to give shorter polypeptide chains (DpCs). The resulting DpC lesions are efficiently bypassed by translesion synthesis (TLS) DNA polymerases like κ, η, δ, *etc.*, although polymerase bypass efficiency as well as correct base insertion depends heavily on size, sequence context, and position of peptides in DpCs. This chapter explores various synthetic methods to generate these lesions including detailed experimental procedures for the construction of DpCs and DPCs *via* reductive amination and oxime ligation. Further we describe biochemical experiments to investigate the effects of these lesions on DNA polymerase activity and fidelity.

1. Introduction

In eukaryotic cells, genomic DNA is wrapped around histone octamers to form a highly compact and dynamic chromatin structure. Reversible interactions between DNA and regulatory proteins are vital for diverse cellular processes such as transcription, replication, cell division and repair. However, cellular proteins can become irreversibly trapped on genomic DNA following exposure to reactive species of both endogenous and exogenous origin, leading to DNA-protein cross-links (DPCs) (Connelly & Leach, 2004; Ide, Shoulkamy, Nakano, Miyamoto-Matsubara, & Salem, 2011; Pachva, Kisselev, Matkarimov, Saparbaev, & Groisman, 2020; Tretyakova, Groehler, & Ji, 2015). Because of their enormous size, DPCs distort DNA helix structure and not only block interaction between other proteins and DNA but also interfere with DNA transactions including transcription, replication, and repair, leading to genomic instability and cell death (Duxin, Dewar, Yardimci, & Walter, 2014; Ide et al., 2011; Izzotti, Cartiglia, Taningher, De Flora, & Balansky, 1999; Larsen et al., 2019; Nakano et al., 2012; Wickramaratne et al., 2016; Wu, Lee, Chen, & Kuo, 2002). Many nuclear proteins including histones, DNA polymerases, DNA repair proteins, and transcription factors have been shown to participate in DPC formation (Groehler, Degner, & Tretyakova, 2017; Lai et al., 2016; Loeber et al., 2009; Michaelson-Richie et al., 2010; Neale, Pan, & Keeney, 2005; Pachva et al., 2020; Spruijt et al., 2013; Vaz et al., 2016; Yang et al., 1996). DPCs can be induced by a various chemical and physical agents including 1,2,3,4-diepoxybutane (DEB)

(Gherezghiher, Ming, Villalta, Campbell, & Tretyakova, 2013), endogenous aldehydes (Ji, Shao, Han, Seiler, & Tretyakova, 2017; Li, Zhang, et al., 2017), ionizing radiation (Dizdaroglu, Gajewski, Reddy, & Margolis, 1989), UV light (Alexander & Moroson, 1962; Hockensmith, Kubasek, Evertsz, Mesner, & von Hippel, 1993; Moss, Dimitrov, & Houde, 1997), transition metals (Mattagajasingh & Misra, 1996; Patierno, Sugiyama, Basilion, & Costa, 1985; Zhitkovich, Voitkun, Kluz, & Costa, 1998), therapeutic agents including nitrogen mustards and platinum drugs (Groehler, Villalta, Campbell, & Tretyakova, 2016; Loeber et al., 2009; Michaelson-Richie et al., 2011), and reactive oxygen species (Gajewski & Dizdaroglu, 1990; Johansen, Muller, Xu, & Burrows, 2005; Xu, Muller, Ye, & Burrows, 2008) (Fig. 1). DPCs are also formed upon exposure to environmental pollutants, cigarette smoke, automotive and diesel exhaust,

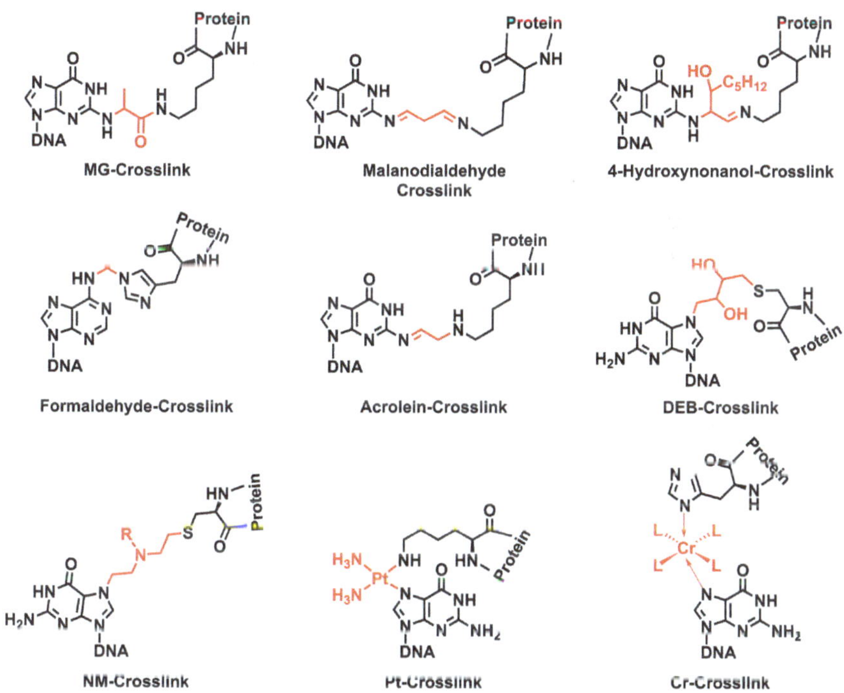

Fig. 1 Chemical structures of DNA-protein cross-links induced by methylglyoxal, acrolein, malondialdehyde, 4-hydroxynonenal, formaldehyde, 1,2,3,4-diepoxybutane, nitrogen mustards, cisplatin and chromium III. *Reproduced with permission from Groehler, A., Degner, A., & Tretyakova, N. Y. (2017). Mass spectrometry-based tools to characterize DNA-protein cross-linking by* bis-electrophiles. Basic & Clinical Pharmacology & Toxicology, 121 (Suppl. 3), *63–77. https://doi.org/10.1111/bcpt.12751.*

occupational chemicals, foodstuffs, and physiological metabolites generated by normal cellular processes and lipid peroxidation (Brambilla et al., 1985; Summerfield & Tappel, 1984). Despite their ubiquitous formation in living cells, structural features and protein composition of DPC lesions are not well established due to their inherent heterogeneity and the difficulty of analyzing macromolecular conjugates containing both proteins and DNA.

Because of their extremely bulky size, DPCs are difficult to repair since access of repair proteins to the lesion site can be hampered by steric hindrance. As a consequence, these lesions accumulate in various organs such as heart and brain tissues with age (Izzotti et al., 1999) and are hypothesized to play a crucial role in aging, cancer and neurodegenerative diseases (Wu et al., 2002).

The conflict between DPC and DNA replication machineries can cause replication errors leading to base substitutions, deletions and insertions around DPC sites (Stingele, Bellelli, & Boulton, 2017; Tretyakova et al., 2015; Vaz, Popovic, & Ramadan, 2017). Obstruction of replication forks by DPCs can cause fork stalling eventually leading to DNA double strand breaks, which are among the most toxic lesions in cells (Sparks et al., 2019). Presence of these lesions in nucleosome core particles significantly alters the gene expression and regulation mechanisms (Li et al., 2017; Raiber et al., 2018). For instance, individuals deficient in SPRTN, a critical gene required for DPC repair, develop the Ruijs–Aalfs syndrome, a human autosomal recessive disorder characterized by accelerated aging, genomic instability, and early-onset of hepatocellular carcinoma (Maskey et al., 2017). DPCs may also contribute to the Fanconi anemia phenotype-recessive hereditary disease characterized by chromosomal instability and hypersensitivity to DNA-DNA cross-links (Klages-Mundt & Li, 2017). Patients with Fanconi anemia eventually develop bone marrow failure and have a substantially heightened risk of acute myeloid leukemia and solid tumors (Klages-Mundt & Li, 2017). Despite significant advancements in the field of DPCs, full understanding of these lesions is still hampered by the lack of efficient methods to synthesize site specific, hydrolytically stable DPCs for *in vitro* and *in vivo* studies. In this chapter we describe experimental procedures for efficient synthesis of DNA-peptide (DpC) as well as DNA-protein cross-links (DPC) and primer extension studies to elucidate their impact on DNA replication.

There are several methods in the literature reporting the chemical synthesis of DpC and DPCs. Previously, our lab reported the construction of DpCs and DPCs *via* Cu(I) catalyzed (3 + 2) Huisgen-Meldal-Sharpless

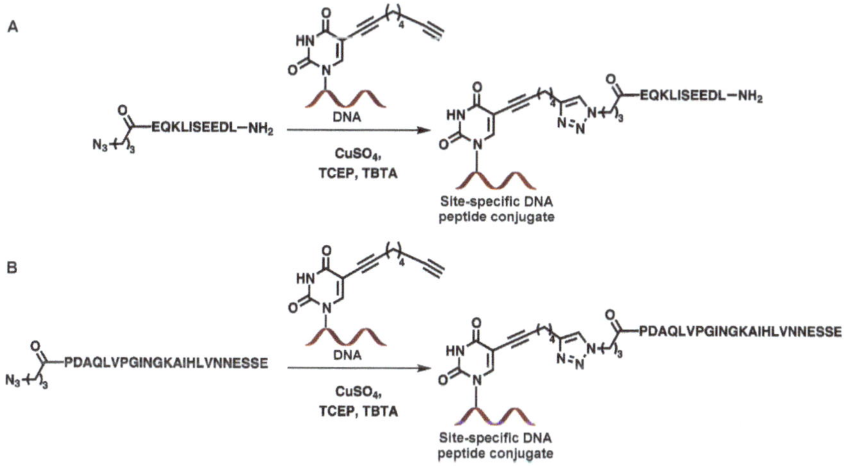

Scheme 1 Synthesis of site-specific DNA-peptide cross-links by (3 + 2) Huisgen-Meldal-Sharpless cycloaddition. Synthetic 10-mer and 23-mer peptides were prepared *via* solid phase peptide synthesis and appended with an N-terminal 4-azidobutanoic acid group. Cross links to peptides of different length (10-mer and 23-mer) were prepared (A and B, respectively). *Reproduced with permission from Yeo J. E., Wickramaratne S., Khatwani S., Wang Y.-C., Vervacke J., & Distefano M. D. (2014). Synthesis of site-specific DNA–protein conjugates and their effects on DNA replication.* ACS Chemical Biology, *9(6) 1860–1868. https://doi.org/10.1021/cb5001795.*

cycloaddition ("click chemistry") (Amblard, Cho, & Schinazi, 2009). The reactive 1,8-octadiynyl side chain was introduced on a pyrimidine nucleoside *via* cross-coupling and further it was converted into phosphoramidite building block and employed in solid phase synthesis to give DNA containing site specific 5-(octa-1,7-diynyl)-uracil (Scheme 1) (Yeo et al., 2014).

The DNA bearing the alkynyl side chain was further reacted with peptides containing an azide at the N terminus in the presence of Cu(I) to afford the DNA-peptide conjugates in quantitative yields (Yeo et al., 2014). In addition, we also reported the synthesis of 7-deaza-dG derived DNA peptide conjugates. These conjugates serve as hydrolytically stable mimics of N7-dG lesions formed in cells treated with *bis*-electrophiles such as nitrogen mustards, platinum compounds and 1,2,3,4-diepoxybutane (Wickramaratne, Mukherjee, Villalta, Schärer, & Tretyakova, 2013). In this case, an aldehyde group was masked as a vicinal diol by introducing 7-deaza-7-(2,3-dihydroxyprop-1-yl)-dG into DNA (Scheme 2). Following solid phase synthesis, DNA was oxidized with $NaIO_4$ to give the corresponding 7-deaza-7-(2-oxoethyl)-dG containing DNA *in-situ* and incubated with lysine containing peptide to give hydrolytically unstable DNA-peptide

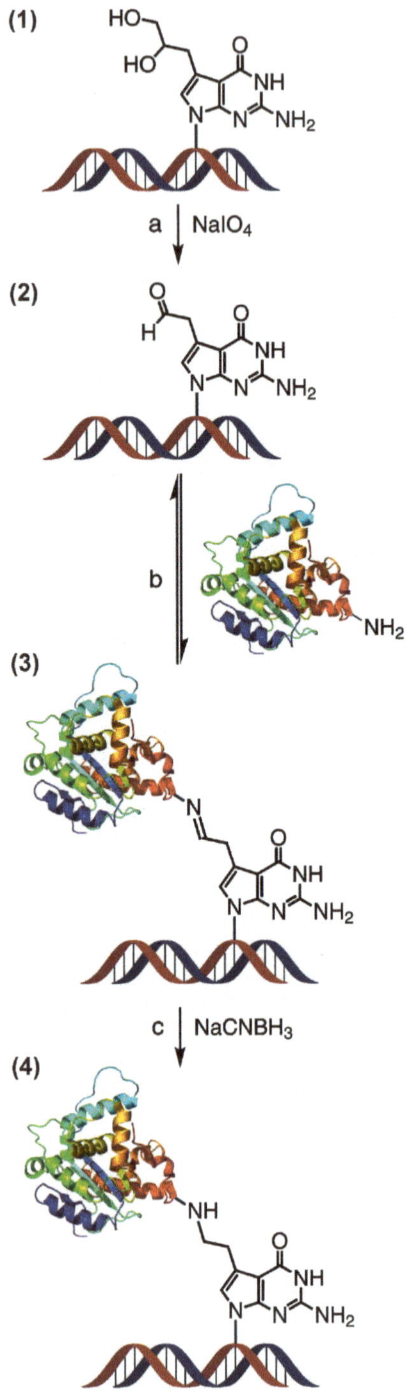

Scheme 2 Formation of DNA-protein cross-links conjugated to C7 position of 7-deazaguanine. *Reproduced with permission from Wickramaratne, S., Mukherjee, S., Villalta, P. W., Schärer, O. D., & Tretyakova, N. Y. (2013). Synthesis of sequence-specific DNA-protein conjugates via a reductive amination strategy.* Bioconjugate Chemistry, 24*(9), 1496–1506. https://doi.org/10.1021/bc400018u.*

conjugate (Scheme 2). The resulting Schiff base conjugate was reduced with $NaCNBH_3$ at room temperature to give a stable amine linkage by reductive amination (Scheme 2). DNA-protein cross-links were also prepared in a similar fashion.

Recently, we and others showed that 5-formylcytosine (5fC), an endogenous oxidation product of the major DNA epigenetic mark 5-methylcytosine, forms reversible Schiff base conjugates with lysine and arginine residues of histone proteins in cells (Scheme 3) (Ji, Shao, et al., 2017;Li, Zhang, et al., 2017; Raiber et al., 2018). Histones H2A, H2B and H4 have been shown to participate in formation of DPCs with 5fC in nucleosomal core particles (Li et al., 2017; Raiber et al., 2018). Such conjugates influence chromatin structure and stability and are thought to mediate gene expression levels in cells. In addition to their effects on gene expression, 5-formylC-histone cross-links are likely to interfere with DNA replication and repair. Reductive amination technology can be used to stabilize the resulting imine conjugates, resulting in stable amino cross-links (Scheme 3).

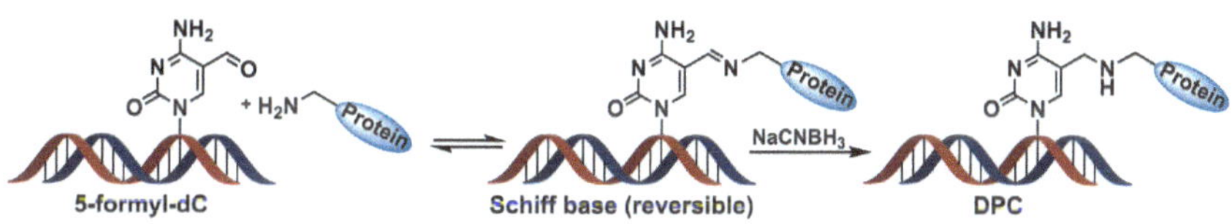

Scheme 3 Reversible histone-DNA cross-linking and their stabilization *via* reduction. From Pujari, S. S., Zhang, Y., Ji, S., Distefano, M. D., & Tretyakova, N. Y. (2018). Site-specific cross-linking of proteins to DNA via a new bioorthogonal approach employing oxime ligation. *Chemical Communications, 54*(49), 6296–6299. https://doi.org/10.1039/C8CC01300D, used with permission.

Synthesis, complete characterization, and effects of DpCs and DPCs on cellular transactions have been reported and discussed below (Ji, Shao, et al., 2017). However, a major limitation of the reductive amination methodology that makes it less favorable is that a mixture of products is often formed due to the participation of multiple Lys or Arg protein side chains in the cross-linking reaction. Also, DPC and DpC yields are in the range of 30–60% depending on DNA and polypeptide sequence. Furthermore, the use of strong reducing agents such as borohydride may lead to protein denaturation, precluding structural studies of the resulting DNA-protein conjugates. In an effort to overcome these limitations, we employed oxime ligation technology for site specific conjugation of proteins to DNA by taking advantage of the ability of substituted alkoxyamines to form stable oxime conjugates with aldehydes under mild conditions (Scheme 4) (Kölmel & Kool, 2017; Pujari, Zhang, Ji, Distefano, & Tretyakova, 2018).

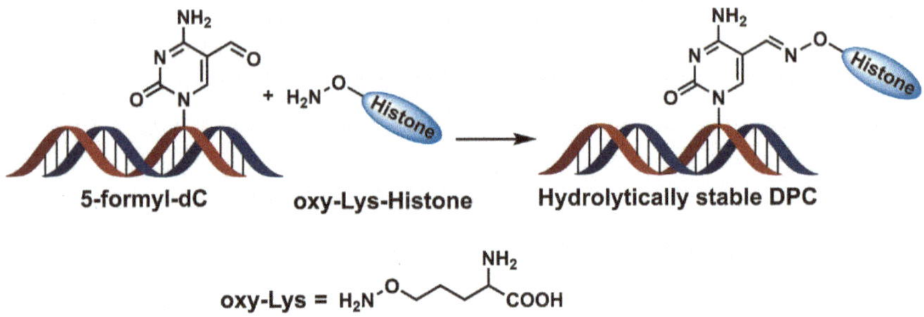

Scheme 4 Oxime ligation to generate DNA-histone H3 cross-links.

A structural analogue of lysine, where the ε-CH_2 group was replaced with oxygen (oxy-Lys), was site-specifically incorporated into peptides using solid phase synthesis. Oxy-Lys containing peptides were ligated with proteins using sortase-A mediated ligation (Dorr, Ham, An, Chaikof, & Liu, 2014). DNA strands containing site-specific 5-formyl-dC (5fC) were conjugated to oxy-Lys-containing peptides and proteins to obtain structurally defined, hydrolytically stable DNA-peptide and DNA-protein cross-links in excellent yields (Scheme 4) (Pujari et al., 2018). These cross-links were characterized by gel electrophoresis and mass spectrometry. Here we describe the synthetic protocols to prepare the DpCs and DPCs *via* reductive amination as well as oxime ligation. Furthermore, primer extension studies to elucidate the impact of these DpCs and DPCs on the DNA replication are discussed. For construction of the corresponding site-specific DPCs, oxy-lysine containing peptides were ligated with protein *via* sortase ligation to yield engineered proteins containing site specific oxy-Lysine (Scheme 5) (Pujari et al., 2018). Further, the Histone H3 DPCs generated

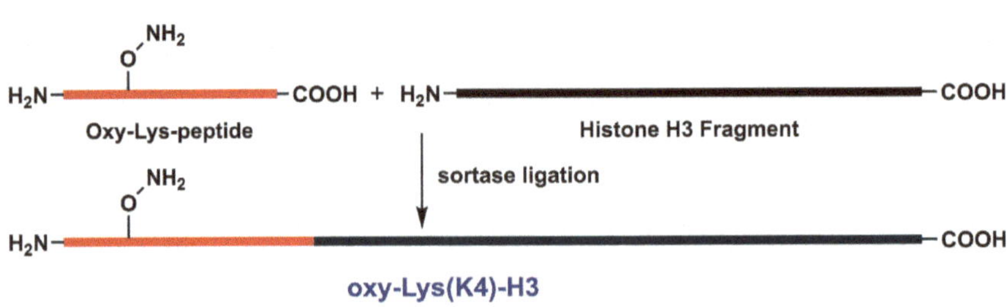

Scheme 5 Construction of oxy-Lys containing histone H3 *via* sortase mediated ligation.

by both oxime ligation as well as reductive amination were used to investigate the effect of these lesions on DNA machineries (Scheme 5).

2. Synthesis of DNA-peptide cross-links by reductive amination

Because of their unusual size and their propensity to block access of DNA repair machinery, DPC lesions require specialized repair mechanisms in order to be removed from the genome. It has been suggested that *in vivo*, the protein component of DPCs is proteolytically degraded by specialized metalloproteases (SPRTN/Wss1) and/or the ubiquitin/proteasome pathway to form smaller polypeptide conjugates (DpCs) (Larsen et al., 2019) which could be removed *via* the nucleotide excision repair (NER) pathway (Stingele et al., 2017). In addition, various enzymatically generated DPCs such as those containing Top1ccs and Top2ccs have been shown to be targets for degradation by the proteasome (Desai, Liu, Vazquez-Abad, & D'Arpa, 1997; Mao, Desai, Ting, Hwang, & Liu, 2001). Following proteasome-mediated degradation, a covalently-bound DNA-peptide adduct remains. This peptide adduct is then processed by the enzyme tyrosyl-DNA-phosphodiesterase 1 or 2 (Tdp1, Tdp2), which hydrolyzes the covalent bond between the DNA and the peptide (Pommier et al., 2006; Pouliot, Yao, Robertson, & Nash, 1999). Hence it is of paramount importance to study the biological effects of the DNA-peptide lesions (DpCs). Here we describe the experimental procedures for preparation of DNA-peptide cross-links by reductive amination as well as oxime ligation.

For the construction of DNA-peptide cross-links (DpCs) *via* reductive amination, the 5fC containing **DNA-1** (Table 1) was incubated with lysine containing peptides (**Pep-1–Pep-3 in** Table 1) in sodium phosphate buffer at 37 °C for 3 h. The reaction mixture was further incubated with $NaCNBH_3$ overnight, and the resulting DpCs were analyzed by denaturing PAGE and mass spectrometry. The representative PAGE and MS data for DpC generated *via* reductive amination between **DNA-1** and **Pep-1** are shown in Fig. 2 (Ji, Park, et al., 2019; Ji, Thomforde, et al., 2019). Similarly, DpCs conjugated to the C-7 position of 7-deazaG were obtained by first oxidizing 7-deaza-7-(2,3-dihydroxypropan-1-yl)-dG containing DNA (**DNA-2**) with $NaIO_4$ in sodium phosphate buffer for 6 h at 4 °C in the dark to generate 7-deaza-7-(2-oxoethyl)-2′-deoxyguanosine. In the subsequent step, **Pep-4** (Ac-EQKLISEEDL-$CONH_2$, Table 1) derived

Table 1 Synthetic oligonucleotides and peptides used for preparation of DpCs and DPCs *via* reductive amination and oxime ligation.

	Sequences[a]
DNA-1	5'-d(ATG GCG TGC TAT)-3'
DNA-2	5'-d(AGG GTT TTC CCA GTC ACG ACG TT)-3' (Pande et al., 2017)
DNA-3	5'-d(AGG GTT TTC CCA GTC ACG ACG TT)-3' (Shaofei Ji et al., 2017)
DNA-4	5'-d(GCC AGT GCC AAG CTT GCA TGC CT)-3'
Pep-1	NH2-RPKPQQFFGLM-COOH
Pep-2	NH2-YGGFMTSEKSQTPLVTLFKNAIIKNAYKKGE-COOH
Pep-3	NH2-YAEGTFISDYSIAMDKIHQQDFVNWLLAQKGKKNDWKHNITQ GKKK KKKSKTKCVIM-COOH (Shaofei Ji et al., 2017)
Pep-4	NHAc-EQKLISEEDL-COOH (Pande et al., 2017)
Pep-5	NH2-K*GG A-COOH (Pujari et al., 2018)
Pep-6	NH2-GGK GLG K*GG A-COOH (Pujari et al., 2018)
Pep-7	NH2-ARTK*QTARK STGGKAPRKQ LATKAARKSA PATG-COOH

[a] where **C** is 5fC, **G** is 7-deaza-7-dihydroxypropyl-2'-deoxyguanosine, **K*** is oxy-Lysine, TG-threonine glycolic acid.

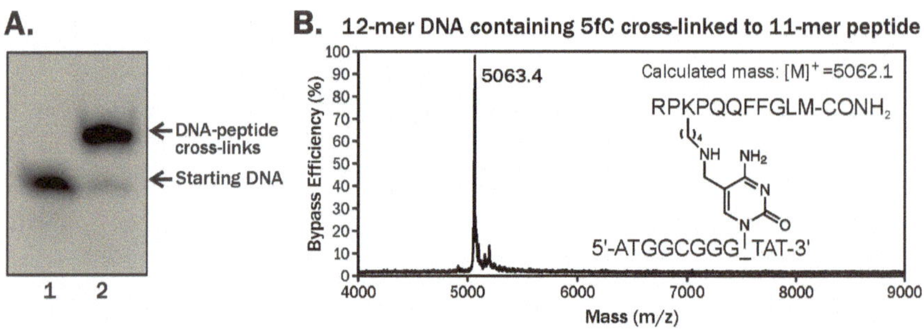

Fig. 2 (A) Denaturing PAGE analysis of reaction mixtures after reductive amination reaction between DNA-1 and Pep-1. The 5′-ends of oligonucleotides were radiolabeled with [γ-^{32}P] ATP and T4 PNK. (B) Representative MALDI-MS spectrum of synthetic DNA-peptide conjugate. *Reproduced with permission from Ji, S., Park, D., Kropachev, K., Kolbanovskiy, M., Fu, I., Broyde, S., et al. (2019). 5-Formylcytosine-induced DNA-peptide cross-links reduce transcription efficiency, but do not cause transcription errors in human cells.* Journal of Biological Chemistry, 294*(48), 18387–18397. https://doi.org/10.1074/jbc.RA119.009834.*

from the c-Myc protein (EQKLISEEDL) was added, followed by 0.5 M $NaCNBH_3$ at 37 °C overnight to form the corresponding DpC (Scheme 2) (Ji, Park, et al., 2019; Ji, Thomforde, et al., 2019).

2.1 Equipment

- Digital dry bath (Labnet)
- Electrophoresis chamber (Bio-Rad)

- Voltage generator (Bio-Rad)
- 394 DNA/RNA synthesizer (Applied Biosystems)
- PS3 Peptide synthesizer (Protein Technologies. Inc.)
- Illustra Micro Biospin G-25 column (GE Healthcare, Pittsburgh, PA)
- Speedvac equipment
- Shaker (Scilogex)
- Typhoon FLA 7000 instrument (GE Healthcare, Pittsburgh, PA)
- MALDI-TOF-MS

2.2 Buffers and reagents

- Oligonucleotide containing 5fC
- Peptide containing lysine residue(s)
- γ-^{32}P ATP (PerkinElmer Life Sciences, Boston, MA)
- T4 PNK buffer (New England Biolabs, Beverly, MA)
- 4.5 mM phosphate buffer pH 7.2
- $NaCNBH_3$
- Sterile H_2O
- Agarose gel (20%, *w/v*) (from agarose powder)
- Urea
- 1 × TBE buffer
- Ammonium persulfate
- *N*, *N*, *N*′, *N*′-Tetramethylethylenediamine (TEMED)
- LDS gel loading buffer (20 mM EDTA in 95% formamide containing 0.05% bromophenol blue and xylene cyanol)
- 3-Hydroxypicolinic acid (3-HPA)

2.3 Procedure

1. Radiolabel the synthetic 12-mer oligodeoxynucleotide containing 5fC (**DNA-1**, Table 1) by incubating it with γ ^{32}P ATP (1 μL) and T4 PNK (20 units) in 20 μL 1 × T4 PNK buffer at 37 °C for 1 h
2. After 1 h, heat the reaction mixture at 65 °C for 10 min to inactivate the enzyme and then filter through an illustra MicroSpin G-25 column to remove excess γ-^{32}P ATP
3. Incubate the ^{32}P-end-labeled 12-mer oligodeoxynucleotide containing 5fC (12.5 pmol) with 160-fold molar excess of synthetic polypeptide **Pep-1** (2 nmol, Table 1) in 16 μL sodium phosphate buffer (4.5 mM, pH 7.4) at 37 °C for 3 h

4. After 3 h, incubate the reaction mixtures in the presence of 25 mM $NaCNBH_3$ (20 μL final total volume) at 37 °C overnight to generate reduced forms of DNA-peptide conjugates. Concentrate the reaction mixture to dryness under vacuum using Speedvac apparatus
5. Reconstitute the reaction mixture in 20 μL of water and add 20 μL of LDS gel loading buffer to it and mix
6. Heat the above mixture at 90 °C for 10 min and spin the reaction vessels for 30 s
7. Load the reaction mixture on denaturing 20% (*w*/*v*) PAGE containing 7 M urea.
8. Run the gel at 300 V in 1 × TBE buffer until the xylene cyanol band migrated to the bottom (~ 2.5–3 h depending on the length of the DNA and the peptide).
9. Take out the gel plate from the chamber and keep it on the glass plate and cover with saran wrap and place the clean radio-imaging plate on top of it and keep it in dark for overnight.
10. Scan the plate next day on a Typhoon FLA 7000 instrument and detect the radiolabeled DNA strands and DPCs, analyze and quantify the bands using ImageQuant TL 8.0.

2.4 Notes

1. All radioactivity procedures including radiolabeling, heating the reaction mixtures, running the gel, radio-plate imaging should be conducted in a specialized radiation room only.
2. To generate non-radiolabeled DPCs for further experiments, the same reaction conditions were used with non-radiolabeled DNA (0.5–1 nmol scale). To purify DPCs, divide the reaction mixture in 3–4 parts and separate on the 20% PAGE as described above.
3. After running the gel, take out the gel from glass plate and wash the gel 3 × times with water.
4. Cut one lane corresponding to reaction mixture and stain it with SYBR green for 1 h and wash 3 × with water and visualize on Typhoon FLA 7000 instrument and mark the DPC band. Align this small piece of gel with the rest of the gel accurately and cut the non-stained bands of DPC.
5. Cut these DPC bands into very small gel pieces and perform freeze and thaw cycle for at least 10 times in 1 × TEA buffer (1 mL) and then incubate them in temperature controlled oscillator at 37 °C for 2 days.

Centrifuge and carefully collect the solution and concentrate. After concentration, reconstitute it in sterile water, desalt using Sep-Pak C18 cartridges and concentrate in half under vacuum. By this method, 30–50% of the DpCs can be extracted from the gel.

6. Measure the concentration of desalted, non-radiolabeled DpCs with a NanoDrop spectrophotometer. A concentration of approximately 0.05–0.1 pmol/μL is desired.
7. For MS analysis, gel extracted and desalted DNA–peptide cross-links were mixed with 3-HPA matrix and spotted on MALDI-TOF plate and characterized by MALDI-TOF-MS.

3. Synthesis of DNA-peptide cross-links by oxime ligation

As discussed earlier in this chapter, the reductive amination method to synthesize DpCs described above typically produces a mixture of products since the aldehyde group in 5fC can react with multiple Lys and Arg residues within the protein/peptide (Ji, Shao, et al., 2017; Li et al., 2017). The yield of this reaction is also in the range of low to medium making it less viable method. Therefore, the oxime ligation technology was developed in our laboratory (Kölmel & Kool, 2017; Pujari et al., 2018). Fmoc-protected-oxy-Lysine was synthesized in our laboratory and this unnatural amino acid was incorporated into a peptide derived from histone H4 by solid phase peptide synthesis. The obtained peptide was reacted with DNA containing 5fC to obtain site-specific DNA-peptide cross-link in excellent yields. A detailed experimental procedure is outlined below. In brief, oxy-Lys containing peptides **Pep-5** or **Pep-6** (Table 1)were reacted with 5fC-containing **DNA-1** (Table 1) in 100 mM ammonium acetate buffer pH 4.5 at 37 °C for overnight. It is worth noting the reaction proceeds smoothly with up to 95% yield when aniline is used as a catalyst. Various reaction conditions like duration, temperature, and pH were explored for the best conjugation yields (Pujari et al., 2018). The best yields for the oxime ligation were obtained when using 100 mM ammonium acetate buffer, pH 4.5 containing having 100 mM aniline at 37 °C for 18 h. The oxime DpCs were purified by HPLC and analyzed by PAGE and MS (Fig. 3).

In a similar fashion, DpCs conjugated to the C7 position of 7-deaza-dG were obtained. Briefly, 23-mer oligonucleotide containing a 7-deaza-7-(2,3-dihydroxypropan-1-yl)-dG was oxidized with $NaIO_4$ as detailed before (Scheme 6).

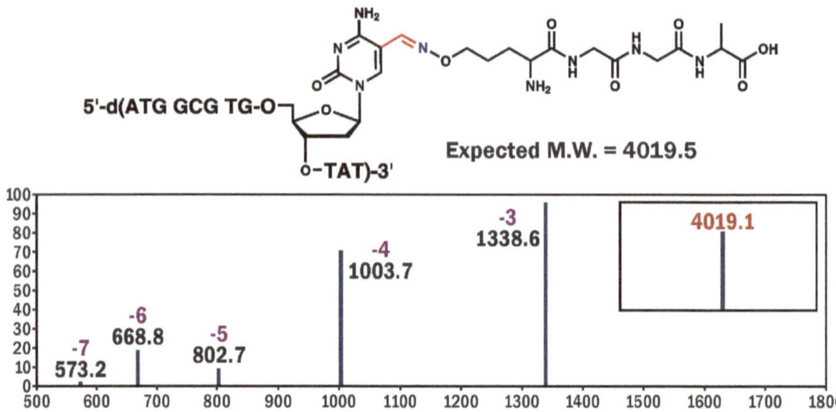

Fig. 3 ESI^{+} mass spectrum and deconvoluted MS (insert) of DpC between 5-formyl-dC containing **DNA-1** and oxy-Lys containing **Pep-5**. *From Pujari, S. S., Zhang, Y., Ji, S., Distefano, M. D., & Tretyakova, N. Y. (2018). Site-specific cross-linking of proteins to DNA via a new bioorthogonal approach employing oxime ligation.* Chemical Communications, *54(49), 6296–6299. https://doi.org/10.1039/C8CC01300D, used with permission.*

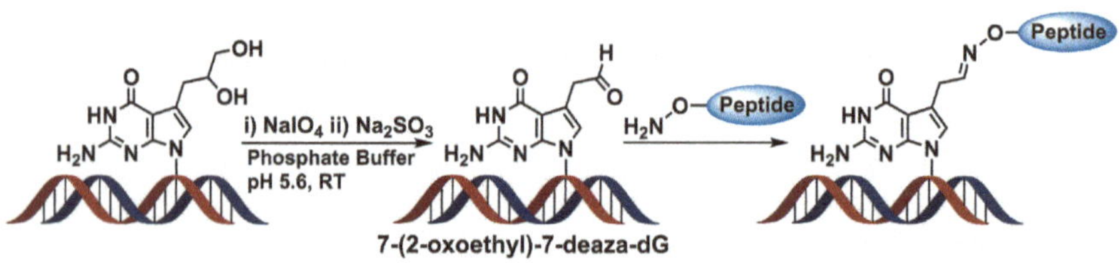

Scheme 6 DpC generation at the C7 position of 7-deaza-dG *via* oxime ligation. *From Pujari, S. S., Zhang, Y., Ji, S., Distefano, M. D., & Tretyakova, N. Y. (2018). Site-specific cross-linking of proteins to DNA via a new bioorthogonal approach employing oxime ligation.* Chemical Communications, *54(49), 6296–6299. https://doi.org/10.1039/C8CC01300D, used with permission.*

The main difference between reductive amination and oxime ligation approaches for synthesis of 7-deaza-dG DpCs is avoiding the addition of $NaCNBH_3$ since the oxime DpCs are hydrolytically stable and do not require reduction. A representative MS is shown below (Fig. 4).

3.1 Equipment

- Digital dry bath (Labnet)
- Electrophoresis chamber (Bio-Rad)
- Voltage generator (Bio-Rad)
- 394 DNA/RNA synthesizer (Applied Biosystems)

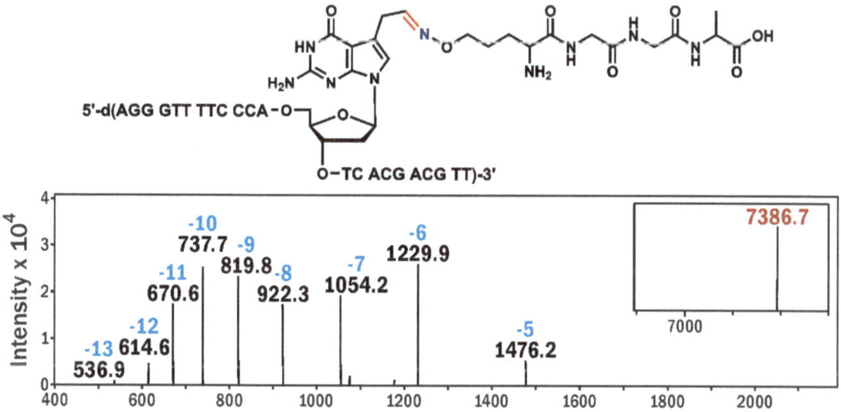

Fig. 4 ESI$^+$ mass spectrum and deconvoluted spectrum (insert) of DPC formed between 7-deaza-7-(2-oxoethyl) dG containing DNA-2 and oxy-Lys containing **Pep-5**. *From Pujari, S. S., Zhang, Y., Ji, S., Distefano, M. D., & Tretyakova, N. Y. (2018). Site-specific cross-linking of proteins to DNA via a new bioorthogonal approach employing oxime ligation.* Chemical Communications, *54(49), 6296–6299. https://doi.org/10.1039/C8CC01300D, used with permission.*

- PS3 Peptide synthesizer (Protein Technologies. Inc.)
- Illustra Micro Biospin G-25 column (GE Healthcare, Pittsburgh, PA)
- Speedvac equipment
- Shaker (Scilogex)
- Agilent 1260 infinity HPLC
- Synergi hydro-RP column
- Typhoon FLA 7000 instrument (GE Healthcare, Pittsburgh, PA)
- Agilent 1100 HPLC-MSD ion trap mass spectrometer
- Zorbax 300 SB-C18 column (150 × 0.5 mm, 5 μm)

3.2 Buffers and reagents

- Oligonucleotide containing 5fC
- Peptide containing oxy lysine residues
- γ-^{32}P ATP (PerkinElmer Life Sciences, Boston, MA)
- T4 PNK buffer (New England Biolabs, Beverly, MA)
- Sterile H_2O
- Agarose gel (20%, *w/v*) (from agarose powder)
- Urea
- 1 × TBE buffer
- Ammonium persulfate
- *N*, *N*, *N'*, *N'*-Tetramethylethylenediamine (TEMED)

- LDS gel loading buffer (20 mM EDTA in 95% formamide containing 0.05% bromophenol blue and xylene cyanol)
- 100 mM ammonium acetate buffer pH 4.5
- Aniline
- 100 mM TEA buffer
- 25 mM Ammonium acetate buffer
- Acetonitrile

3.3 Procedure

1. Radiolabel the synthetic 5fC-containing 12-mer oligodeoxynucleotide (**DNA-1**, Table 1) by incubating it with γ-^{32}P-ATP (1 μL) and T4 PNK (20 units) in 20 μL 1 × T4 PNK buffer at 37 °C for 1 h.
2. Heat the reaction mixture at 65 °C for 10 min to inactivate the enzyme and then filter through an Illustra Micro Biospin G-25 column to remove excess γ-^{32}P ATP.
3. Incubate the resulting ^{32}P-end-labeled 12-mer (3 pmol) with a 100 × molar excess of synthetic **Pep-5** or **Pep-6** (300 pmol, Table 1) in 20 μL ammonium acetate buffer (100 mM, pH 4.5) containing 100 mM aniline at 37 °C overnight (~18 h).
4. After 18 h, concentrate the reaction mixture to dryness under vacuum.
5. Reconstitute the reaction mixture in 20 μL of water, add 20 μL LDS gel loading buffer to it and mix.
6. Heat the above mixture at 90 °C for 10 min and spin the reaction vessels for 30 s.
7. After heating the reaction mixture to 90 °C for 10 min, load it on denaturing 20% (*w*/*v*) PAGE containing 7 M urea.
8. Run the gel at 300 V in 1 × TBE buffer until the xylene cyanol band migrated to the bottom (~2.5-3 h depending on the length of the DNA and peptide).
9. Take out the gel plate from the chamber and keep it on the glass plate and cover with saran wrap and place the clean radio-imaging plate on top of it and keep it in dark for overnight.
10. Scan the plate on a Typhoon FLA 7000 instrument and detect the radiolabeled DNA strands and DPCs, analyze and quantify using ImageQuant TL 8.0.

3.4 Notes

1. All radioactivity involving procedures like radiolabeling, heating the reaction mixtures, running the gel, radio-plate imaging should be conducted in a radiation room only.

2. To obtain the non-radiolabeled DpCs for further experiments, the same reaction conditions were employed with non-radiolabeled DNA (0.5–1 nmol scale), and the reaction mixtures were purified by RP-HPLC. HPLC purification of DpCs was performed on an Agilent 1260 infinity HPLC system using a Synergi 4u, hydro-RP 80 Å column with buffers A: 100 mM triethylammonium acetate buffer, pH 7.2; B: ACN.
3. The major HPLC peak corresponding to DpC was collected and concentrated using a speedvac apparatus. After concentration, reconstitute it in sterile water (1 mL), desalt using NAP-5 columns and concentrate in half.
4. Measure the concentration of desalted, non-radiolabeled DpCs with a NanoDrop spectrophotometer. A concentration of approximately 0.05–0.1 pmol/μL is desired.
5. For MS analysis, HPLC purified and NAP-5 desalted DNA-peptide cross-links were analyzed by an Agilent 1100 HPLC-MSD ion trap mass spectrometer using Zorbax 300 SB-C18 column (150 × 0.5 mm, 5 μm) using 25 mM ammonium acetate buffer pH 7.2 and ACN in the negative ion mode. A Zorbax 300 SB-C18 column (150 × 0.5 mm, 5 μm) was eluted at a flow rate of 15 μL/min using 25 mM ammonium acetate, pH 7.2 (A) and acetonitrile (B). Solvent composition was linearly changed from 2% to 50% B in 25 min. The mass spectrometer was operated in the negative ion mode with the mass range of *m/z* 400–1800. MS^2 fragmentation was carried out by collision-induced dissociation with an isolation width of 4.0 m/z and fragmentation amplitude of 0.95 V.
6. The DpCs can be isolated either by gel extraction (employing freeze-thaw cycles) or by RP-HPLC purification. The RP-HPLC purification method is more advantageous over gel extraction since it is less laborious, requires less time, and no loss of DpC occurs during purification.

4. Synthesis of DNA-protein cross-links by reductive amination

Due to their enormous size as compared with other nucleobase lesions and their ability to disrupt DNA duplex structure and DNA-protein interactions, DPCs are anticipated to interfere with DNA replication and transcription. However, the structural and biological consequences of DPCs have not been fully elucidated due to the lack of convenient methodologies to generate these DPCs in laboratory as model hydrolytically stable lesions.

Accessing such DPC substrates has been limited due to the synthetic challenge of covalently linking two complex biomolecules (DNA and proteins) in a site-specific manner. Few methods have been developed for the synthesis of site specific DNA-protein cross-links. Our laboratory has been developing novel methodologies to generate synthetic DPCs structurally analogous to DPC adducts found in cells. A bioorthogonal approach employing click chemistry (Amblard et al., 2009) between azide-functionalized proteins and alkyne-containing DNA was used to generate structurally defined DPC conjugates. The azide groups were incorporated *via* synthetic methods for short peptides and enzymatically for a larger protein, while alkyne-containing DNA (5-octadiynyl-dU) was generated by solid phase synthesis (Yeo et al., 2014). Cu(I) catalyzed "click" reaction between the protein and the DNA was performed, resulting in cross-links that are site-specific in regard to both biomolecules (Scheme 2) (Yeo et al., 2014).

The most common site of DNA involved in DPC formation following treatment with *bis*-electrophiles is the N7 of guanine. However, N7 guanine alkylation destabilizes the β-glycosidic bond of the modified nucleoside, leading to spontaneous depurination (Gates, 2009; Gates, Nooner, & Dutta, 2004). Our laboratory developed a new methodology to create hydrolytically stable structural mimics of N7 guanine conjugated DPCs by reductive amination reactions between the Lys and Arg side chains of proteins and acetaldehyde functionalities of the modified 7-deazaguanine residues within DNA (Scheme 1) (Pujari et al., 2018). The resulting model DPCs are structurally analogous to the N7 guanine adducts generated by antitumor nitrogen mustards, platinum compounds, 1,2,3,4-diepoxybutane, and chlorooxirane (Scheme 6). Recently we and others showed that 5fC mediated DNA-protein cross-linking can be achieved in moderate yields *via* reductive amination. In this approach, DNA containing 5fCis incubated with proteins containing lysine and arginine residues such as histones to give Schiff base conjugates (Scheme 3). The resulting imine adducts are stabilized by treatment of $NaCNBH_3$ to give hydrolytically stable model DPCs. Briefly, DNA strands bearing 5fC (12-mer **DNA-1** or 23-mer **DNA-3**, Table 1) were incubated with Histone H2A, H3, or H4 protein individually in sodium phosphate buffer to generate Schiff base cross-links, which were subsequently stabilized *via* reduction (Ji, Shao, et al., 2017). These DPCs were further fully characterized by PAGE and mass spectrometry (Figs. 5 and 6).

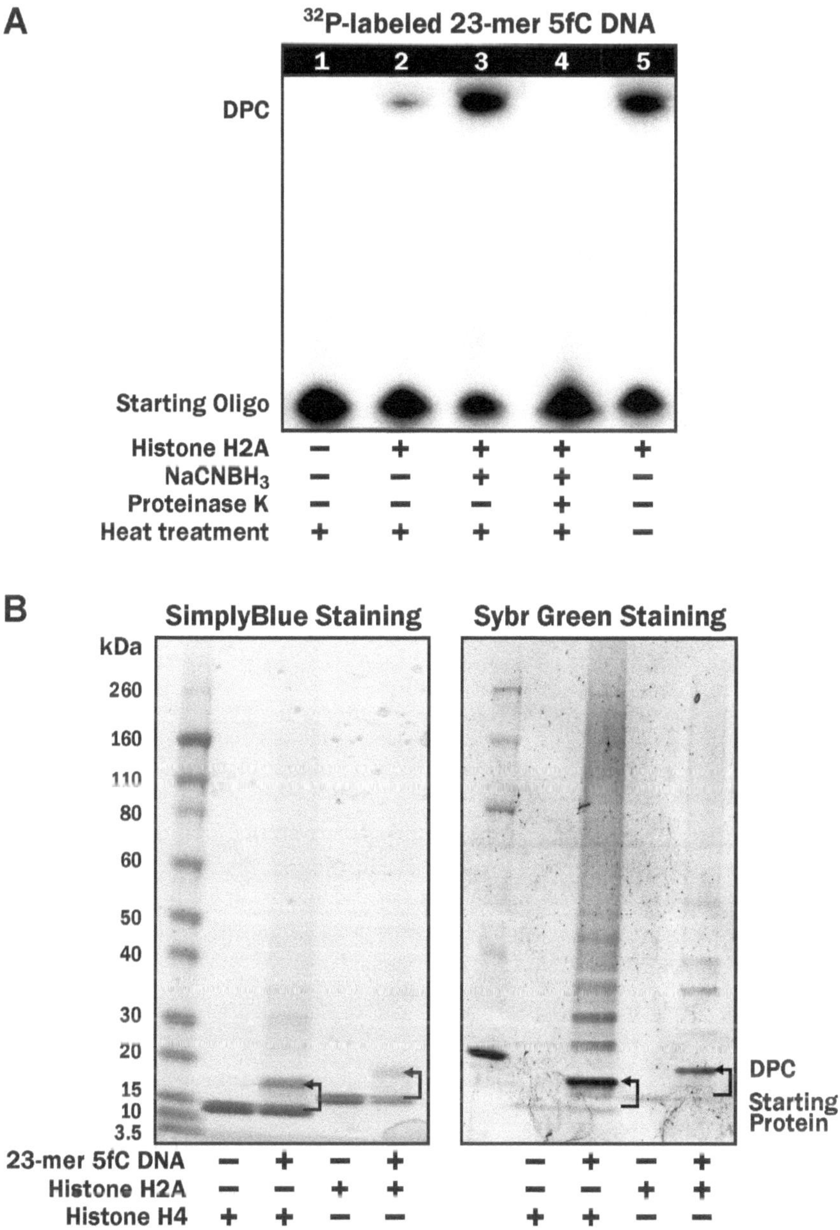

Fig. 5 Denaturing SDS-PAGE analysis of DNA–protein cross-links using (A) ^{32}P-labeled 5fC **DNA-3** with Histone H2A and (B) SimplyBlue Protein staining (left panel) and SYBR Green DNA staining (right panel) of **DNA-3** with Histone H2A and H4. *From Ji, S., Shao, H., Han, Q., Seiler, C. L., & Tretyakova, N. Y. (2017). Reversible DNA–protein cross-linking at epigenetic DNA marks.* Angewandte Chemie International Edition, *56(45), 14130–14134. https://doi.org/10.1002/anie.201708286, used with permission.*

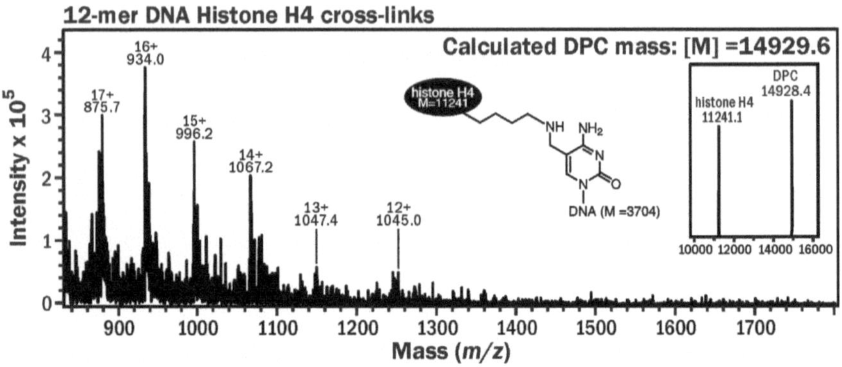

Fig. 6 ESI^+ mass spectrum of histone H4 (M = 11,241) cross-linked to 5fC-containing **DNA-1** (Table 1, M = 3704.4). The cross-links were stabilized by $NaCNBH_3$ treatment. The signal at *m/z* 14,928.6 corresponds to DPC structure shown on top. *From Ji, S., Shao, H., Han, Q., Seiler, C. L., & Tretyakova, N. Y. (2017). Reversible DNA–protein cross-linking at epigenetic DNA marks.* Angewandte Chemie International Edition, 56*(45), 14130–14134. https://doi.org/10.1002/anie.201708286, used with permission.*

4.1 Equipment

- Digital dry bath (Labnet)
- Electrophoresis chamber (Bio-Rad)
- Voltage generator (Bio-Rad)
- Illustra Micro Biospin G-25 column (GE Healthcare, Pittsburgh, PA)
- Speedvac equipment
- 394 DNA/RNA synthesizer (Applied Biosystems)
- PS3 Peptide synthesizer (Protein Technologies. Inc.)
- Shaker (Scilogex)
- Typhoon FLA 7000 instrument (GE Healthcare, Pittsburgh, PA)
- MALDI-TOF-MS mass spectrometer

4.2 Buffers and reagents

- DNA oligodeoxynucleotide containing 5fdC
- Histone proteins
- γ-^{32}P ATP (PerkinElmer Life Sciences, Boston, MA)
- T4 PNK buffer (New England Biolabs, Beverly, MA)
- 4.5 mM phosphate buffer pH 7.2
- $NaCNBH_3$
- Sterile H_2O
- Agarose gel (20%, *w*/*v*) (from agarose powder)
- Urea

- 1 × TBE buffer
- Ammonium persulfate
- *N*, *N*, *N'*, *N'*-Tetramethylethylenediamine (TEMED)
- LDS gel loading buffer (20 mM EDTA in 95% formamide containing 0.05% bromophenol blue and xylene cyanol)
- SYBR green
- Simply Blue
- 3-Hydroxypicolinic acid (3-HPA)

4.3 Procedure

1. Radiolabel synthetic 12-mer oligodeoxynucleotide containing 5fC (**DNA-3** in Table 1) by incubating it with γ-^{32}P-ATP (1 μL) and T4 PNK (20 units) in 20 μL 1 × T4 PNK buffer at 37 °C for 1 h.
2. After 1 h, heat the reaction mixture at 65 °C for 10 min to inactivate the enzyme and then filter through an Illustra Micro Biospin G-25 column to remove excess γ-^{32}P ATP.
3. Incubate ^{32}P-end-labeled 23-mer oligodeoxynucleotide containing 5fC (50 pmol) with 2-fold molar excess of each individual recombinant protein (Histone H2A or H4) in 16 μL sodium phosphate buffer (4.5 mM, pH 7.4) at 37 °C for 3 h.
4. The reaction mixture is then incubated in the presence of 25 mM $NaCNBH_3$ (20 μL final total volume) at 37 °C overnight to generate reduced forms of DPCs.
5. Concentrate the reaction mixture to half volume under reduced pressure (speedvac apparatus).
6. Reconstitute the reaction mixture in 10 μL sterile water and add 20 μL LDS gel loading buffer to it and mix.
7. Heat the above mixture at 90 °C for 10 min and spin the reaction vessels for 30 s in a tabletop centrifuge.
8. After heating the reaction mixture to 90 °C for 10 min, load it on denaturing 20% (*w*/*v*) PAGE containing 7 M urea.
9. Run the gel at 300 V in 1 × TBE buffer until the xylene cyanol band migrates to the bottom (~2.5–3 h).
10. Take out the gel plate from the chamber, keep it on the glass plate and cover with saran wrap. Place the clean radio-imaging plate on top of it and keep it in dark for overnight.
11. Scan the plate on a Typhoon FLA 7000 instrument and detect the radiolabeled DNA strands and DPCs, analyze and quantify using ImageQuant TL 8.0.

4.4 Notes

1. DPCs generated by this method are a mixture of products since the aldehyde group of 5fC can react with multiple lysine or arginine residues of the protein.
2. All the procedures like radiolabeling, heating the reaction mixtures, running the gel, and radioplate imaging should be conducted in a radiation room only.
3. To prepare non-radiolabeled DPCs for further experiments, conduct the same reaction with nonradiolabeled DNA (0.5–1 nmol scale), divide the reaction mixture into 3–4 parts, and run on the 20% PAGE as described above.
4. After running the gel, take out the gel from glass plate and wash the gel 3 times with water.
5. Cut one lane from the loaded portion of the reaction mixture and stain it with SYBR green for 1 h and wash 3 × with water (to visualize DNA).
6. Further, stain the gel with Simply Blue dye overnight upon gentle shaking. Remove excess dye from the gel by repetitive washes with water. Align this small piece of gel with the rest of the gel to locate unstained DPC in the gel and cut it out with a razor blade.
7. Cut the DPC-containing gel band into small pieces and perform freeze and thaw cycle in 1 × TEA buffer at least 10 times and then incubate in a temperature controlled oscillator at 37 °C for 2 days. Centrifugate and carefully collect the supernatant and concentrate to reduce the volume in half. Reconstitute it in sterile water (1 mL), desalt using an Amicon 3 K filter and concentrate to half the volume. By this method, approximately 20–30% of the DPCs can be extracted. An alternative method to extract DPCs from the gels is described in the next section.
8. Measure the concentration of desalted, non-radiolabeled DPCs with a NanoDrop spectrophotometer. A concentration of approximately 0.05–0.1 pmol/μL is desired.
9. For MS analysis, gel-extracted and desalted DPCs are mixed with 3-HPA matrix and spotted on a MALDI-TOF plate and characterized by MALDI-TOF-MS.

5. Synthesis of DNA-protein cross-links by oxime ligation

As discussed earlier, in the reductive amination approach, Schiff base conjugates between basic side chains of proteins and DNA are reduced with

$NaCNBH_3$ to produce a stable amino conjugate (Scheme 3) (Ji, Shao, et al., 2017). A major limitation of the reductive amination methodology is that a mixture of products is often formed due to the participation of multiple Lys or Arg protein side chains in the cross-linking reaction, lacking site-specificity and the reactions are often low yielding (low-moderate yields) (Ji, Shao, et al., 2017). Furthermore, the use of strong reducing agents may lead to protein denaturation, precluding structural studies of the resulting DNA-protein conjugates. In an effort to develop feasible methodologies to synthesize model site-specific DPCs in quantitative yields, we developed a new methodology for site specific conjugation of proteins to DNA by taking advantage of the ability of substituted alkoxyamines to form stable oxime conjugates with aldehydes under mild conditions (Scheme 4) (Kölmel & Kool, 2017; Pujari et al., 2018). A structural analogue of lysine, where the ε-CH_2 group was replaced with oxygen (oxy-Lys), was site-specifically incorporated into peptides using solid phase peptide synthesis. Oxy-Lys containing peptides or native Lys controls were ligated with VHH-7 protein at the C-terminus using sortase A mediated ligation to give the corresponding fusion proteins containing oxy-Lysine or Lys at a specified position (Scheme 5) (Pujari et al., 2018). Oligonucleotides bearing 5-formyl-dC or 7-hydroxypropyl-7-deaza-dG were conjugated to oxy-Lys-containing proteins to obtain structurally defined DNA-protein cross-links (Schemes 4 and 6). The reaction products were resolved by denaturing SDS-PAGE (Pujari et al., 2018). The reaction of **VHH7–2** with 5-formyl-dC **DNA-1** was performed in NH_4OAc buffer at pH 4.5 in the presence of aniline, leading to the formation of DPC conjugates (Fig. 7).

Protein bands with reduced mobility (M ~ 19 kDa) were identified as DPC because they could be visualized by both DNA (Fig. 7A) and protein stains (Fig. 7B). Similarly, VHH7 proteins were conjugated to 7-deaza-7-(2-oxyethyl)-dG containing DNA (**DNA-2**) resulting in covalent DPCs in ~60% yield for **VHH7–2**. To identify amino acid residue(s) of VHH7 participating in cross-linking, the DNA portion of gel purified DPC was digested to nucleosides, and the resulting protein-nucleoside conjugate was then subjected to tryptic digestion. Tryptic peptides were detected by nanoLC-ESI^+-MS/MS analysis using an Orbitrap Elite mass spectrometer (Pujari et al., 2018). Only the oxy-Lys residue was found to be cross-linked to 5-formyl-dC (Fig. 8), while four other Lys and Arg residues were intact. Therefore, oxime ligation is site-specific for oxy-Lys within **VHH7–2** and 5-formyl-dC in DNA.

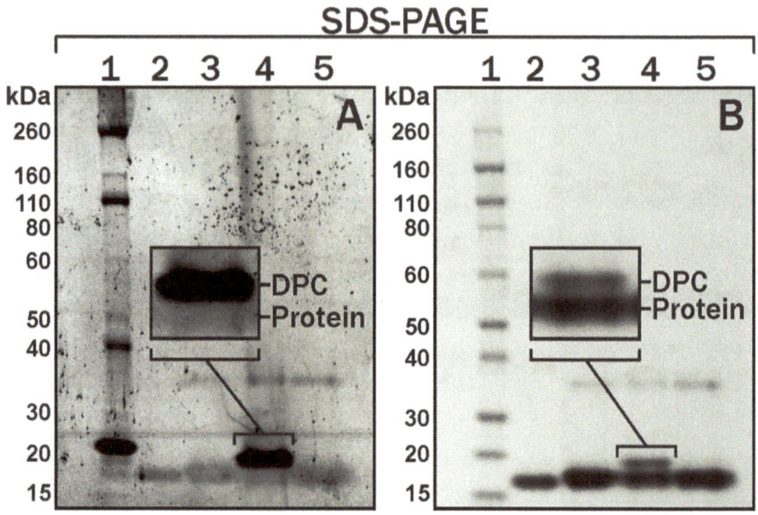

Fig. 7 Denaturing SDS-PAGE analyses of 5-formyl-dC-mediated DNA-VHH7 protein cross-links (A) SYBR green staining of DNA; (B) Simply blue staining of proteins. Lane 1. Protein ladder. 2. Unmodified **VHH7-1** protein. 3. **VHH7-2** containing oxy-Lys. 4. **VHH7-2**+**DNA-1**. 5. **VHH7-2**+DNA containing no aldehyde group. *From Pujari, S. S., Zhang, Y., Ji, S., Distefano, M. D., & Tretyakova, N. Y. (2018). Site-specific cross-linking of proteins to DNA via a new bioorthogonal approach employing oxime ligation.* Chemical Communications, *54(49), 6296–6299. https://doi.org/10.1039/C8CC01300D, used with permission.*

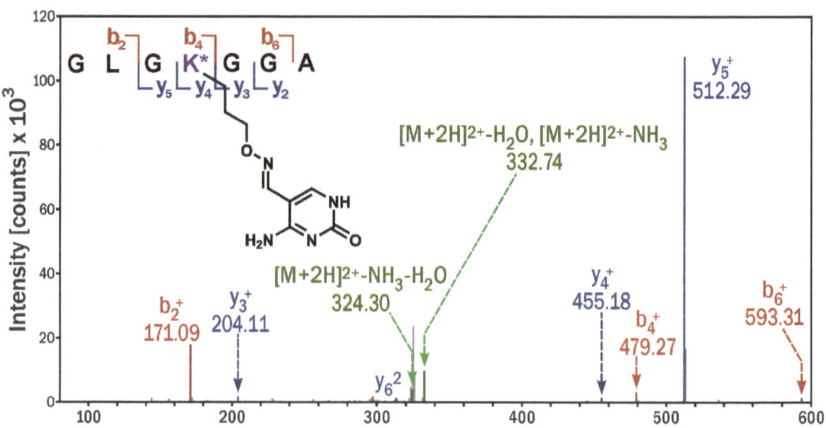

Fig. 8 NanoLC-NSI$^+$-MS/MS spectrum of **VHH7–2** tryptic peptide containing 5-formyl-dC cross-link to oxy-Lys. *From Pujari, S. S., Zhang, Y., Ji, S., Distefano, M. D., & Tretyakova, N. Y. (2018). Site-specific cross-linking of proteins to DNA via a new bioorthogonal approach employing oxime ligation.* Chemical Communications, *54(49), 6296–6299. https://doi.org/10.1039/C8CC01300D, used with permission.*

5.1 Equipment

- Digital dry bath (Labnet)
- Electrophoresis chamber (Bio-Rad)
- Voltage generator (Bio-Rad)
- 394 DNA/RNA synthesizer (Applied Biosystems)
- Illustra Micro Bio-spin G-25 column (GE Healthcare, Pittsburgh, PA)
- Speedvac concentrator
- Shaker (Scilogex)
- Typhoon FLA 7000 instrument (GE Healthcare, Pittsburgh, PA)
- LiCor Odyssey
- Orbitrap Elite mass spectrometer

5.2 Buffer and reagents

- 5fC containing synthetic oligodeoxynucleotide
- Oxy-Lys containing VHH7 protein generated by sortase mediated ligation
- γ-^{32}P ATP (PerkinElmer Life Sciences, Boston, MA)
- T4 PNK buffer (New England Biolabs, Beverly, MA)
- 100 mM ammonium acetate buffer pH 4.5
- Aniline (Sigma Aldrich)
- Sterile H_2O (Invitrogen)
- Agarose gel (20%, *w/v*) (from agarose powder, Biorad)
- Urea (Sigma Aldrich)
- 1 × TBE buffer
- Ammonium persulfate (Across Organics)
- *N*, *N*, *N*′, *N*′-Tetramethylethylenediamine (TEMED) (Sigma)
- LDS gel loading buffer (20 mM EDTA in 95% formamide containing 0.05% bromophenol blue and xylene cyanol)
- SYBR green
- Simply Blue

5.3 Procedure

1. Incubate synthetic 12-mer oligodeoxynucleotides containing 5fC (**DNA-1** in Table 1, 300 pmol) with γ-^{32}P-ATP (2 μL, PerkinElmer Life Sciences, Boston, MA) and T4 PNK (20 units) in 20 μL 1 × T4 PNK buffer at 37 °C for 1 h.
2. After 1 h, heat the reaction mixture at 65 °C for 10 min and filter through an Illustra Micro Bio-spin G-25 column to remove excess γ-^{32}P ATP.

3. Incubate radiolabeled 5fC-containing DNA 12-mer (**DNA-1** in Table 1, 3 pmol) with 100-fold molar excess of VHH-7 protein containing oxy-Lys at its C-terminus (300 pmol) in 20 μL ammonium acetate buffer containing aniline (100 mM, pH 4.5) at 37 °C overnight (~18 h).
4. Reconstitute the reaction mixture in 20 μL gel loading buffer (20 mM EDTA in 95% formamide containing 0.05% bromophenol blue and xylene cyanol) and heat it at 90 °C for 10 min and spin the reaction vessels for 30 s.
5. After heating, load the reaction mixture on denaturing SDS-PAGE.
6. Run the gel at 300 V in 1 × TBE buffer until the xylene cyanol band migrated to the bottom (~ 2.5–3 h).
7. Take out the gel plate from the chamber, keep it on the glass plate and cover with Saran wrap. Place the clean radio-imaging plate on top of it and keep it in dark for overnight.
8. Scan the plate on Typhoon FLA 7000 instrument and detect the radiolabeled DNA strands and DPCs, analyze and quantify using ImageQuant TL 8.0.

5.4 Notes

1. DPCs generated by this method are site-specific in regard to both the protein and the DNA.
2. All the procedures including radiolabeled DNA—end-labeling, conjugation reactions, running the gel, and radioimaging—should be conducted in a radiation room only.
3. In a similar fashion, histone H3 DPCs can be also generated *via* oxime ligation. Modified histone H3 containing oxy-Lys at K4 position was constructed as shown in Scheme 5 (for peptide sequence see Table 1, **Pep-7**) and conjugated with either 12-mer **DNA-1** or 23-mer **DNA-4** and the DPCs were purified by 4–12% Bis-tris SDS-PAGE separation
4. To extract non-radiolabeled DPCs from the gel, the same reaction is conducted with non-radiolabeled DNA (0.5–1 nmol), and the reaction mixtures are separated by gel. SDS-PAGE gel was stained with SYBR green for 30 min (DNA part of DPC will stain) and washed three times with water and visualized on LiCor Odyssey.
5. After washing the gel with water for another 30 min, SimplyBlue stain (for staining protein part of DPC) was added and the gel was stained

for overnight and the next day it was destained by repeated washing with water to observe a clean band on the gel.

6. Excise the DPC bands from the gel with a clean sharp razor, cut the gel into small pieces, and transfer into microcentrifuge tubes. DPC-containing gel pieces were further destained by incubating in 500 μL ACN:water (1:1) multiple times until complete decoloration of the gel pieces was observed. The gel pieces were then macerated with a plastic pestle on ice for 30 s and 500 μL of freshly prepared 100 mM ammonium bicarbonate buffer pH 8.0 containing 0.1% SDS was added and vigorously vortexed on a benchtop vortexer for 10 min, and the mixture was centrifuged at 14,000 × *g* for 5 min (Takemori et al., 2020). Collect the supernatant carefully and reduce the volume in half by centrifuging it in room temperature. Avoid complete drying of the DPC since it may stick to the microcentrifuge tube walls and lose the DPC. Desalt the DPC further by an Amicon 3K filter and concentrate the DPCs and store in freezer for further use. By this method, around 85% of the DPCs can be extracted. This method offers an alternative to the freeze-thaw method discussed above.
7. Measure the concentration of desalted, non-radiolabeled DPCs with a NanoDrop spectrophotometer. A concentration of approximately 0.05–0.1 pmol/μL is desired.

6. Polymerase bypass studies of DNA-protein and DNA-peptide conjugates

DNA-protein cross-links (DPCs) are highly cytotoxic lesions that obstruct essential DNA transactions and whose resolution is critical for cell and organismal fitness. Due to their unusually bulky size, DPCs pose challenges to replicative polymerases. We have shown that DNA-protein cross-links containing 7-deaza-7-dihydroxypropyl-dG conjugated to GFP, AlkB and histone-H4 proteins prepared by reductive amination (Scheme 2) completely blocked human TLS polymerases η and κ (Wickramaratne et al., 2016). No primer extension products were observed in either standing-start or running-start polymerase bypass assay (Wickramaratne et al., 2016). Similar outcomes were also observed for the DPCs generated by CuAAC (Scheme 1). 5-Octadinyl-dU containing oligonucleotides were conjugated to GFP under click conditions, and the resulting DPCs completely blocked human TLS polymerases η and κ (Yeo et al., 2014). Recently, we reported the ability of human TLS polymerases η and κ to catalyze DNA synthesis past 5fC-mediated conjugates to short peptides (Ji et al., 2018). A combination of gel electrophoresis and mass spectrometry techniques were employed to characterize the efficiency of polymerase bypass past 5fC-protein and polypeptide lesions. Both reversible Schiff base 5fC-histone H2A and H4 cross-links as well reductively stabilized irreversible DNA-histone cross-links were prepared (Ji et al., 2018). Reversible Schiff base DPCs were used directly, while reductively stabilized DPCs were purified by denaturing PAGE, and used for the polymerase bypass study. To investigate the effects of C5-dC conjugated DNA-protein cross-links on DNA replication, primer-template complexes containing unmodified dC, 5fC, or DNA-protein cross-links were subjected to *in vitro* replication in the presence of human DNA polymerases η and κ, and the primer extension products were analyzed by denaturing PAGE (Fig. 9). These experiments revealed that the presence of reversible 5fC-histone cross-links dramatically reduced the amount of full length primer extension products in reactions catalyzed by human lesion bypass polymerases η and κ. Further, primer extension experiments were repeated using gel purified DPC with stable amino linkages (Scheme 3). We found that C5-dC DPC made up of full size histone proteins completely inhibited DNA replication as no full length products were observed for either histone H2A or H4 DPCs (Fig. 9C and D) (Ji et al., 2018). However, when the protein component of the histone DPC was digested to single amino acids using proteinase K, the ability of polymerase η to bypass the lesion was restored (Ji et al., 2018).

We also observed that the 5fC mediated DNA-histone H3 cross-links prepared both by either reductive amination and/or oxime ligation blocked the human TLS polymerases η and κ completely (Ji et al., 2018).

As all the above results indicate, DPCs pose a great risk to cellular machinery. However, the mechanisms by which cells respond to and resolve DPCs in cells remain incompletely understood. Because of the diversity of DPCs in structure, size and context, cells employ a variety of repair mechanisms to resolve different DPCs. One of the most efficient ways to bypass these lesions is to degrade the protein part of the DPC to much smaller polypeptides so that the DNA polymerases can easily bypass these lesions or they can be removed *via* the nucleotide excision repair pathway (NER). Proteolytic degradation of the protein can take place by variety of mechanisms and some are listed below.

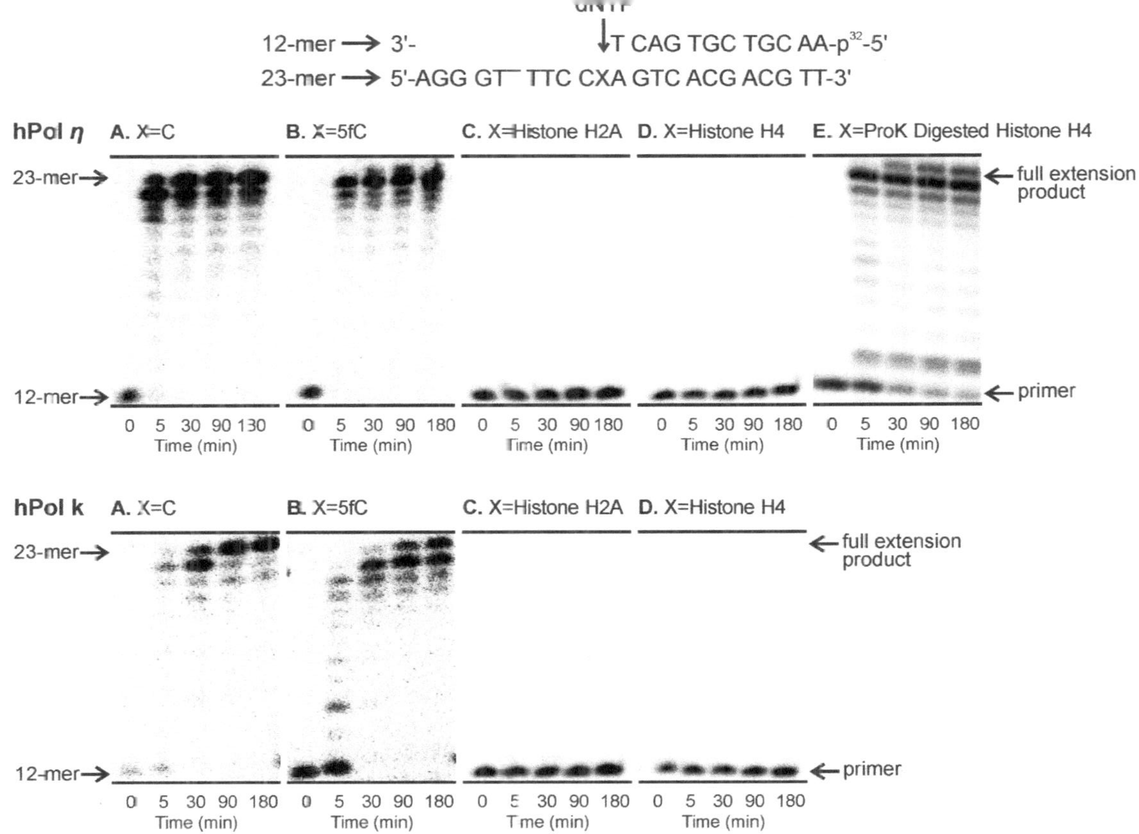

Fig. 9 Primer extension assays for replication bypass of reductively stabilized DNA–protein cross-links by hPol η (top panel) and hPol κ (bottom panel). ^{32}P-labeled 12-mer primer was annealed with 23-mer template containing unmodified dC (A), 5-formyl-dC (B), or C5-dC cross-links to full size histone H2A (E) or histone H4 (F). Polymerase reactions were initiated by addition of DNA polymerases and a mixture of dNTP and quenched at pre-selected time points prior to loading onto 20% denaturing PAGE. *From Ji, S., Fu, I., Naldiga, S., Shao, H., Basu, A. K., Broyde, S., et al. (2018). 5-Formylcytosine mediated DNA-protein cross-links block DNA replication and induce mutations in human cells.* Nucleic Acids Research, *46(13), 6455–6469. https://doi.org/10.1093/nar/gky444, used with permission.*

(A) Cell cycle independent *proteolytic degradation of DPCs*—where cells employ various proteases to reduce the size of the protein component of DPC and allow the DNA replication to proceed.

(B) *Replication coupled proteolysis by metalloproteases*: DNA dependent metalloproteases such as Wss1 in yeasts and its counterpart SPRTN in higher eukaryotes degrade DPC proteins in a replication coupled manner to mitigate the effect of DPCs on replication forks.

(C) *DPC proteolysis by the ubiquitin-proteasome pathway*: The ubiquitin-proteasome system can be involved in degradation of DPC proteins. Many known DPCs such as TOP1ccs, TOP2ccs as well as DPCs containing Polβ or PARP1 have been shown to be degraded by the proteasome (Desai et al., 1997; Mao et al., 2001; Mohni et al., 2019; Prasad, Horton, Dai, & Wilson, 2019; Quiñones et al., 2015; Sun, Saha, Saha, Jo, & Pommier, 2020).

As discussed above, translesion synthesis DNA polymerases η and κ—which are specialized polymerases known to bypass other bulky DNA lesions—were completely blocked by covalent DNA-protein lesions. Unlike replicative polymerases, whose tight active sites undergo conformational changes upon binding the correct dNTP, TLS polymerases of the Y family have more open active sites that enable them to bypass bulky DNA lesions. These specialized polymerases are recruited to stalled replication forks and traverse the damaged sites (Sale, Lehmann, & Woodgate, 2012; Vaisman & Woodgate, 2017). In addition, we have shown that large DPCs completely inhibit T7 RNA polymerase as well (Ji, Park, et al., 2019; Ji, Thomforde, et al., 2019). Although proteolytic cleavage of DPC removes the replication block and allows the polymerases to proceed, the resulting DNA-peptide lesions are potentially mutagenic. Previous investigations have revealed that the ability of DNA polymerases to bypass DNA-peptide conjugates is dependent on the lesion size, the attachment site within the DNA, and polymerase identity (Cho & Guengerich, 2013; Duxin et al., 2014; Kuo, Griffith, & Kreuzer, 2007; Minko et al., 2008; Nakano et al., 2007; Novakova, Kasparkova, Malina, Natile, & Brabec, 2003; Permana & Snapka, 1994; Yamanaka et al., 2010). Lloyd and co-workers (Minko et al., 2008) have reported that human polymerase (hpol) κ and its *Escherichia coli* orthologue pol IV were able to catalyze error-free primer extension past DNA templates containing tetra- and dodecapeptides conjugated to the N^2 position of guanine *via* a trimethylene linker. Polymerase hpol *ν* was capable of bypassing these peptides conjugated to N^6 of adenine *via* a trimethylene linkage, with the correct base (dT) inserted opposite the adduct as well (Yamanaka et al., 2010). In contrast, analogous adducts at the N2 of guanine completely blocked DNA replication (Yamanaka et al., 2010). Guengerich and co-workers (Cho & Guengerich, 2013) reported that human polymerases and κ and ι as well as bacterial polymerases pol T7 and DPO4, were capable of replicating DpCs. Replication of 10-mer peptide (EQKLISEEDL-NH2) conjugated to the C5 position of thymidine generated *via* click chemistry proceeded in error-prone manner. It was found that both hpol κ and η can incorporate either of the four deoxynucleotides opposite the DNA-peptide conjugate, suggesting the possibility for both transition and transversion mutations (Susith Wickramaratne et al., 2015). Further, standing start primer extension past DpCs of dG- and deaza-DHP-dG containing DNA with 10-mer peptide Ac-EQKLISEEDL-NH2 (prepared *via* reductive amination, Scheme 2) resulted in complete extension of primer (20-mer) by hPols η and κ albeit with low efficiencies. In contrast, the 23-mer peptide cross-link completely blocked primer indicating that the ability of TLS polymerases to bypass DpC lesions is a function of peptide size (Susith Wickramaratne et al., 2015).

Basu and others showed the efficiency of translesion synthesis (TLS) past these DpC was 76% compared to that of the unmodified control (Pande et al., 2017). The DpC induced 20% targeted G→A and G→T plus 15% semitargeted mutations, notably at a guanine (G5) five bases 3′ to the lesion site (Pande et al., 2017). We observed targeted C→T transitions and deletion mutations when human translesion synthesis polymerase η (hPol η) bypassed a 5fC-11-mer peptide conjugate in the CXA sequence context (Thomforde et al., 2021). TLS polymerases such as Y-family Pols η, ι, κ, and Rev 1 and B-family Pol ζ lack normal proofreading mechanisms and are known to induce mutations during DNA replication (Thomforde et al., 2021). We also showed that the presence of 5fC DNA–peptide cross-link significantly inhibits transcription in human HEK293T cells but does not induce transcription errors (Ji, Park, et al., 2019; Ji, Thomforde, et al., 2019). Furthermore, local DNA sequence context can have a profound effect on polymerase bypass of nucleobase lesions which has been previously demonstrated (Akagi et al., 2020; Gahlon, Schweizer, & Sturla, 2013; Shriber, Leitner-Dagan, Geacintov, Paz-Elizur, & Livneh, 2015). However, the ability of TLS polymerases to bypass DpC lesions located in different sequence contexts could be different. Previously we reported that, hPol η and κ knockdown in human cells influences replication bypass fidelity of DPCs conjugated at the C7 position of 7-deaza-dG (Pande et al., 2017). To study the effect of DpCs on the polymerase bypass, 5fC mediated

DNA-peptide cross-links were generated *via* reductive amination with DNA containing 5fC and peptides of different lengths (Ji et al., 2018). A combination of gel electrophoresis and mass spectrometry techniques was employed to characterize the efficiency and the fidelity of polymerase bypass past 5fC-polypeptide lesions. Stable DNA-peptide cross-links containing 11-mer peptide RPKPQQFFGLM-CONH2 and 31-mer peptide YGGFMTSEKSQ TPLVTLFKNAIIKNAYKK GE were prepared. In contrast to DNA-histone cross-links, hPol η and κ were able to extend the primer opposite smaller cross-links containing 11-mer peptide (RPKPQQFFGLM-CONH2 and a 31-mer peptide (YGGFMTSEKSQTPLVTL FKNAIIKN AYKKGE) (Fig. 10) (Ji et al., 2018).

Single nucleotide incorporation assays with individual dNTPs revealed that while hPol η inserted either one of the four dNTPs opposite 11-mer peptide-dC conjugates, pol κ predominantly incorporated an incorrect base, dA. Similar results were obtained for substrates containing a larger cross-link to 31-mer peptide, although incorporation efficiency was reduced (Fig. 11) confirming 5fC mediated DNA-peptide lesions interfere with fidelity of TLS DNA polymerases.

To assess the effects of local DNA sequence context on replication, we investigated hPol η-catalyzed *in vitro* replication in the presence of 5fC DpC lesions placed in seven different local DNA sequence contexts (CXC, CXG, CXT, CXA, AXA, GXA, and TXA). Polymerase η was chosen on the basis of previous studies demonstrating its ability to bypass DNA-polypeptide cross-links, while the peptide sequence containing oxy-Lys was derived from the N- terminal region of histone H4 (Thomforde et al., 2021). The DNA-peptide conjugates were prepared by oxime ligation, and primer extension as well as steady-state kinetics experiments were performed in the presence of hPol η, to examine the effects of DNA sequence context on the ability of hPol η to add correct (dGMP) and incorrect nucleotides (dAMP) opposite the DNA-peptide cross-link (Thomforde et al., 2021) (Fig. 12).

We observed that the neighboring bases significantly affected the efficiency of incorporation of dAMP opposite the DpC adduct, while the addition of the correct nucleotide (dGMP) is uniform across the seven sequences examined (Fig. 12). The highest bypass efficiency was seen for a CXT sequence, where the thymine is on the 3′-side of the template base. Incorporation of dAMP opposite DpC was preferred over the correct nucleotide (dGMP), with a misinsertion frequency (f) of 1.92 in a CXA sequence context (Thomforde et al., 2021).

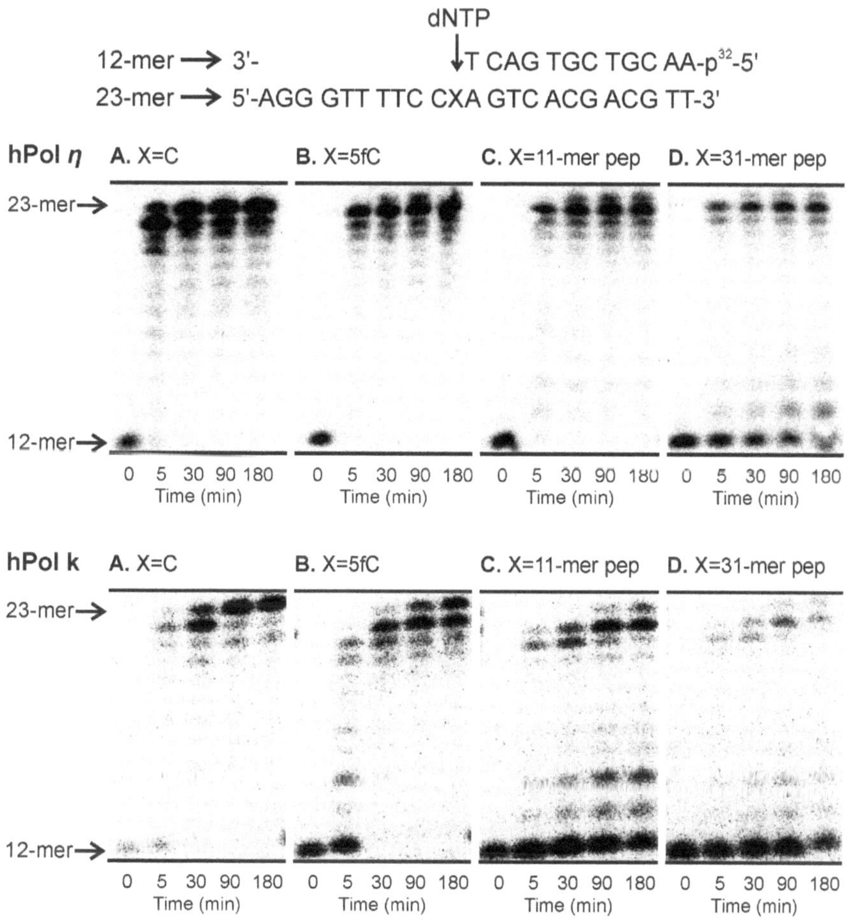

Fig. 10 Primer extension assays for replication bypass of DNA–peptide cross-links by hPol η (top panel) and hPol κ (bottom panel). ^{32}P-labeled 12-mer primer was annealed with 23-mer template containing unmodified dC (A), 5-formyl-dC (B), or C5-dC cross-links to 11-mer peptide RPKPQQFFGLM-CONH2 (C) 31-mer peptide YGGFMTSEKSQTPLVTLFKNAIIKNA YKKGE (D). Polymerase reactions were initiated by addition of DNA polymerases and a mixture of dNTP and quenched at pre-selected time points prior to loading onto 20% denaturing PAGE. *From Ji, S., Fu, I., Naldiga, S., Shao, H., Basu, A. K., Broyde, S., et al. (2018). 5-Formylcytosine mediated DNA-protein cross-links block DNA replication and induce mutations in human cells.* Nucleic Acids Research, *46(13), 6455–6469. https://doi.org/10.1093/nar/gky444, used with permission.*

6.1 Equipment

- Digital dry bath (Labnet)
- Sequencing gel electrophoresis chamber (Thermo Fisher)
- Voltage generator (Bio-Rad)

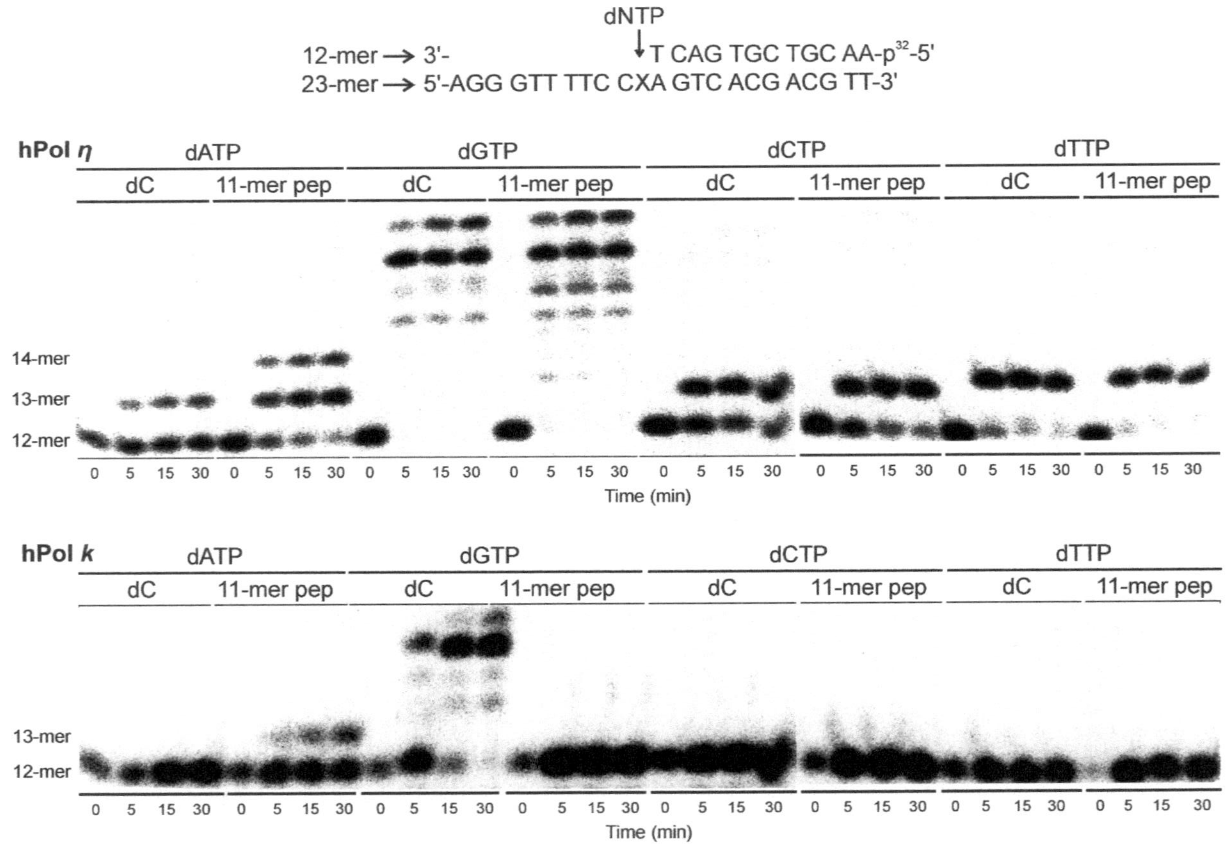

Fig. 11 Single nucleotide insertion opposite unmodified dC (control), and 5fC-11-mer peptide cross-links (X = RPKPQQFFGLM-CONH2). ^{32}P-ATP labeled primer-template complexes were incubated with hPol η (top panel) or hPol κ in the presence of 50uM individual dNTP and then quenched at pre-selected times. *From Ji, S., Fu, I., Naldiga, S., Shao, H., Basu, A. K., Broyde, S., et al. (2018). 5-Formylcytosine mediated DNA-protein cross-links block DNA replication and induce mutations in human cells.* Nucleic Acids Research, *46(13), 6455–6469. https://doi.org/10.1093/nar/gky444, used with permission.*

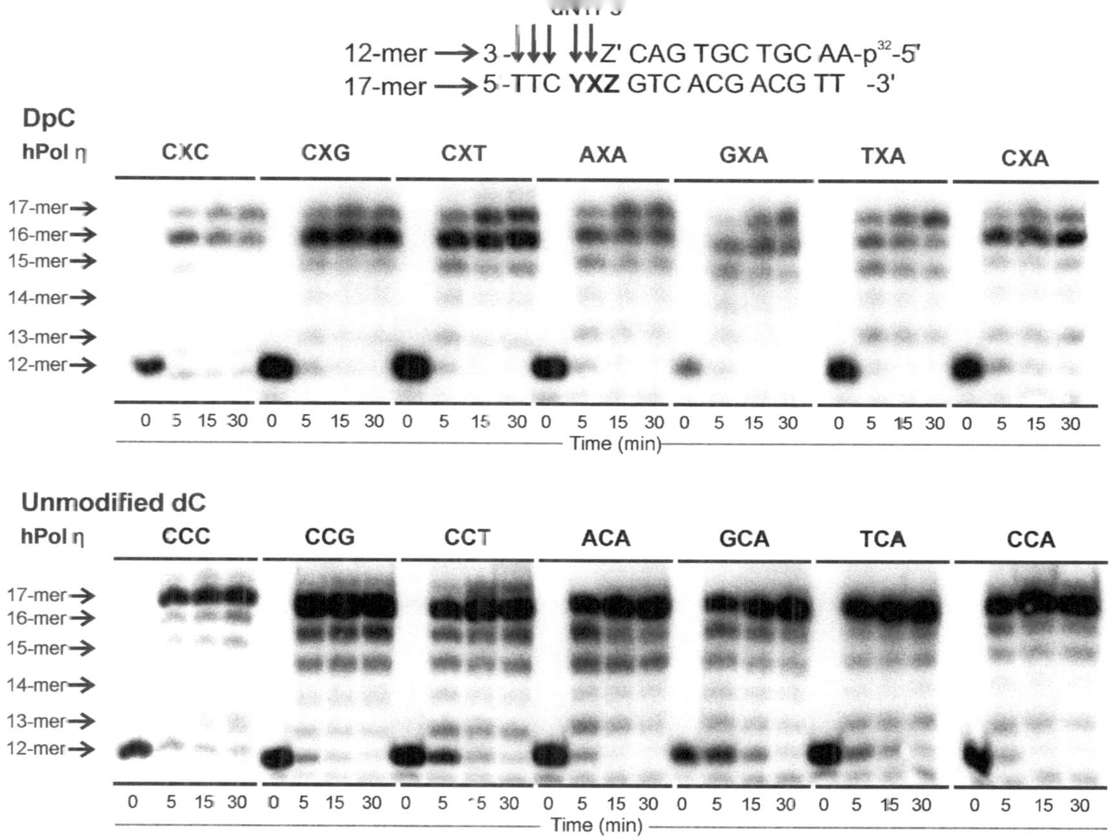

Fig. 12 Primer extension assays for replication bypass of covalent DNA-peptide cross-links by hPol η. The ^{32}P-labeled 12-mer primer was annealed with 17-mer DNA containing 5fC conjugated to an 11-mer peptide (X = DpC) (top) or unmodified dC (X = dC) (bottom) *via* oxime ligation. YXZ corresponds to the sequence context used, as listed above the X-axis. Z′ corresponds to the nucleotide complementary to Z in each sequence context. Primer extension reactions were initiated by the addition of hPol η and a mixture of dNTPs. Reactions were quenched at specific time points (0, 5, 15, and 30 min), and mixtures loaded onto a 20% PAGE gel containing 7 M urea. *From Thomforde, J., Fu, I., Rodriguez, F., Pujari, S. S., Broyde, S., & Tretyakova, N. (2021). Translesion synthesis past 5-formylcytosine-mediated DNA–peptide cross-links by hpolη is dependent on the local DNA sequence.* Biochemistry*, 60(23), 1797–1807. https://doi.org/10.1021/acs.biochem.1c00130, used with permission.*

- Illustra Micro Bio-spin G-25 column (GE Healthcare, Pittsburgh, PA)
- Speedvac equipment
- Phosphorimager plate
- Typhoon FLA 7000 instrument (GE Healthcare, Pittsburgh, PA)

6.2 Buffers and reagents

- Oligodeoxynucleotide template strands (23-mers) containing C, 5fC, DpC, or DPC at position X, 5′-d(AGG GTT TTC CXA GTC ACG ACG TT-3′)
- Oligodeoxynucleotide primer (12-mer complimentary to the 3′-end of the template, 5′-d(AAC GTC GTG ACT-3′)
- γ-^{32}P ATP (PerkinElmer Life Sciences, Boston, MA)
- T4 PNK buffer (New England Biolabs, Beverly, MA)
- Standard dNTP mixture containing ATP, GTP, CTP and TTP
- Recombinant human polymerase η
- Recombinant human polymerase κ
- 50 mM Tris-HCl (pH7.5) 5 mM DTT, 5 mM $MgCl_2$, 50 mM NaCl, 100 μg/mL BSA and 10% glycerol (*v*/v)
- Sterile H_2O (Invitrogen)
- 20% PAGE
- Urea (Sigma Aldrich)
- 1 × TBE buffer
- Ammonium persulfate (Across organics)
- *N*, *N*, *N*′, *N*′-Tetramethylethylenediamine (TEMED) (Sigma)
- LDS gel loading buffer (20 mM EDTA in 95% formamide containing 0.05% bromophenol blue and xylene cyanol)

6.3 Procedure

1. Incubate the synthetic DNA 12-mer primer (500 pmol, 5′-d(AAC GTC GTG ACT-3′) with γ-^{32}P-ATP (2 μL, PerkinElmer Life Sciences, Boston, MA) and T4 PNK (20 units) in 20 μL 1 × T4 PNK buffer at 37 °C for 1 h.
2. After 1 h, heat the reaction mixture at 65 °C for 10 min and filter through an Illustra Micro Bio-spin G-25 column to remove excess of γ-^{32}P ATP.
3. Mix radiolabeled 12-mer primer (30 pmol) with 1.5 eq. of 23-mer DNA template strands containing either site-specific DNA-protein or DNA-peptide cross-links 5′-d(AGG GTT TTC CXA GTC

ACG ACG TT-3′, where X is either dC, 5fC, peptide adducted to DNA or protein adducted to DNA) in 10 mM Tris-HCl buffer (pH 8.0) and 50 mM NaCl.

4. Heat the mixtures at 90 °C for 10 min and allow to slowly cool down to room temperature overnight to generate primer–template complexes for primer extension assays.
5. Next day, incubate the ^{32}P-labeled primer-template duplexes containing unmodified dC, 5-formyl-dC, DNA-peptide or DNA-protein cross-links (1 pmol) with human DNA polymerase η and κ (2 pmol) in the presence of 50 mM Tris-HCl (pH 7.5), 5 mM DTT, 5 mM MgC12, 50 mM NaCl, 100 μg/mL BSA and 10% glycerol (v/v) (total volume, 30 μL) at 37 °C.
6. Prepare the 20% PAGE sequencing gel and keep pre-running the gel at 1900 V for 2 h at room temperature.
7. Initiate the polymerization by addition of all four dNTPs at 1 mM final concentration at 37 °C
8. Take 4 μL aliquots of the reaction mixture at pre-selected time points (0, 5, 30, 90, 180 min) and quench with 4 μL of gel loading buffer (20 mM EDTA in 95% formamide containing 0.05% bromophenol blue and xylene cyanol).
9. Heat the reaction aliquots at 90 °C for 5 min, spin down for 30 s and load them onto 20% denaturing PAGE containing 7 M urea.
10. Run the gel at 1900 V for 3 h in 1 × TBE buffer. After the gel run is finished (by visualizing the xylene cyanol band migrated to the lower middle part of the gel), take out the gel plate keeping gel on one plate.
11. Wrap the gel with saran wrap completely and place a clean phosphorimager plate on the gel and place the gel in dark to image for overnight.
12. Next day visualize the phosphorimager plate on a Typhoon FLA 7000 and detect the radiolabeled extension products, analyze and quantify using ImageQuant TL 8.0.

6.4 Notes

1. Disassemble the gel only when the gel plate is warm, otherwise the gel will stick to the glass plate making it hard to separate and gel may break.
2. Prior running the gel, siliconize either one or both glass plates with silicone oil.
3. All the experiments should be conducted in a radiation room.

4. The polymerase reaction can be initiated either by adding the dNTPs or by adding polymerase depending on the convenience of the assay. Kinetic parameters for nucleotide addition opposite the lesion can be obtained by conducting steady state kinetics experiments with individual dNTPs.

7. Summary and conclusions

In conclusion, robust methodologies to prepare DNA-protein and DNA-peptide cross-links have been developed. The more advantageous oxime ligation method was employed to prepare site-specific cross-links with high yields, and to avoid the addition of external agents to stabilize the cross-links. In contrast, in reductive amination method, $NaCNBH_3$ is used to stabilize the cross-links and yields are low to moderate lacking site-specificity in regard to protein/peptide. We and other laboratories reported that 5fC-mediated DPCs have a profound effect on DNA replication. Mostly, nearly all of the DNA-protein cross-links completely block the TLS human polymerases. The shorter DNA-polypeptide conjugates which are remnants of proteolytic degradation of the protein part of DPC have been effectively bypassed irrespective of the sequence composition and methods of DpC construction. Human DNA polymerase η catalyzes replication bypass of 5fC-mediated DNA-peptide cross-links attached to the C-5 position of dC and the N-7 of dG, but the outcomes vary considerably depending on the length of the peptide attached to DNA and local DNA sequence on either side of the DpC lesion. The neighboring bases significantly affect the efficiency of incorporation of dAMP opposite the DpC adducts.

Acknowledgments

We would like to thank Robert Carlson (University of Minnesota Cancer Center) for his help preparing the figures and acquiring the reprint permissions. This work is supported by an R01 grant from the NEIHS (ES-023350).

References

Akagi, J.-I., Hashimoto, K., Suzuki, K., Yokoi, M., de Wind, N., Iwai, S., et al. (2020). Effect of sequence context on Polζ-dependent error-prone extension past (6-4) photoproducts. *DNA Repair*, *87*, 102771. https://doi.org/10.1016/j.dnarep.2019.102771.

Alexander, P., & Moroson, H. (1962). Cross-linking of deoxyribonucleic acid to protein following ultra-violet irradiation of different cells. *Nature*, *194*(4831), 882–883. https://doi.org/10.1038/194882a0.

Amblard, F., Cho, J. H., & Schinazi, R. F. (2009). Cu(I)-catalyzed Huisgen azide-alkyne 1,3-dipolar cycloaddition reaction in nucleoside, nucleotide, and oligonucleotide chemistry. *Chemical Reviews*, *109*(9), 4207–4220. https://doi.org/10.1021/cr9001462.

Brambilla, G., Sciabà, L., Faggin, P., Finollo, R., Bassi, A. M., Ferro, M., et al. (1985). Methylglyoxal-induced DNA-protein cross-links and cytotoxicity in Chinese hamster ovary cells. *Carcinogenesis*, *6*(5), 683–686. https://doi.org/10.1093/carcin/6.5.683.

Cho, S.-H., & Guengerich, F. P. (2013). Replication past the butadiene diepoxide-derived DNA adduct S-[4-(N(6)-deoxyadenosinyl)-2,3-dihydroxybutyl]glutathione by DNA polymerases. *Chemical Research in Toxicology*, *26*(6), 1005–1013. https://doi.org/10.1021/tx400145e.

Connelly, J. C., & Leach, D. R. (2004). Repair of DNA covalently linked to protein. *Molecular Cell*, *13*(3), 307–316. https://doi.org/10.1016/s1097-2765(04)00056-5.

Desai, S. D., Liu, L. F., Vazquez-Abad, D., & D'Arpa, P. (1997). Ubiquitin-dependent destruction of topoisomerase I is stimulated by the antitumor drug camptothecin. *Journal of Biological Chemistry*, *272*(39), 24159–24164. https://doi.org/10.1074/jbc.272.39.24159.

Dizdaroglu, M., Gajewski, E., Reddy, P., & Margolis, S. A. (1989). Structure of a hydroxyl radical-induced DNA-protein crosslink involving thymine and tyrosine in nucleohistone. *Biochemistry*, *28*(8), 3625–3628. https://doi.org/10.1021/bi00434a071.

Dorr, B. M., Ham, H. O., An, C., Chaikof, E. L., & Liu, D. R. (2014). Reprogramming the specificity of sortase enzymes. *Proceedings National Academy of Sciences United States of America*, *111*(37), 13343–13348. https://doi.org/10.1073/pnas.1411179111.

Duxin, J. P., Dewar, J. M., Yardimci, H., & Walter, J. C. (2014). Repair of a DNA-protein crosslink by replication-coupled proteolysis. *Cell*, *159*(2), 346–357. https://doi.org/10.1016/j.cell.2014.09.024.

Gahlon, H. L., Schweizer, W. B., & Sturla, S. J. (2013). Tolerance of base pair size and shape in postlesion DNA synthesis. *Journal of the American Chemical Society*, *135*(17), 6384–6387. https://doi.org/10.1021/ja311434s.

Gajewski, E., & Dizdaroglu, M. (1990). Hydroxyl radical induced cross-linking of cytosine and tyrosine in nucleohistone. *Biochemistry*, *29*(4), 977–980. https://doi.org/10.1021/bi00456a020.

Gates, K. S. (2009). An overview of chemical processes that damage cellular DNA: Spontaneous hydrolysis, alkylation, and reactions with radicals. *Chemical Research in Toxicology*, *22*(11), 1747–1760. https://doi.org/10.1021/tx900242k.

Gates, K. S., Nooner, T., & Dutta, S. (2004). Biologically relevant chemical reactions of N7-alkylguanine residues in DNA. *Chemical Research in Toxicology*, *17*(7), 839–856. https://doi.org/10.1021/tx049965c.

Gherezghiher, T. B., Ming, X., Villalta, P. W., Campbell, C., & Tretyakova, N. Y. (2013). 1,2,3,4-Diepoxybutane-induced DNA–protein cross-linking in human fibrosarcoma (HT1080) cells. *Journal of Proteome Research*, *12*(5), 2151–2164. https://doi.org/10.1021/pr3011974.

Groehler, A., Degner, A., & Tretyakova, N. Y. (2017). Mass spectrometry-based tools to characterize DNA protein cross-linking by bis-electrophiles. *Basic & Clinical Pharmacology & Toxicology*, *121*(Suppl. 3), 63–77. https://doi.org/10.1111/bcpt.12751.

Groehler, A., Villalta, P. W., Campbell, C., & Tretyakova, N. (2016). Covalent DNA–protein cross-linking by phosphoramide mustard and nornitrogen mustard in human cells. *Chemical Research in Toxicology*, *29*(2), 190–202. https://doi.org/10.1021/acs.chemrestox.5b00430.

Hockensmith, J. W., Kubasek, W. L., Evertsz, E. M., Mesner, L. D., & von Hippel, P. H. (1993). Laser cross-linking of proteins to nucleic acids. II. Interactions of the bacteriophage T4 DNA replication polymerase accessory proteins complex with DNA. *Journal of Biological Chemistry*, *268*(21), 15721–15730.

Ide, H., Shoulkamy, M. I., Nakano, T., Miyamoto-Matsubara, M., & Salem, A. M. (2011). Repair and biochemical effects of DNA-protein crosslinks. *Mutation Research*, *711*(1–2), 113–122. https://doi.org/10.1016/j.mrfmmm.2010.12.007.

Izzotti, A., Cartiglia, C., Taningher, M., De Flora, S., & Balansky, R. (1999). Age-related increases of 8-hydroxy-2′-deoxyguanosine and DNA-protein crosslinks in mouse organs. *Mutation Research*, *446*(2), 215–223. https://doi.org/10.1016/s1383-5718(99)00189-8.

Ji, S., Fu, I., Naldiga, S., Shao, H., Basu, A. K., Broyde, S., et al. (2018). 5-Formylcytosine mediated DNA-protein cross-links block DNA replication and induce mutations in human cells. *Nucleic Acids Research*, *46*(13), 6455–6469. https://doi.org/10.1093/nar/gky444.

Ji, S., Park, D., Kropachev, K., Kolbanovskiy, M., Fu, I., Broyde, S., et al. (2019). 5-Formylcytosine-induced DNA-peptide cross-links reduce transcription efficiency, but do not cause transcription errors in human cells. *Journal of Biological Chemistry*, *294*(48), 18387–18397. https://doi.org/10.1074/jbc.RA119.009834.

Ji, S., Shao, H., Han, Q., Seiler, C. L., & Tretyakova, N. Y. (2017). Reversible DNA–protein cross-linking at epigenetic DNA marks. *Angewandte Chemie International Edition*, *56*(45), 14130–14134. https://doi.org/10.1002/anie.201708286.

Ji, S., Thomforde, J., Rogers, C., Fu, I., Broyde, S., & Tretyakova, N. Y. (2019). Transcriptional bypass of DNA–protein and DNA–peptide conjugates by T7 rna polymerase. *ACS Chemical Biology*, *14*(12), 2564–2575. https://doi.org/10.1021/acschembio.9b00365.

Johansen, M. E., Muller, J. G., Xu, X., & Burrows, C. J. (2005). Oxidatively induced DNA-protein cross-linking between single-stranded binding protein and oligodeoxynucleotides containing 8-oxo-7,8-dihydro-2′-deoxyguanosine. *Biochemistry*, *44*(15), 5660–5671. https://doi.org/10.1021/bi047580n.

Klages-Mundt, N. L., & Li, L. (2017). Formation and repair of DNA-protein crosslink damage. *Science China. Life Sciences*, *60*(10), 1065–1076. https://doi.org/10.1007/s11427-017-9183-4.

Kölmel, D. K., & Kool, E. T. (2017). Oximes and hydrazones in bioconjugation: Mechanism and catalysis. *Chemical Reviews*, *117*(15), 10358–10376. https://doi.org/10.1021/acs.chemrev.7b00090.

Kuo, H. K., Griffith, J. D., & Kreuzer, K. N. (2007). 5-Azacytidine induced methyltransferase-DNA adducts block DNA replication in vivo. *Cancer Research*, *67*(17), 8248–8254. https://doi.org/10.1158/0008-5472.Can-07-1038.

Lai, Y., Yu, R., Hartwell, H. J., Moeller, B. C., Bodnar, W. M., & Swenberg, J. A. (2016). Measurement of endogenous versus exogenous formaldehyde-induced DNA-protein crosslinks in animal tissues by stable isotope labeling and ultrasensitive mass spectrometry. *Cancer Research*, *76*(9), 2652–2661. https://doi.org/10.1158/0008-5472.CAN-15-2527.

Larsen, N. B., Gao, A. O., Sparks, J. L., Gallina, I., Wu, R. A., Mann, M., et al. (2019). Replication-coupled DNA-protein crosslink repair by SPRTN and the proteasome in xenopus egg extracts. *Molecular Cell*, *73*(3), 574–588. e577 https://doi.org/10.1016/j.molcel.2018.11.024.

Li, F., Zhang, Y., Bai, J., Greenberg, M. M., Xi, Z., & Zhou, C. (2017). 5-Formylcytosine yields DNA–protein cross-links in nucleosome core particles. *Journal of the American Chemical Society*, *139*(31), 10617–10620. https://doi.org/10.1021/jacs.7b05495.

Loeber, R. L., Michaelson-Richie, E. D., Codreanu, S. G., Liebler, D. C., Campbell, C. R., & Tretyakova, N. Y. (2009). Proteomic analysis of DNA – protein cross-linking by antitumor nitrogen mustards. *Chemical Research in Toxicology*, *22*(6), 1151–1162. https://doi.org/10.1021/tx900078y.

Mao, Y., Desai, S. D., Ting, C. Y., Hwang, J., & Liu, L. F. (2001). 26 S proteasome-mediated degradation of topoisomerase II cleavable complexes. *Journal of Biological Chemistry*, *276*(44), 40652–40658. https://doi.org/10.1074/jbc.M104009200.

Maskey, R. S., Flatten, K. S., Sieben, C. J., Peterson, K. L., Baker, D. J., Nam, H. J., et al. (2017). Spartan deficiency causes accumulation of topoisomerase 1 cleavage complexes and tumorigenesis. *Nucleic Acids Research*, *45*(8), 4564–4576. https://doi.org/10.1093/nar/gkx107.

Mattagajasingh, S. N., & Misra, H. P. (1996). Mechanisms of the carcinogenic chromium(VI)-induced DNA-protein cross-linking and their characterization in cultured intact human cells. *Journal of Biological Chemistry*, *271*(52), 33550–33560. https://doi.org/10.1074/jbc.271.52.33550.

Michaelson-Richie, E. D., Loeber, R. L., Codreanu, S. G., Ming, X., Liebler, D. C., Campbell, C., et al. (2010). DNA−protein cross-linking by 1,2,3,4-diepoxybutane. *Journal of Proteome Research*, *9*(9), 4356–4367. https://doi.org/10.1021/pr1000835.

Michaelson-Richie, E. D., Ming, X., Codreanu, S. G., Loeber, R. L., Liebler, D. C., Campbell, C., et al. (2011). Mechlorethamine-induced DNA–protein cross-linking in human fibrosarcoma (ht1080) cells. *Journal of Proteome Research*, *10*(6), 2785–2796. https://doi.org/10.1021/pr200042u.

Minko, I. G., Yamanaka, K., Kozekov, I. D., Kozekova, A., Indiani, C., O'Donnell, M. E., et al. (2008). Replication bypass of the acrolein-mediated deoxyguanine DNA-peptide cross-links by DNA polymerases of the dinb family. *Chemical Research in Toxicology*, *21*(10), 1983–1990. https://doi.org/10.1021/tx800174a.

Mohni, K. N., Wessel, S. R., Zhao, R., Wojciechowski, A. C., Luzwick, J. W., Layden, H., et al. (2019). HMCES maintains genome integrity by shielding abasic sites in single-strand DNA. *Cell*, *176*(1–2), 144–153. e113 https://doi.org/10.1016/j.cell.2018.10.055.

Moss, T., Dimitrov, S. I., & Houde, D. (1997). UV-laser crosslinking of proteins to DNA. *Methods*, *11*(2), 225–234. https://doi.org/10.1006/meth.1996.0409.

Nakano, T., Morishita, S., Katafuchi, A., Matsubara, M., Horikawa, Y., Terato, H., et al. (2007). Nucleotide excision repair and homologous recombination systems commit differentially to the repair of DNA-protein crosslinks. *Molecular Cell*, *28*(1), 147–158. https://doi.org/10.1016/j.molcel.2007.07.029.

Nakano, T., Ouchi, R., Kawazoe, J., Pack, S. P., Makino, K., & Ide, H. (2012). T7 RNA polymerases backed up by covalently trapped proteins catalyze highly error prone transcription. *Journal of Biological Chemistry*, *287*(9), 6562–6572. https://doi.org/10.1074/jbc.M111.318410.

Neale, M. J., Pan, J., & Keeney, S. (2005). Endonucleolytic processing of covalent protein-linked DNA double-strand breaks. *Nature*, *436*(7053), 1053–1057. https://doi.org/10.1038/nature03872.

Novakova, O., Kasparkova, J., Malina, J., Natile, G., & Brabec, V. (2003). DNA-protein cross-linking by trans-[PtCl(2)(E-iminoether)(2)]. A concept for activation of the trans geometry in platinum antitumor complexes. *Nucleic Acids Research*, *31*(22), 6450–6460. https://doi.org/10.1093/nar/gkg863.

Pachva, M. C., Kisselev, A. F., Matkarimov, B. T., Saparbaev, M., & Groisman, R. (2020). DNA-histone cross-links: Formation and repair. *Frontiers in Cell and Development Biology*, *8*(1627). https://doi.org/10.3389/fcell.2020.607045.

Pande, P., Ji, S., Mukherjee, S., Schärer, O. D., Tretyakova, N. Y., & Basu, A. K. (2017). Mutagenicity of a model DNA-peptide cross link in human cells: Roles of translesion synthesis DNA polymerases. *Chemical Research in Toxicology*, *30*(2), 669–677. https://doi.org/10.1021/acs.chemrestox.6b00397.

Patierno, S. R., Sugiyama, M., Basilion, J. P., & Costa, M. (1985). Preferential DNA-protein cross-linking by NiCl2 in magnesium-insoluble regions of fractionated Chinese hamster ovary cell chromatin. *Cancer Research*, *45*(11 Pt 2), 5787–5794.

Permana, P. A., & Snapka, R. M. (1994). Aldehyde-induced protein-DNA crosslinks disrupt specific stages of SV40 DNA replication. *Carcinogenesis*, *15*(5), 1031–1036. https://doi.org/10.1093/carcin/15.5.1031.

Pommier, Y., Barcelo, J. M., Rao, V. A., Sordet, O., Jobson, A. G., Thibaut, L., et al. (2006). Repair of topoisomerase I-mediated DNA damage. *Progress in Nucleic Acid Research and Molecular Biology, 81*, 179–229. https://doi.org/10.1016/s0079-6603(06)81005-6.

Pouliot, J. J., Yao, K. C., Robertson, C. A., & Nash, H. A. (1999). Yeast gene for a Tyr-DNA phosphodiesterase that repairs topoisomerase I complexes. *Science, 286*(5439), 552–555. https://doi.org/10.1126/science.286.5439.552.

Prasad, R., Horton, J. K., Dai, D.-P., & Wilson, S. H. (2019). Repair pathway for PARP-1 DNA-protein crosslinks. *DNA Repair, 73*, 71–77. https://doi.org/10.1016/j.dnarep.2018.11.004.

Pujari, S. S., Zhang, Y., Ji, S., Distefano, M. D., & Tretyakova, N. Y. (2018). Site-specific cross-linking of proteins to DNA via a new bioorthogonal approach employing oxime ligation. *Chemical Communications, 54*(49), 6296–6299. https://doi.org/10.1039/C8CC01300D.

Quiñones, J. L., Thapar, U., Yu, K., Fang, Q., Sobol, R. W., & Demple, B. (2015). Enzyme mechanism-based, oxidative DNA–protein cross-links formed with DNA polymerase β in vivo. *Proceedings of the National Academy of Sciences of the United States of America, 112*(28), 8602–8607. https://doi.org/10.1073/pnas.1501101112.

Raiber, E.-A., Portella, G., Martínez Cuesta, S., Hardisty, R., Murat, P., Li, Z., et al. (2018). 5-Formylcytosine organizes nucleosomes and forms Schiff base interactions with histones in mouse embryonic stem cells. *Nature Chemistry, 10*(12), 1258–1266. https://doi.org/10.1038/s41557-018-0149-x.

Sale, J. E., Lehmann, A. R., & Woodgate, R. (2012). Y-family DNA polymerases and their role in tolerance of cellular DNA damage. *Nature Reviews Molecular Cell Biology, 13*(3), 141–152. https://doi.org/10.1038/nrm3289.

Shriber, P., Leitner-Dagan, Y., Geacintov, N., Paz-Elizur, T., & Livneh, Z. (2015). DNA sequence context greatly affects the accuracy of bypass across an ultraviolet light 6-4 photoproduct in mammalian cells. *Mutation Research, 780*, 71–76. https://doi.org/10.1016/j.mrfmmm.2015.08.002.

Sparks, J. L., Chistol, G., Gao, A. O., Räschle, M., Larsen, N. B., Mann, M., et al. (2019). The CMG helicase bypasses DNA-protein cross-links to facilitate their repair. *Cell, 176*(1–2), 167–181. e121 https://doi.org/10.1016/j.cell.2018.10.053.

Spruijt, C. G., Gnerlich, F., Smits, A. H., Pfaffeneder, T., Jansen, P. W. T. C., Bauer, C., et al. (2013). Dynamic readers for 5-(hydroxy)methylcytosine and its oxidized derivatives. *Cell, 152*(5), 1146–1159. https://doi.org/10.1016/j.cell.2013.02.004.

Stingele, J., Bellelli, R., & Boulton, S. J. (2017). Mechanisms of DNA–protein crosslink repair. *Nature Reviews Molecular Cell Biology, 18*(9), 563–573. https://doi.org/10.1038/nrm.2017.56.

Summerfield, F. W., & Tappel, A. L. (1984). Cross-linking of DNA in liver and testes of rats fed 1,3-propanediol. *Chemico-Biological Interactions, 50*(1), 87–96. https://doi.org/10.1016/0009-2797(84)90134-0.

Sun, Y., Saha, L. K., Saha, S., Jo, U., & Pommier, Y. (2020). Debulking of topoisomerase DNA-protein crosslinks (TOP-DPC) by the proteasome, non-proteasomal and non-proteolytic pathways. *DNA Repair, 94*, 102926. https://doi.org/10.1016/j.dnarep.2020.102926.

Takemori, A., Butcher, D. S., Harman, V. M., Brownridge, P., Shima, K., Higo, D., et al. (2020). PEPPI-MS: Polyacrylamide-gel-based prefractionation for analysis of intact proteoforms and protein complexes by mass spectrometry. *Journal of Proteome Research, 19*(9), 3779–3791. https://doi.org/10.1021/acs.jproteome.0c00303.

Thomforde, J., Fu, I., Rodriguez, F., Pujari, S. S., Broyde, S., & Tretyakova, N. (2021). Translesion synthesis past 5-formylcytosine-mediated DNA–peptide cross-links by hpolη is dependent on the local DNA sequence. *Biochemistry, 60*(23), 1797–1807. https://doi.org/10.1021/acs.biochem.1c00130.

Tretyakova, N. Y., Groehler, A.t., & Ji, S. (2015). DNA-protein cross-links: Formation, structural identities, and biological outcomes. *Accounts of Chemical Research*, *48*(6), 1631–1644. https://doi.org/10.1021/acs.accounts.5b00056.

Vaisman, A., & Woodgate, R. (2017). Translesion DNA polymerases in eukaryotes: What makes them tick? *Critical Reviews in Biochemistry and Molecular Biology*, *52*(3), 274–303. https://doi.org/10.1080/10409238.2017.1291576.

Vaz, B., Popovic, M., Newman, J. A., Fielden, J., Aitkenhead, H., Halder, S., et al. (2016). Metalloprotease SPRTN/DVC1 orchestrates replication-coupled DNA-protein crosslink repair. *Molecular Cell*, *64*(4), 704–719. https://doi.org/10.1016/j.molcel.2016.09.032.

Vaz, B., Popovic, M., & Ramadan, K. (2017). DNA-protein crosslink proteolysis repair. *Trends in Biochemical Sciences*, *42*(6), 483–495. https://doi.org/10.1016/j.tibs.2017.03.005.

Wickramaratne, S., Boldry, E. J., Buehler, C., Wang, Y.-C., Distefano, M. D., & Tretyakova, N. Y. (2015). Error-prone translesion synthesis past DNA-peptide cross-links conjugated to the major groove of DNA via C5 of thymidine. *Journal of Biological Chemistry*, *290*(2), 775–787. https://doi.org/10.1074/jbc.M114.613638.

Wickramaratne, S., Ji, S., Mukherjee, S., Su, Y., Pence, M. G., Lior-Hoffmann, L., et al. (2016). Bypass of DNA-protein cross-links conjugated to the 7-deazaguanine position of DNA by translesion synthesis polymerases. *Journal of Biological Chemistry*, *291*(45), 23589–23603. https://doi.org/10.1074/jbc.M116.745257.

Wickramaratne, S., Mukherjee, S., Villalta, P. W., Schärer, O. D., & Tretyakova, N. Y. (2013). Synthesis of sequence-specific DNA-protein conjugates via a reductive amination strategy. *Bioconjugate Chemistry*, *24*(9), 1496–1506. https://doi.org/10.1021/bc400018u.

Wu, F. Y., Lee, Y. J., Chen, D. R., & Kuo, H. W. (2002). Association of DNA-protein crosslinks and breast cancer. *Mutation Research*, *501*(1–2), 69–78. https://doi.org/10.1016/s0027-5107(02)00006-4.

Xu, X., Muller, J. G., Ye, Y., & Burrows, C. J. (2008). DNA–protein cross-links between guanine and lysine depend on the mechanism of oxidation for formation of C5 vs C8 guanosine adducts. *Journal of the American Chemical Society*, *130*(2), 703–709. https://doi.org/10.1021/ja077102a.

Yamanaka, K., Minko, I. G., Takata, K.-I., Kolbanovskiy, A., Kozekov, I. D., Wood, R. D., et al. (2010). Novel enzymatic function of DNA polymerase ν in translesion DNA synthesis past major groove DNA–peptide and DNA–DNA cross-links. *Chemical Research in Toxicology*, *23*(3), 689–695. https://doi.org/10.1021/tx900449u.

Yang, S. W., Burgin, A. B., Huizenga, B. N., Robertson, C. A., Yao, K. C., & Nash, H. A. (1996). A eukaryotic enzyme that can disjoin dead-end covalent complexes between DNA and type I topoisomerases. *Proceedings of the National Academy of Sciences of the United States of America*, *93*(21), 11534–11539. https://doi.org/10.1073/pnas.93.21.11534.

Yeo, J. E., Wickramaratne, S., Khatwani, S., Wang, Y.-C., Vervacke, J., Distefano, M. D., et al. (2014). Synthesis of site-specific DNA–protein conjugates and their effects on DNA replication. *ACS Chemical Biology*, *9*(8), 1860–1868. https://doi.org/10.1021/cb5001795.

Zhitkovich, A., Voitkun, V., Kluz, T., & Costa, M. (1998). Utilization of DNA-protein cross-links as a biomarker of chromium exposure. *Environmental Health Perspectives*, *106*(Suppl. 4), 969–974. https://doi.org/10.1289/ehp.98106s4969.

CHAPTER EIGHTEEN

An efficient chemical screening method for structure-based inhibitors to nucleic acid enzymes targeting the DNA repair-replication interface and SARS CoV-2

Davide Moiani[a,b], Todd M. Link[a,b], Chris A. Brosey[a,b], Panagiotis Katsonis[c], Olivier Lichtarge[c], Youngchang Kim[d,e], Andrzej Joachimiak[d,e], Zhijun Ma[f], In-Kwon Kim[f], Zamal Ahmed[a,b], Darin E. Jones[g], Susan E. Tsutakawa[h,*], and John A. Tainer[a,b,h,*]

[a]Department of Cancer Biology, University of Texas MD Anderson Cancer Center, Houston, TX, United States
[b]Department of Molecular & Cellular Oncology, University of Texas MD Anderson Cancer Center, Houston, TX, United States
[c]Department of Molecular and Human Genetics, Baylor College of Medicine, Houston, TX, United States
[d]Center for Structural Genomics of Infectious Diseases, Consortium for Advanced Science and Engineering, University of Chicago, Chicago, IL, United States
[e]Department of Biochemistry and Molecular Biology, University of Chicago, Chicago, IL, United States
[f]Department of Chemistry, University of Cincinnati, Cincinnati, OH, United States
[g]Department of Pharmaceutical Sciences, The University of Arkansas for Medical Sciences, Little Rock, AR, United States
[h]Biophysics and Integrated Bioimaging, Lawrence Berkeley National Laboratory, Berkeley, CA, United States
*Corresponding authors: e-mail address: setsutakawa@lbl.gov; jtainer@mdanderson.org

Contents

Methods in Enzymology, Volume 661
ISSN 0076-6879
https://doi.org/10.1016/bs.mie.2021.09.003

Abstract

We present a Chemistry and Structure Screen Integrated Efficiently (CASSIE) approach (named for Greek prophet Cassandra) to design inhibitors for cancer biology and pathogenesis. CASSIE provides an effective path to target master keys to control the repair-replication interface for cancer cells and SARS CoV-2 pathogenesis as exemplified here by specific targeting of Poly(ADP-ribose) glycohydrolase (PARG) and ADP-ribose glycohydrolase ARH3 macrodomains plus SARS CoV-2 nonstructural protein 3 (Nsp3) Macrodomain 1 (Mac1) and Nsp15 nuclease. As opposed to the classical massive effort employing libraries with large numbers of compounds against single proteins, we make inhibitor design for multiple targets efficient. Our compact, chemically diverse, 5000 compound Goldilocks (GL) library has an intermediate number of compounds sized between fragments and drugs with predicted favorable ADME (absorption, distribution, metabolism, and excretion) and toxicological profiles. Amalgamating our core GL library with an approved drug (AD) library, we employ a combined GLAD library virtual screen, enabling an effective and efficient design cycle of ranked computer docking, top hit biophysical and cell validations, and defined bound structures using human proteins or their avatars. As new drug design is increasingly pathway directed as well as molecular and mechanism based, our CASSIE approach facilitates testing multiple related targets by efficiently turning a set of interacting drug discovery problems into a tractable medicinal chemistry engineering problem of optimizing affinity and ADME properties based upon early co-crystal structures. Optimization efforts are made efficient by a computationally-focused iterative chemistry and structure screen. Thus, we herein describe and apply CASSIE to define prototypic, specific inhibitors for PARG vs distinct inhibitors for the related macrodomains of ARH3 and SARS CoV-2 Nsp3 plus the SARS CoV-2 Nsp15 RNA nuclease.

1. Introduction

Because of the high failure rate and extremely high cost of clinical trials, many drugs going into clinical trials are often modifications of approved drugs. For example, we identified the antibiotic Novobiocin (NVB) in a high throughput screen, as a DNA polymerase theta (POLθ) synthetically lethal with Homologous Recombination (HR) deficiency and thus a candidate target for HR-deficient cancers (Zhou et al., 2021). However, it is difficult to alter the complex structure of approved drugs and natural products for a new target (Atanasov et al., 2021; Omanakuttan et al., 2012). On the other side of the size range, fragment screening where crystals are probed with small molecules was designed to identify novel chemical structures (Murray & Blundell, 2010; Wilson 3rd et al., 2021), but obtaining larger compounds with higher affinity and selectivity plus with the proper orientation as indicated by the fragment-bound structures has proved problematic for drug discovery. Conceptually, an ideal process would be to start drug discovery using a Goldilocks (GL) compound library that is pre-selected for ideal drug characteristics and that is sized bigger than fragments but smaller than approved drugs. This intermediate size enables chemical elaboration to improve molecule properties, either for binding or for drug characteristics. To exemplify the versatility of our approach, we present our GLAD library (our GL library combined with an Approved Drug or AD library) and our Chemistry and Structure Screen Integrated Efficiently (CASSIE) pipeline and then their application to define lead inhibitors for two distinct target families: ADP ribosylase macrodomains and SARS CoV-2 Nsp15 endoribonuclease.

Controlling repair-replication and innate immune responses, macrodomains are an important drug target family: they counteract ADP-ribosylation in DNA repair and in viral infection, and thus are a fundamental cellular response to stress. In the context of DNA damage, they regulate efficient targeting and kinetics for the DNA damage response and are a target for synthetic lethality in DNA repair deficient cancer cells. Trapped PARP1 and DNA adducts facilitate MRE11 and RAD51 recruitment (Moiani et al., 2018; Syed & Tainer, 2018) and stall replication forks. Their repair involves removal of poly-ADP-ribosylation (PARylation) by macrodomain-containing glycohydrolases such as poly(ADP-ribose) glycohydrolase (PARG) (Houl et al., 2019). In the host antiviral environment, viral pathogens such as SARS CoV-2 employ a related macrodomain, Nsp3 Mac1, to remove ADP-ribosylation to dysregulate host immune response (Brosey et al., 2021). These features make macrodomains a focus for structure-based drug discovery as promising therapeutic targets both for cancer and antiviral agents. A key issue for DNA repair and virus macrodomain targets concerns homology of viral and human macrodomains and its impact on inhibitor binding pockets.

As part of an effort to develop inhibitors for additional SARS CoV-2 non-structural proteins (Nsp), we also applied our CASSIE approach for the SARS CoV-2 endoribonuclease Nsp15. Nsp15 is implicated in viral pathogenicity and evading host innate immune response as defective Nsp15 led to reduced viral pathogenicity in CoV models (Ancar et al., 2020; Deng et al., 2019; Kindler et al., 2017; Yuen et al., 2020). Its nidoviral uridylate-specific endoribonuclease (NendoU) function may regulate viral RNA synthesis, limit recognition of viral RNA by cellular sensors, and block early, protective antiviral host cell response. In particular, Nsp15 may help CoV evade host detection by trimming long 5′-polyuridine (polyU) tracts, a pathogen-associated molecular pattern (PAMP) that cells use to detect viruses (Hackbart, Deng, & Baker, 2020). Nsp15 NendoU activity is necessary and sufficient for inhibiting eIF2α-dependent and -independent stress granules (SGs) that function as an antiviral hub and whose suppression is needed for CoV replication (Gao et al., 2021). Thus, an Nsp15i could be of value by critically combining a direct anti-viral effect by enabling the early, protective immune response to viruses that thereby averts later excessive dysregulated damaging immune activation (Channappanavar et al., 2016; Channappanavar & Perlman, 2017); consistently, an Nsp15 natural product inhibitor (Nsp15i) efficiently neutralized SARS-CoV-2 (Hong et al., 2021). Yet, Nsp15 is understudied and virus-specific with no viable clinical drug, and the few cell-active inhibitors identified lack specificity (Choi et al., 2021; Hong et al., 2021). Thus, we apply CASSIE to target Nsp15, as it is essential in CoV biology (Kim et al., 2020; Xu et al., 2021) and as it could be an independent or anti-CoV cocktail target.

As an overview of our methodology, we begin by harnessing Evolutionary Trace (ET) methods to distinguish conserved and variable sequence areas of the active site pocket as shown for PARG and Nsp3 Mac1 (Brosey et al., 2021). Based upon an evolutionary perspective of surface sequence conservation, we focus on sites offering optimal specificity. Next, we use our CASSIE approach of in silico screening of our selective GLAD library against available protein structures and experimental validation by binding assays (Fig. 1). By employing a chemically-diverse, tractably-sized 5000 compound GL library of compounds larger than fragments but smaller than drugs, we collapse potentially immense chemical library space into a sparse set of cell-friendly chemotypes. Our tactical approach for efficient computational screens is to identify the top 15–100 candidates for binding measurements. Top binding candidates are employed for crystallization experiments to obtain X-ray crystal structures of inhibitor candidates with micromolar or better binding. When human structures are not available, useful crystallographic feedback is obtained by employing avatars that incorporate essential target site features (Moiani et al., 2018). We use structural results to provide a rational basis to build focused chemical libraries to improve binding affinity plus favorable ADME (absorption, distribution, metabolism, and excretion) properties and toxicological chemotypes. We show here how our GLAD library and CASSIE pipeline can provide inhibitor tools for cell biology and leads for preclinical drug discovery.

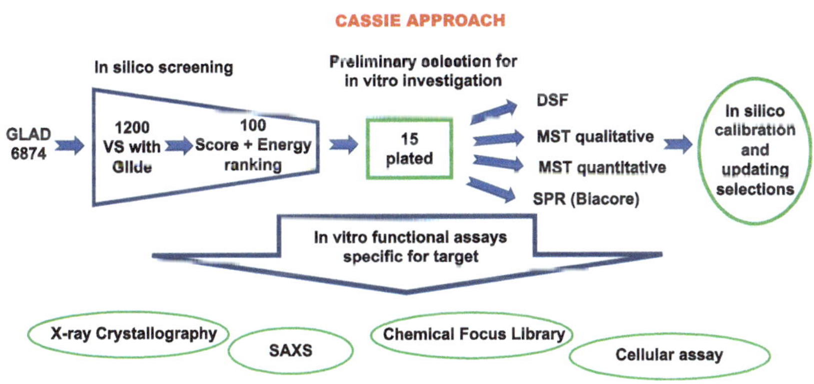

Fig. 1 Schematic flow-chart representation of the CASSIE approach.

2. Targeting evolutionary conservation and identifying functional sites

The multiple interfaces, that DNA repair proteins have with their substrate and with other DNA repair and replication proteins and that viral proteins have for pathogen–host interactions, pose a challenge for in silico screening. It is key to define a target functional site and to identify residues relevant to the ligand docking volume. While structural and biological data may help define target sites, evolutionary sequence conservation complements the identification of novel regions of functional relevance and helps to prioritize sites. ET analysis combines phylogenetic relationships and sequence information to identify residues and regions of functional importance (Lichtarge, Bourne, & Cohen, 1996; Mihalek, Res, & Lichtarge, 2004). Importantly, ET scores protein residues according to the phylogenetic distances of the homologous sequences that vary at that residue and outputs these scores in a range of 0 (most important) to 100 (least important). These scores can be subsequently mapped to available structures or protein structure predictions (Lua & Lichtarge, 2010).

Clusters of high-ranking residues signify functionally important regions, such as active sites or critical allosteric sites (Lees-Miller et al., 2020), with confidence in these clusters made stronger by the independence of ET calculation from available structural information. Notably, these identified residues can be input into the in silico screening programs to define Virtual High-Throughput Screening (VHTS) receptor space. See Section 5 for VHTS input into the docking program, GLIDE. Here, we outline how to use the ET server (http://evolution.lichtargelab.org/) (Lichtarge et al., 1996; Mihalek et al., 2004).

2.1 Identify homologous sequences by BLAST searches

The ET server uses the Blastall 2.2.15 program (Altschul et al., 1997) and the NCBI non-redundant (nr) UniRef90, and Uniref100 databases (Maglott, Ostell, Pruitt, & Tatusova, 2007; Suzek et al., 2015). Typically, at least 50 homologous sequences in an alignment are used to calculate the ET scores. The exact number depends on the availability of homologous sequences and the sequence selection procedure. Technically, we can run ET with as many sequences as we like. The more sequences we use, the slower it is to align them and score (these relationships are exponential rather than linear). In the ET server, the user can define the maximum number of sequences (the default is 500 from the BLAST search). We typically start with 5000 sequences from BLAST.

2.2 Select sequences for the multiple sequences alignment (MSA)

Synthetic constructs, mutant sequences, and fragments of sequences should not be included in the MSA. Therefore, ET embodies filtering algorithms to remove such BLAST hits. Furthermore, redundant sequences may not be necessary or may bias the ET outcome to functions specific to the over-represented phylogenetic branches, so redundancy should be avoided. Scripts embedded to the ET server may select representative homologs at various phylogenetic distances that have the fewest alignment gaps compared to the query sequence. Typically, this process chooses approximately 160 sequences.

2.3 Align the sequences using MUSCLE alignment program

In the extremely rare cases that technical errors appear using MUSCLE alignment program (Edgar, 2004), the ET pipeline may turn to ClustalW alignment program (Thompson, Higgins, & Gibson, 1994). The steps 2.1–2.3 may be skipped if users prefer to input manually curated MSA.

2.4 Run ET analysis

We used the option "position-specific gap-reducing real-valued trace." Although the ET algorithm may proceed with less homologous sequences, it is recommended to use MSA with at least five homologous sequences, although such small numbers of sequences may result in low resolution for ET scores (many ties in the residue ranks). Low resolution of ET ranks may also be obtained for larger number of homologous sequences, when they have high percentage identity compared to the query sequence. The ET analysis outputs scores for each residue in the range of 0 (most important) to 100 (least important)

2.5 Color available structures according to ET scores

This can be done with the PyMOL program and the PyETV plugin (Lua & Lichtarge, 2010). Fig. 2 shows the PDB structures: 4B1G (human PARG),

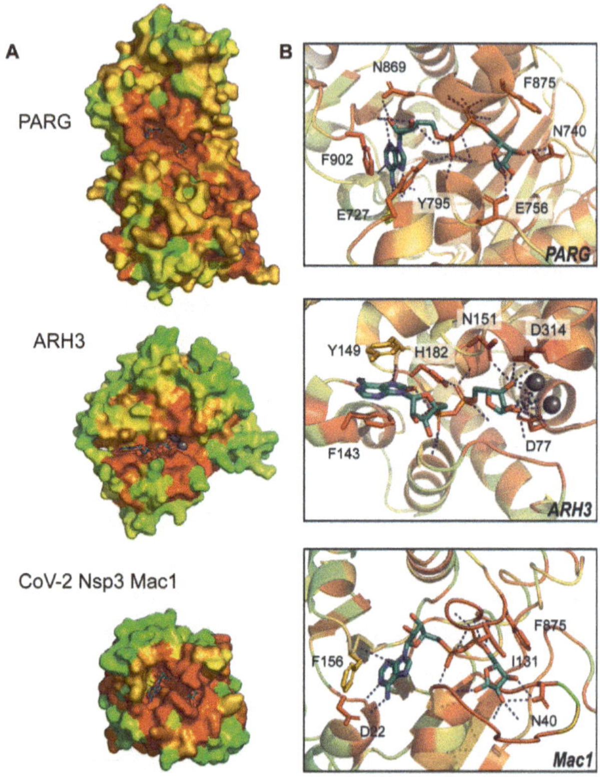

Fig. 2 Evolutionary trace analysis identifies functional residues within PARG, ARH3, and CoV-2 Nsp3 Mac1 active sites. High-ranking ET values (red-orange) cluster at the active sites of each enzyme. Inspection of residues within the active sites (right panel) shows greatest conservation among residues coordinating the ADPr pyrophosphate linker and terminal ribose and increased variation within the adenine pocket.

6D36 (human ARH3), and 6W02 (CoV-2 Mac1), colored by prioritization scores (red notes the most important and green notes the least important residues). The PyETV plugin offers additional schemes to represent the important residues and additional functions, such as measuring the z-score for the non-random clustering of the top ET residues.

2.6 Structural analysis of ET results on PARG, ARH3, and SARS CoV-2 Nsp3 Mac1

Applying this approach to PARG as a key repair-replication regulator and for comparative exemplary macrodomains ARH3 and SARS CoV-2 Nsp3 Mac1, highlights the active site as a shared primary region of functional significance in all three enzymes (Fig 2). Inspection of the active site reveals greater importance among residues coordinating the ADP ribose (ADPr) pyrophosphate linker and terminal ribose group relative to those stabilizing the adenine moiety. This is underscored by the unique binding conformations assumed by aromatic residues within the adenine pocket (c.f. PARG—phenylalanine 902, tyrosine 795; ARH3—phenylalanine 143, tyrosine 149; SARS CoV-2 Mac1—phenylalanine 156). These structural analyses by ET generally identify the active sites and other targets for screening. In this example, they suggest that selective targeting of the adenine pocket can be leveraged to identify ligands specific to each enzyme.

2.7 ET analysis of SARS CoV-2 proteins

We note that ET analyses for all SARS CoV-2 proteins, are available through an interactive GUI (http://cov.lichtargelab.org/) (Wang et al., 2021).

3. GLAD library

3.1 GLAD library description

The GLAD library has been assembled based upon experimental feedback over the last 5 years to consist of 6874 compounds and is divided into two major families as discussed in the introduction GL (Goldilocks) and AD (Approved Drugs).

The GL library is subdivided into compound families (Fig. 3). The GL library was first assembled with 2500 compounds and then expanded based upon experimental results in two different steps to 4763. The initial GL compounds were selected to ensure chemical diversity using calculated two-dimensional fingerprints and the Tanimoto similarity index (Bajusz, Racz, & Heberger, 2015). Additionally, compounds were selected to minimize potential compound promiscuity (Bruns & Watson, 2012, Dahlin et al., 2015), assay interference (e.g., fluorescence), and with predicted favorable physicochemical properties (size, solubility, cellular permeability, etc.) (Lipinski, Lombardo, Dominy, & Feeney, 2001; Veber et al., 2002). The GL library includes compounds that are brominated, fluorinated, Protein–Protein Interaction disruptors (PPI), Fsp3 (containing sp3 hybridized carbon units), Superior (compounds optimized for solubility, low toxicity and cell permeability), peptidomimetic, and/or are targeting proteins involved in apoptosis and DNA Repair. The GL library was optimized for the percentage of each compound subfamily to its current composition (Fig. 3).

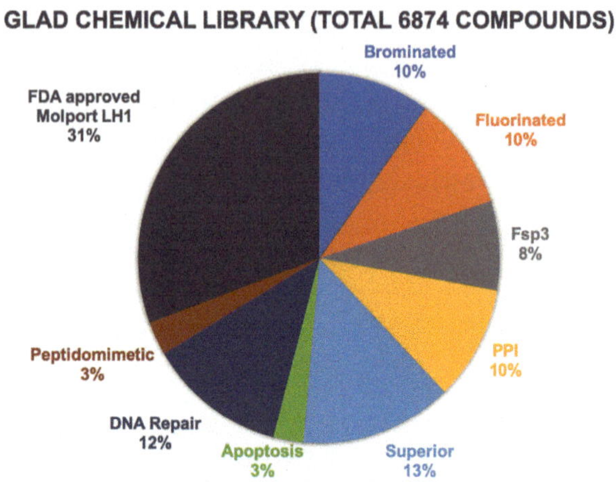

Fig. 3 Pie chart of GLAD library and its subfamilies distribution. Schematic representation of the distribution of 6874 GLAD compounds in its subfamilies with explicit percentage: AD (FDA approved library and others), Brominated, Fluorinated, Fsp3 carbon hybridization, PPI (Protein–Protein Interaction disruptors), Superior (optimized for Cell assay), compounds targeting Apoptosis and DNA Repair proteins (ligands involved or tested in our in-house DNA Repair system), Peptidomimetic.

The last expansion (2111 compounds), called AD (Approved Drugs), provides larger soluble compounds including Vitamins, lipids, natural compounds and additional human metabolites representing circa 31% of the full GLAD library. A complete virtual entity of the library has been assembled for in silico study. For experimental measurements, copies of the library are stored both in powder form and as DMSO solutions for in vitro and in vivo experiments.

3.2 Representation of the virtual GLAD library for in silico virtual screening

The material library is accompanied by a full collection of structure data format (.sdf) files representing all compounds with related information such as chemical identification, calculated properties, molecular weight, and SMILES. These are provided by our supplier (https://lifechemicals.com/).

3.3 Ligand files preparation

Upload the .sdf files for the full GLAD library to Maestro.

3.4 Optimize the ligands for in silico study

Run LigPrep program (Chen & Foloppe, 2010). In the program, we use the force field OPLS4 (Lu et al., 2021), which generates multiple conformers for each ligand to be incorporated in the GLAD in silico library for Virtual Screening. This collection of compounds is supplemented with additional compound controls that are known to bind to the target. Several human metabolites are already present in the AD subdivision.

4. Protein target and grid preparation

To reduce the computational complexity, we generate a ligand receptor grid onto the protein target site of interest. We use a rigid docking site with flexible ligands (Schnecke, Swanson, Getzoff, Tainer, & Kuhn, 1998). The PDB structure selected for in silico study needs to be prepared, optimized and minimized (Nguyen et al., 2021). If a protein target has an allosteric mechanism and the active site where the grid is centered has more than one preferential conformation, then multiple target structures need to be prepared in parallel to run multiple virtual screens (Moiani et al., 2018).

4.1 Prepare the protein structure

In the Protein Preparation Wizard (Sastry, Adzhigirey, Day, Annabhimoju, & Sherman, 2013), we strip waters and non-metal ligands from the structure. We keep metals and assign with proper charges. We change non-standard amino acids to standard amino acids, like selenomethionine to methionine. We either complete missing loops or cap termini resulting from missing loops with N-terminal acetyl and C-terminal amide capping groups and which gives the proper zwitterionic charge.

4.2 Generate a grid in the target area

This is typically an active site or protein-protein interaction surface. In the Receptor grid generation, we pick residues in the active site. The program will build a cube, typically with a range of dimensions of 10–15 Å on each side. If using a large grid (e.g., double the size), the calculations may take 1–2 days or longer, depending on computing power. Multiple grids can be generated and targeted at the same time. Results from multiple calculations and multiple grids requires a more accurate evaluation to generate the proper ranking of results.

4.3 Grids for macrodomains

For PARG, we used three structures (PDB ID 6OAK and PDB ID 4B1H with two conformations of phenylalanine 902) with the cubic dimensions (12 Å on each side) around phenylalanine 902, tyrosine 795, and phenylalanine 875 (based on the human structure). The Receptor grid generation program calculated the midpoint between the three c-alpha and centered the grid around the midpoint. We also generated grids for MAC1 (PDB ID 7KG3) and for ARH3 (PDB ID 6D36).

4.4 Grids for Nsp15 trimeric assembly

For Nsp15 (PDB ID 6WLC), we centered our grid on only Tyrosine 343 as it interacts with tipiracil and UMP (Kim et al., 2020, 2021). Notably, we generated the grid on a Nsp15 homotrimer, as the active site is near the subunit interface.

5. Virtual screen and ranking parameter selections

Next, we dock our GLAD library into the selected cubic grids of the protein using Glide program (Friesner et al., 2006). Glide allows the ligand to adopt multiple conformations but the protein is kept rigid. The virtual screening parameters are optimized using a workflow implemented in the Schrödinger package

5.1 Run a virtual screen calculation

In the Glide docking program we use "standard precision" and a parallelizing multiprocessor cluster and enable the MMGBSA scoring calculation in Prime. MM-GBSA energies (ΔG) are calculated from binding energies including covalent, van der Waals (VDW), coulombic, solvation, hydrogen bonding as well as from packing and energy differences related to strain in protein and ligand (Genheden & Ryde, 2015). As needed for speed, we use VHTS receptor space, which offers a lower precision but faster calculation.

5.2 Identify the top 10% compounds based on the docking score

Some ligands may occur multiple times because they bind in different conformations.

5.3 Rescore these top 10% compounds with the MMGBSA energy score

We have the most leads when we combine the docking score (weighted 70%) with MMGBSA energy (weighted 30%) (Moiani et al., 2009; Moiani, Cavallotti, Famulari, & Schmuck, 2008). We find both top binders but also more chemically diverse compounds in the list.

5.4 Remove compound redundancies

When the compound appears multiple times in the list because of different conformations, we typically remove all but the top scoring conformation.

5.5 Pick top scoring ligands with unique chemical features

We analyze the subfamily type of the GLAD library in the full list of results, view the ligands docked on available structures, and select about 15 top *in silico* leads, which contain the highest number of unique chemical features, also known as chemical singletons. Fig. 4 (A–E) presents our selection for the

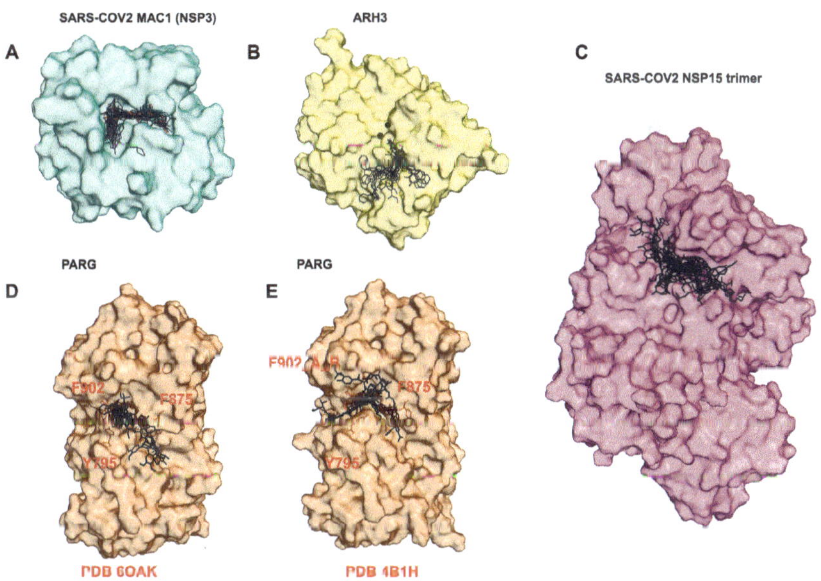

Fig. 4 Surface presentation of target proteins in complex with top 15 in silico leads. (A) MAC1 from PDB ID 7KG3 surface representation in complex with top 15 in silico results. (B) ARH3 from PDB ID 6D36 surface representation, magnesium ions black sphere, in complex with top 15 in silico results. (C) Nsp15 from PDB ID 6WLC) surface representation, in complex with top 15 in silico results. (D and E) PARG from three structures PDB ID 6OAK and PDB ID 4B1H with dual conformation **a** and **b** for amino acid F902 surface representation with key residue in stick in complex with top 15 results.

macrodomain targets and Nsp15 for each of them a top 15 compounds were selected per ranking, with chemical singleton redundancies cleaned. The target is presented in surface representation to better display the population of top *in silico* leads.

5.6 Final selection for in vitro experiments

The final selection of 15 is plated and moved to in vitro structural, binding and functional assays (see Sections 6 and 7).

5.7 PARG in silico leads

We merged the top compounds from in silico virtual screens performed on three different conformational structure of human PARG. Although we used different structures, some compounds were found in both docking analyses (Fig. 4D and E). After merging, the top ranked were 90% AD, and only two compounds were from the GL, a PPI and a DNA Repair uracil derivative. Interestingly, atracurium, in the top 15, has chemical similarity to a small unit of branched PAR.

5.8 ARH3 in silico leads

We merged the results obtained from targeting two grids simultaneously. To target the adenosyl clamp between phenylalanine 127 and tyrosine 133, the two grids were centered close to metal active site, specifically at residue asparagine 135. The results were grouped together and properly re-ranked as described in Section 5.3. The top 15 in silico leads' distribution shows 50/50 ratio between GL and AD. Specifically the highest scoring AD are drugs approved for infection and for targeting the immune system, which is interesting in light of the immune evasion role of this viral enzyme. The GL top in silico leads are peptidomimetics, PPI, Superior, and molecules targeting DNA Repair proteins.

5.9 SARS CoV-2 MAC1 (Nsp3) in silico leads

After applying our ranking methods on the results from Glide, we analyzed the chemotype of top 15 in silico leads. The top two compounds are part of the DNA Repair subfamily of GL, and specifically are uracil derivatives. ADPr scored as third in the ranking, matching the high affinity for MAC1, proven by in vitro measurements (in Section 7) and shown in Fig. 5A. Top in silico leads include peptidomimetics and PPI disruptors. More than

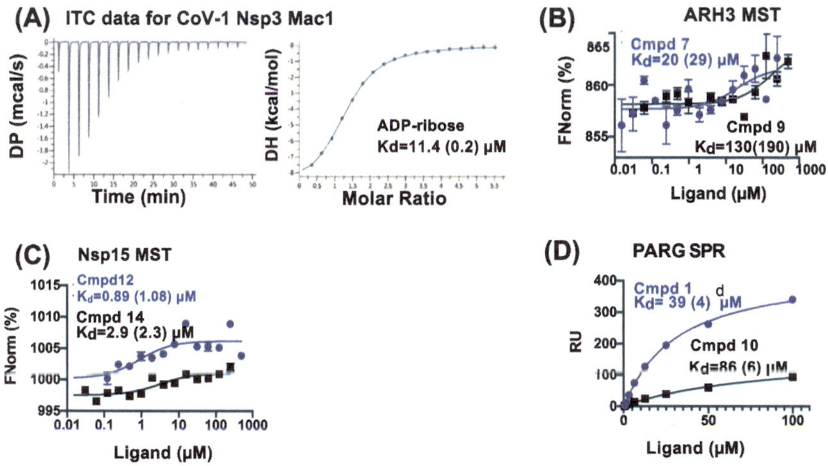

Fig. 5 Representative binding curves for in vitro binding experiments for CoV-1 Nsp3 Mac1, ARH3, Nsp15 and PARG. (A) ITC data for CoV-1 Nsp3 Mac1 binding to ADP-ribose. (B) MST data for ARH3 binding to two of the top scored compounds. Binding was detected for 3 of the 15 compounds with binding constants ranging from 20 to 400 μM. (C) MST data for Nsp15 binding to two of the top scored compounds. Binding was detected for 12 of the 15 compounds with binding constants ranging from 180 nM to 400 μM. (D) SPR data for PARG binding to two of the top scored compounds. Binding was detected for 6 of the 16 compounds with binding constants ranging from 23 to 1000 μM. *Panel (A): Reproduced from Brosey, C. A., Houl, J. H., Katsonis, P., Balapiti-Modarage, L. P. F., Bommagani, S., Arvai, A., et al. (2021). Targeting SARS-CoV-2 Nsp3 macrodomain structure with insights from human poly(ADP ribose) glycohydrolase (PARG) structures with inhibitors.* Progress in Biophysics and Molecular Biology. 163, *171–186.*

50% of top results were AD, including an anticancer drug (Daunorubicin) and an antiviral drug (Valganiclovir). None of the compounds bound MAC1, based on Microscale Thermophoresis (MST) binding results except ADPr. The graphic model of top silico leads for Mac1 is shown in Fig. 4A.

5.10 SARS CoV-2 Nsp15 in silico leads

The analysis of top in silico leads shows 70% are AD, while the rest include a PPI and three DNA Repair subfamily uracil derivatives. Among the AD results, we found certain peptides like Goserelin and Desmopressin or complex molecules like Anidulafungin (an antifungal drug). Interestingly, we found some high molecular weight compounds as top in silico leads with an extended binding surface as shown in Fig. 4C.

6. Fast qualitative and quantitative in vitro binding assays of top in silico results

As part of CASSIE (Fig. 1), we validate our in silico hits with binding assays. The top in silico selections are physically plated at suitable concentrations for qualitative binding. Compounds are prepared from the mother solution (directly plated by supplier) and made uniform at 10 mM concentration in DMSO. Typically, the in vitro techniques selected for binding studies are Microscale thermophoresis (MST) or Surface Plasmon Resonance (SPR, Biacore). Measuring ligand-protein affinity constants ranging from nM to mM, MST tracks thermophoresis, the movement of fluorescently labeled molecules in and out of a small temperature gradient. When a protein binds a ligand, there may be changes in protein conformation or even the protein hydration shell. These changes can alter the rate of movement (thermophoresis) of the protein–ligand complex as compared to the control, protein without added ligand. This is monitored by changes in fluorescence, with varying ligand concentrations to generate a binding curve from which affinity is calculated. The deflections in the binding curves can be either positive or negative as the deflection is dependent on the relative rates of the protein without ligand vs the protein ligand complex. However, if there is no change in the thermophoresis rates upon a binding event, this event will be silent or undetected using MST. SPR tracks ligand binding by measuring changes in polarized light striking the target protein or protein complex attached to a sensor surface. The protein is usually covalently attached to a sensor surface or captured using an affinity tag such as his_6 or streptavidin. These two binding assays, one detecting changes in thermophoresis of the protein by binding ligand and the other detecting changes in target mass upon binding, provide complementary information and validation of the small molecule in silico lead. The selection of top compounds from in silico are run in parallel to a well-known binder for the enzyme and compared. Compounds identified with adequate affinity move to a quantitative test, such as isothermal calorimetry (ITC) which usually requires added material to ensure reproducibility of multiple measurements.

6.1 MST assay

We typically measure MST on a Monolith NT1.15. Proteins are labeled with Atto488 NHS-ester (ATTO-TEC, GBH) according to the manufacturer's protocol with labeling efficiency 1:1 protein-to-dye ratio. Ligands are serially diluted into solution of PBS, 0.5 mM fresh TCEP, 0.01% Tween-20 with the Atto488-labeled protein. Samples are loaded into standard capillaries and each set analyzed in triplicate. Data fitting with the MO Affinity Analysis software for a series of compounds is performed at the lowest fixed power and time which gives the largest signal changed based on the either known binders or the most consistent change for that series. For ARH3 analysis, ADPr was used as a control and the settings were at high power with thermophoresis at 20 s. For Nsp15, the most consistent change for the series was with medium power with thermophoresis at 10 s. Capillaries with measure intensities greater than 1% DMSO or with intensities (±) 5% of the overall average are excluded from the analysis. Binding was labeled as not detected (ND) if the data could not be fit to the K_d model, or the calculated response amplitude was less than 2.

For ARH3, binding was detected for 3 of the 15 compounds with binding constants ranging from 20 to 400 μM. For Nsp15 binding was detected for 12 of the 15 compounds with binding constants ranging from 180 nM to 400 μM. For PARG there was no detectable change in the thermophoresis based on the above criteria and as a result, SPR was used.

6.2 SPR assay

We typically measure SPR on a Biacore T200 with proteins attached to the sensor surface. For PARG, 1 mg/ml protein was covalently attached to a Biacore CM5 chip using amine coupling via an EDC/NHS (N-(3-dimethylaminopropyl)-N'-ethylcarbodiimide hydrochloride/ N-hydroxysuccinimide) reaction. A control channel is treated similarly, except using ethanolamine instead of protein. The final relative response unit (RU) was 899.7. Serially dilute ligand starting at 100 μM was passed over the surface at 30 μL/min with a contact time of 120 s. The data was fit to a steady state affinity model with the Biacore Evaluation 3.0 software using blank subtracted RU values at 4 s before injection stop for a single run. For PARG, binding was detected for 6 of the 16 compounds with binding constants ranging from 23 to 1000 μM.

6.3 Limitation of these two assays and data evaluation for top in silico leads

MST in qualitative mode is a relatively quick method and needs a low quantity of material (including protein) but the data are not extremely reliable, particularly if the binding data extrapolated fluctuation range is over half millimolar. MST in quantitative mode would be more appropriate and needs multiple measurements in the titration, which requires more time and material with more reproducibility but the final binding K_d is still not precise.

SPR offers better data if the system is properly calibrated. The protein needs to be anchored to the chip and a calibration test implemented. Following these steps, the system can efficiently produce multiple, reliable, and accurate affinity binding measurements. ITC is the best but requires time for calibration and optimization plus an abundant quantity of material.

If micromolar or better binders are not discovered within the top 15, then we recommend testing compounds in multiple binding assays and to test all of the top 100 compounds. In silico docking scores do not directly correlate to experimental affinities.

7. Structural analysis of top in silico compounds

After in vitro binding assay validations of the in silico data, we proceed to structure-based optimizations. Crystallography is an optimal structural validation for precision; however, co-crystal structures may be difficult to obtain with preliminary candidates. With the CASSIE approach, inhibition of function and low micromolar binding are more effectively evaluated with minimal effort and protein.

7.1 What to look for in initial co-crystal structures

In binding pockets, we use crystal structures strategically to identify two different kind of subsites: anchor points and induced plasticity. This approach is based upon our discovery of anchored plasticity to gain 20,000-fold between active site pockets formed by invariant sequences in nitric oxide synthase (NOS) and the ability to selectively inhibit MRE11 endonuclease and exonuclease activities (Garcin et al., 2008; Shibata et al., 2014). Tactically, the larger sized AD library components can identify inducible plasticity regions within binding sites. The intermediate sized GL library components are better at identifying potential anchor points in binding sites that can be used for focused libraries to optimize affinity and drug-like properties by growing compounds into surface subsites with plasticity. Thus, this combination of data enables focused compound libraries to employ a systematic strategy of retaining anchor interactions while growing compounds in directions defined by a vector from the anchor site to flexible subsites. Focused libraries are designed from lead compounds by adding new chemical moieties that can bind in neighboring regions. Target flexibility or plasticity, if identified by structural analysis, can also be harnessed to provide additional regions for growing compounds to increase specificity. By employing this anchored plasticity approach CASSIE provides a systematic approach to enlarging binding pock subsites for increased affinity. This can span from creating new pockets for inhibitors as we did for Map kinase (Perry, Harris, Moiani, Olson, & Tainer, 2009) to building upon substrate analogue structures (Daniels et al., 2000; Doi et al., 2006) and learning how inhibitors displace bound water molecules (Putnam, Arvai, Bourne, & Tainer, 2000). Although we currently use crystal structures and temperature factors, as employed for EXO5 nuclease (Hambarde et al., 2021) to identify target plasticity, we expect to be able to use atomic structures combined with SAXS to identify key plasticity associated with allosteric mechanisms (Brosey & Tainer, 2019; Hammel et al., 2021; Hammel & Tainer, 2021). Observed plasticity can even be used to apply inhibitor doorstoppers to mimic a protein inhibitor and block an enzyme's active site closing as needed for strong DNA binding, by uracil-DNA glycosylase (Nguyen et al., 2021; Putnam et al., 1999).

8. Summary and prospects for advances

Urgent demands for efficient drug design stem from the SARS CoV-2 pandemic (200 million cases and 4.4 million deaths as of August 24, 2021, Johns Hopkins Center for Systems Science and Engineering, https://coronavirus.jhu.edu/map.html) and cancer (19 million new cases and 10 million deaths) (Sung et al., 2021). These diseases underscore the merit of new approaches that reduce bottlenecks and improve efficiency for drug discovery. Yet, meeting this need is challenging as both cancer cells and pathogens have a mutator phenotype that can promote selection of resistance to both the host immune system and drug therapies. For cancer drugs, relevant new mechanistic insights have come from targeting poly (ADP-ribose) polymerase (PARP). PARP inhibitors (PARPi) are successful for ovarian cancer and breast cancer indications, but resistance often develops (Li et al., 2020), suggesting a need to also be able to inhibit additional pathway proteins or complementary pathways.

Interestingly, the clinical success of PARPi's can depend upon trapping PARP1 on the damaged DNA, and this allosteric capacity has been structurally characterized for different PARPi (Zandarashvili et al., 2020). Indeed, we also know from biology that catalytically inactive proteins that bind and hold damaged DNA can control damage outcomes, as seen for ATL that forms a complex with alkylated base damage and thereby alters repair pathway outcome from base repair to the nucleotide excision repair pathway (Tubbs et al., 2009). Analogously, XPG impacts multiple repair pathways by binding and sculpting DNA junctions and protein partners in addition to its nuclease activity (Tsutakawa et al., 2020), FEN1 sculpts 5' flaps to avoid template switching at replication forks (Perry et al., 2006; Trego et al., 2011; Tsutakawa et al., 2017), GRB2 adaptor protein efficiently brings MRE11 nuclease to DNA breaks (Ye et al., 2021), XRCC1 links MRE11 and PolQ to promote alternative end joining of DNA breaks (Eckelmann et al., 2020), SLX4IP binds and maintains SLX4-XPF-ERCC1 complex for inter-strand crosslink repair (Zhang et al., 2019), and acetylation targets oxidative base repair initiation to open chromatin (Bacolla et al., 2021). In extreme examples, transient polyvalent binding by flexible proteins and RNA can create phase separated condensates of functional proteins for processes such as repair, as seen for DNA break repair protein KU and its complex with long non-coding RNA (Thapar et al., 2021). Furthermore, combining structural and computational methods is proving enabling to identify and target molecular mechanisms as well as to understand the mechanisms where point mutations can cause different human diseases (Yan et al., 2019).

All of the above noted observations support the value of molecular mechanistic knowledge for drug discovery. In fact, most successful new drugs result from molecular and mechanistic targeting that requires chemical, biophysical, biochemical and structural knowledge. It therefore becomes important to develop systematic strategies to attain this information efficiently. As a result of this reasoning, we developed the CASSIE pipeline and GLAD library to these potential overcome rate-limiting steps in developing inhibitor tools and preclinical drug candidates.

In Greek myth, Cassandra consistently made accurate predictions that were unfortunately not believed. CASSIE predictions, however, do not require faith as they can be tested and validated by practical levels of time and effort. In contrast, high throughput screens with overly large libraries can in practice be high input with challenges to effectively handle output. The CASSIE pipeline and GLAD library approach provide a manageable number of compounds, hits, and validation measurements within weeks. Getting compounds into cells and crystal structures in minimal time frames enables molecular targets and mechanisms to be efficiently tested directly and thereby provides a critical enabling path for preclinical drug discovery.

Acknowledgments

We thank Chi-Lin Tsai for comments. We acknowledge National Institute of Health (NIH) grants (P01 CA092584, R35CA220430), R01 GM141226 (to I.K.K.), a Robert A. Welch Chemistry Chair, and Cancer Prevention and Research Institute of Texas (CPRIT) RP180813. X-ray diffraction data was collected at SSRL beamline 12-2 and ALS beamlines 12.3.1 [supported by NIH project ALS-ENABLE (P30 GM124169) and the Integrated Diffraction Analysis Technologies (IDAT) program] and 8.3.1 (supported by NIH R01 GM124149 and P30 GM124169). Nsp15 research was supported by the U.S. Department of Energy (DOE) Office of Science through the National Virtual Biotechnology Laboratory, a consortium of DOE national laboratories focused on response to coronavirus disease 2019, with funding provided by the Coronavirus CARES Act.

References

Altschul, S. F., Madden, T. L., Schaffer, A. A., Zhang, J., Zhang, Z., Miller, W., et al. (1997). Gapped BLAST and PSI-BLAST: A new generation of protein database search programs. *Nucleic Acids Research*, *25*, 3389–3402.

Ancar, R., Li, Y., Kindler, E., Cooper, D. A., Ransom, M., Thiel, V., et al. (2020). Physiologic RNA targets and refined sequence specificity of coronavirus EndoU. *RNA*, *26*, 1976–1999.

Atanasov, A. G., Zotchev, S. B., Dirsch, V. M., The International Natural Product Sciences Taskforce, & Supuran, C. T. (2021). Natural products in drug discovery: Advances and opportunities. *Nature Reviews. Drug Discovery*, *20*, 200–216.

Bacolla, A., Sengupta, S., Ye, Z., Yang, C., Mitra, J., De-Paula, R. B., et al. (2021). Heritable pattern of oxidized DNA base repair coincides with pre-targeting of repair complexes to open chromatin. *Nucleic Acids Research*, *49*, 221–243.

Bajusz, D., Racz, A., & Heberger, K. (2015). Why is Tanimoto index an appropriate choice for fingerprint-based similarity calculations? *Journal of Cheminformatics*, *7*, 20.

Brosey, C. A., Houl, J. H., Katsonis, P., Balapiti-Modarage, L. P. F., Bommagani, S., Arvai, A., et al. (2021). Targeting SARS-CoV-2 Nsp3 macrodomain structure with insights from human poly(ADP-ribose) glycohydrolase (PARG) structures with inhibitors. *Progress in Biophysics and Molecular Biology*, *163*, 171–186.

Brosey, C. A., & Tainer, J. A. (2019). Evolving SAXS versatility: Solution X-ray scattering for macromolecular architecture, functional landscapes, and integrative structural biology. *Current Opinion in Structural Biology*, *58*, 197–213.

Bruns, R. F., & Watson, I. A. (2012). Rules for identifying potentially reactive or promiscuous compounds. *Journal of Medicinal Chemistry*, *55*, 9763–9772.

Channappanavar, R., Fehr, A. R., Vijay, R., Mack, M., Zhao, J., Meyerholz, D. K., et al. (2016). Dysregulated type I interferon and inflammatory monocyte-macrophage responses cause lethal pneumonia in SARS-CoV-infected mice. *Cell Host & Microbe*, *19*, 181–193.

Channappanavar, R., & Perlman, S. (2017). Pathogenic human coronavirus infections: Causes and consequences of cytokine storm and immunopathology. *Seminars in Immunopathology*, *39*, 529–539.

Chen, I. J., & Foloppe, N. (2010). Drug-like bioactive structures and conformational coverage with the LigPrep/ConfGen suite: Comparison to programs MOE and catalyst. *Journal of Chemical Information and Modeling*, *50*, 822–839.

Choi, R., Zhou, M., Shek, R., Wilson, J. W., Tillery, L., Craig, J. K., et al. (2021). High-throughput screening of the ReFRAME, Pandemic Box, and COVID Box drug repurposing libraries against SARS-CoV-2 nsp15 endoribonuclease to identify small-molecule inhibitors of viral activity. *PLoS One*, *16*, e0250019.

Dahlin, J. L., Nissink, J. W., Strasser, J. M., Francis, S., Higgins, L., Zhou, H., et al. (2015). PAINS in the assay: Chemical mechanisms of assay interference and promiscuous enzymatic inhibition observed during a sulfhydryl-scavenging HTS. *Journal of Medicinal Chemistry*, *58*, 2091–2113.

Daniels, D. S., Mol, C. D., Arvai, A. S., Kanugula, S., Pegg, A. E., & Tainer, J. A. (2000). Active and alkylated human AGT structures: A novel zinc site, inhibitor and extrahelical base binding. *The EMBO Journal*, *19*, 1719–1730.

Deng, X., van Geelen, A., Buckley, A. C., O'Brien, A., Pillatzki, A., Lager, K. M., et al. (2019). Coronavirus endoribonuclease activity in porcine epidemic diarrhea virus suppresses type I and type III interferon responses. *Journal of Virology*, *93*, e02000–e02018.

Doi, Y., Katafuchi, A., Fujiwara, Y., Hitomi, K., Tainer, J. A., Ide, H., et al. (2006). Synthesis and characterization of oligonucleotides containing 2'-fluorinated thymidine glycol as inhibitors of the endonuclease III reaction. *Nucleic Acids Research*, *34*, 1540–1551.

Eckelmann, B. J., Bacolla, A., Wang, H., Ye, Z., Guerrero, E. N., Jiang, W., et al. (2020). XRCC1 promotes replication restart, nascent fork degradation and mutagenic DNA repair in BRCA2-deficient cells. *NAR Cancer*, *2*, zcaa013.

Edgar, R. C. (2004). MUSCLE: A multiple sequence alignment method with reduced time and space complexity. *BMC Bioinformatics*, *5*, 113.

Friesner, R. A., Murphy, R. B., Repasky, M. P., Frye, L. L., Greenwood, J. R., Halgren, T. A., et al. (2006). Extra precision glide: Docking and scoring incorporating a model of hydrophobic enclosure for protein-ligand complexes. *Journal of Medicinal Chemistry*, *49*, 6177–6196.

Gao, B., Gong, X., Fang, S., Weng, W., Wang, H., Chu, H., et al. (2021). Inhibition of anti-viral stress granule formation by coronavirus endoribonuclease nsp15 ensures efficient virus replication. *PLoS Pathogens*, *17*, e1008690.

Garcin, E. D., Arvai, A. S., Rosenfeld, R. J., Kroeger, M. D., Crane, B. R., Andersson, G., et al. (2008). Anchored plasticity opens doors for selective inhibitor design in nitric oxide synthase. *Nature Chemical Biology*, *4*, 700–707.

Genheden, S., & Ryde, U. (2015). The MM/PBSA and MM/GBSA methods to estimate ligand-binding affinities. *Expert Opinion on Drug Discovery*, *10*, 449–461.

Hackbart, M., Deng, X., & Baker, S. C. (2020). Coronavirus endoribonuclease targets viral polyuridine sequences to evade activating host sensors. *Proceedings of the National Academy of Sciences of the United States of America*, *117*, 8094–8103.

Hambarde, S., Tsai, C. L., Pandita, R. K., Bacolla, A., Maitra, A., Charaka, V., et al. (2021). EXO5-DNA structure and BLM interactions direct DNA resection critical for ATR-dependent replication restart. *Molecular Cell*, *81*, 2989–3006.

Hammel, M., Rashid, I., Sverzhinsky, A., Pourfarjam, Y., Tsai, M. S., Ellenberger, T., et al. (2021). An atypical BRCT-BRCT interaction with the XRCC1 scaffold protein compacts human DNA Ligase IIIalpha within a flexible DNA repair complex. *Nucleic Acids Research*, *49*, 306–321.

Hammel, M., & Tainer, J. A. (2021). X-ray scattering reveals disordered linkers and dynamic interfaces in complexes and mechanisms for DNA double-strand break repair impacting cell and cancer biology. *Protein Science*, *30*, 1735–1756.

Hong, S., Seo, S. H., Woo, S. J., Kwon, Y., Song, M., & Ha, N. C. (2021). Epigallocatechin gallate inhibits the uridylate-specific endoribonuclease Nsp15 and efficiently neutralizes the SARS-CoV-2 strain. *Journal of Agricultural and Food Chemistry*, *69*, 5948–5954.

Houl, J. H., Ye, Z., Brosey, C. A., Balapiti-Modarage, L. P. F., Namjoshi, S., Bacolla, A., et al. (2019). Selective small molecule PARG inhibitor causes replication fork stalling and cancer cell death. *Nature Communications*, *10*, 5654.

Kim, Y., Jedrzejczak, R., Maltseva, N. I., Wilamowski, M., Endres, M., Godzik, A., et al. (2020). Crystal structure of Nsp15 endoribonuclease NendoU from SARS-CoV-2. *Protein Science*, *29*, 1596–1605.

Kim, Y., Wower, J., Maltseva, N., Chang, C., Jedrzejczak, R., Wilamowski, M., et al. (2021). Tipiracil binds to uridine site and inhibits Nsp15 endoribonuclease NendoU from SARS-CoV-2. *Communications Biology*, *4*, 193.

Kindler, E., Gil-Cruz, C., Spanier, J., Li, Y., Wilhelm, J., Rabouw, H. H., et al. (2017). Early endonuclease-mediated evasion of RNA sensing ensures efficient coronavirus replication. *PLoS Pathogens*, *13*, e1006195.

Lees-Miller, J. P., Cobban, A., Katsonis, P., Bacolla, A., Tsutakawa, S. E., Hammel, M., et al. (2020). Uncovering DNA-PKcs ancient phylogeny, unique sequence motifs and insights for human disease. *Progress in Biophysics and Molecular Biology*, *163*, 87–108.

Li, H., Liu, Z. Y., Wu, N., Chen, Y. C., Cheng, Q., & Wang, J. (2020). PARP inhibitor resistance: The underlying mechanisms and clinical implications. *Molecular Cancer*, *19*, 107.

Lichtarge, O., Bourne, H. R., & Cohen, F. E. (1996). An evolutionary trace method defines binding surfaces common to protein families. *Journal of Molecular Biology*, *257*, 342–358.

Lipinski, C. A., Lombardo, F., Dominy, B. W., & Feeney, P. J. (2001). Experimental and computational approaches to estimate solubility and permeability in drug discovery and development settings. *Advanced Drug Delivery Reviews*, *46*, 3–26.

Lu, C., Wu, C., Ghoreishi, D., Chen, W., Wang, L., Damm, W., et al. (2021). OPLS4: Improving force field accuracy on challenging regimes of chemical space. *Journal of Chemical Theory and Computation*, *17*, 4291–4300.

Lua, R. C., & Lichtarge, O. (2010). PyETV: A PyMOL evolutionary trace viewer to analyze functional site predictions in protein complexes. *Bioinformatics*, *26*, 2981–2982.

Maglott, D., Ostell, J., Pruitt, K. D., & Tatusova, T. (2007). Entrez gene: Gene-centered information at NCBI. *Nucleic Acids Research*, *35*, D26–D31.

Mihalek, I., Res, I., & Lichtarge, O. (2004). A family of evolution-entropy hybrid methods for ranking protein residues by importance. *Journal of Molecular Biology*, *336*, 1265–1282.

Moiani, D., Cavallotti, C., Famulari, A., & Schmuck, C. (2008). Oxoanion binding by guanidiniocarbonylpyrrole cations in water: A combined DFT and MD investigation. *Chemistry*, *14*, 5207–5219.

Moiani, D., Ronato, D. A., Brosey, C. A., Arvai, A. S., Syed, A., Masson, J. Y., et al. (2018). Targeting allostery with avatars to design inhibitors assessed by cell activity: Dissecting MRE11 endo- and exonuclease activities. *Methods in Enzymology*, *601*, 205–241.

Moiani, D., Salvalaglio, M., Cavallotti, C., Bujacz, A., Redzynia, I., Bujacz, G., et al. (2009). Structural characterization of a Protein A mimetic peptide dendrimer bound to human IgG. *The Journal of Physical Chemistry*, *113*, 16268–16275.

Murray, C. W., & Blundell, T. L. (2010). Structural biology in fragment-based drug design. *Current Opinion in Structural Biology*, *20*, 497–507.

Nguyen, M. T., Moiani, D., Ahmed, Z., Arvai, A. S., Namjoshi, S., Shin, D. S., et al. (2021). An effective human uracil-DNA glycosylase inhibitor targets the open pre-catalytic active site conformation. *Progress in Biophysics and Molecular Biology*, *163*, 143–159.

Omanakuttan, A., Nambiar, J., Harris, R. M., Bose, C., Pandurangan, N., Varghese, R. K., et al. (2012). Anacardic acid inhibits the catalytic activity of matrix metalloproteinase-2 and matrix metalloproteinase-9. *Molecular Pharmacology*, *82*, 614–622.

Perry, J. J., Harris, R. M., Moiani, D., Olson, A. J., & Tainer, J. A. (2009). p38alpha MAP kinase C-terminal domain binding pocket characterized by crystallographic and computational analyses. *Journal of Molecular Biology*, *391*, 1–11.

Perry, J. J., et al. (2006). WRN exonuclease structure and molecular mechanism imply an editing role in DNA end processing. *Nature Structure & Molecular Biology*, *13*(5), 414–422, 1088. https://doi.org/10.1038/nsmb1088.

Putnam, C. D., Arvai, A. S., Bourne, Y., & Tainer, J. A. (2000). Active and inhibited human catalase structures: Ligand and NADPH binding and catalytic mechanism. *Journal of Molecular Biology*, *296*, 295–309.

Putnam, C. D., Shroyer, M. J., Lundquist, A. J., Mol, C. D., Arvai, A. S., Mosbaugh, D. W., et al. (1999). Protein mimicry of DNA from crystal structures of the uracil-DNA glycosylase inhibitor protein and its complex with *Escherichia coli* uracil-DNA glycosylase. *Journal of Molecular Biology*, *287*, 331–346.

Sastry, G. M., Adzhigirey, M., Day, T., Annabhimoju, R., & Sherman, W. (2013). Protein and ligand preparation: Parameters, protocols, and influence on virtual screening enrichments. *Journal of Computer-Aided Molecular Design*, *27*, 221–234.

Schnecke, V., Swanson, C. A., Getzoff, E. D., Tainer, J. A., & Kuhn, L. A. (1998). Screening a peptidyl database for potential ligands to proteins with side-chain flexibility. *Proteins*, *33*, 74–87.

Shibata, A., Moiani, D., Arvai, A. S., Perry, J., Harding, S. M., Genois, M. M., et al. (2014). DNA double-strand break repair pathway choice is directed by distinct MRE11 nuclease activities. *Molecular Cell*, *53*, 7–18.

Sung, H., Ferlay, J., Siegel, R. L., Laversanne, M., Soerjomataram, I., Jemal, A., et al. (2021). Global cancer statistics 2020: GLOBOCAN estimates of incidence and mortality worldwide for 36 cancers in 185 countries. *CA: A Cancer Journal for Clinicians*, *71*, 209–249.

Suzek, B. E., Wang, Y., Huang, H., McGarvey, P. B., Wu, C. H., & UniProt, C. (2015). UniRef clusters: A comprehensive and scalable alternative for improving sequence similarity searches. *Bioinformatics*, *31*, 926–932.

Syed, A., & Tainer, J. A. (2018). The MRE11-RAD50-NBS1 complex conducts the orchestration of damage signaling and outcomes to stress in dna replication and repair. *Annual Review of Biochemistry*, *87*, 263–294.

Thapar, R., Wang, J. L., Hammel, M., Ye, R., Liang, K., Sun, C., et al. (2021). Mechanism of efficient double-strand break repair by a long non-coding RNA. *Nucleic Acids Research*, *49*, 1199–1200.

Thompson, J. D., Higgins, D. G., & Gibson, T. J. (1994). CLUSTAL W: Improving the sensitivity of progressive multiple sequence alignment through sequence weighting, position-specific gap penalties and weight matrix choice. *Nucleic Acids Research*, *22*, 4673–4680.

Trego, K. S., et al. (2011). The DNA repair endonuclease XPG interacts directly and functionally with the WRN helicase defective in Werner syndrome. *Cell Cycle*, *10*(12), 1998–2007, 15878. https://doi.org/10.4161/cc.10.12.15878.

Tsutakawa, S. E., Sarker, A. H., Ng, C., Arvai, A. S., Shin, D. S., Shih, B., et al. (2020). Human XPG nuclease structure, assembly, and activities with insights for neurodegeneration and cancer from pathogenic mutations. *Proceedings of the National Academy of Sciences of the United States of America*, *117*, 14127–14138.

Tsutakawa, S. E., et al. (2017). Phosphate steering by Flap Endonuclease 1 promotes 5'-flap specificity and incision to prevent genome instability. *Nature Communication*, 15855. https://doi.org/10.1038/ncomms15855.

Tubbs, J. L., Latypov, V., Kanugula, S., Butt, A., Melikishvili, M., Kraehenbuehl, R., et al. (2009). Flipping of alkylated DNA damage bridges base and nucleotide excision repair. *Nature*, *459*, 808–813.

Veber, D. F., Johnson, S. R., Cheng, H.-Y., Smith, B. R., Ward, K. W., & Kopple, K. D. (2002). Molecular properties that influence the oral bioavailability of drug candidates. *Journal of Medicinal Chemistry*, *45*, 2615–2623.

Wang, C., Konecki, D. M., Marciano, D. C., Govindarajan, H., Williams, A. M., Wastuwidyaningtyas, B., et al. (2021). Identification of evolutionarily stable functional and immunogenic sites across the SARS-CoV-2 proteome and greater coronavirus family. *Bioinformatics*, btab406.

Wilson, D. M., 3rd, Deacon, A. M., Duncton, M. A. J., Pellicena, P., Georgiadis, M. M., Yeh, A. P., et al. (2021). Fragment- and structure-based drug discovery for developing therapeutic agents targeting the DNA damage response. *Progress in Biophysics and Molecular Biology*, *163*, 130–142.

Xu, W., Pei, G., Liu, H., Ju, X., Wang, J., Ding, Q., et al. (2021). Compartmentalization-aided interaction screening reveals extensive high-order complexes within the SARS-CoV-2 proteome. *Cell Reports*, *36*, 109482.

Yan, C., Dodd, T., He, Y., Tainer, J. A., Tsutakawa, S. E., & Ivanov, I. (2019). Transcription preinitiation complex structure and dynamics provide insight into genetic diseases. *Nature Structural & Molecular Biology*, *26*, 397–406.

Ye, Z., Xu, S., Shi, Y., Bacolla, A., Syed, A., Moiani, D., et al. (2021). GRB2 enforces homology-directed repair initiation by MRE11. *Science Advances*, *7*, eabe9254.

Yuen, C. K., Lam, J. Y., Wong, W. M., Mak, L. F., Wang, X., Chu, H., et al. (2020). SARS-CoV-2 nsp13, nsp14, nsp15 and orf6 function as potent interferon antagonists. *Emerging Microbes & Infections*, *9*, 1418–1428.

Zandarashvili, L., Langelier, M. F., Velagapudi, U. K., Hancock, M. A., Steffen, J. D., Billur, R., et al. (2020). Structural basis for allosteric PARP-1 retention on DNA breaks. *Science*, *368*, eaax6367.

Zhang, H., Chen, Z., Ye, Y., Ye, Z., Cao, D., Xiong, Y., et al. (2019). SLX4IP acts with SLX4 and XPF-ERCC1 to promote interstrand crosslink repair. *Nucleic Acids Research*, *47*, 10181–10201.

Zhou, J., Gelot, C., Pantelidou, C., Li, A., Yucel, H., Davis, R. E., et al. (2021). A first-in--class polymerase theta inhibitor selectively targets homologous-recombination-deficient tumors. *Nature Cancer*, *2*, 598–610.

FSC
www.fsc.org
MIX
Papier | Fördert gute Waldnutzung
FSC® C083411

Kolibri 360 GmbH
Ferdinand-Jühlke-Straße 7
99095 Erfurt, Deutschland
produktsicherheit@kolibri360.de